The birth of particle physics

The birth of
particle physics

EDITORS

LAURIE M. BROWN
Department of Physics and Astronomy
Northwestern University

LILLIAN HODDESON
Department of Physics
University of Illinois at Urbana-Champaign
and Fermilab

CAMBRIDGE UNIVERSITY PRESS

Cambridge

London   New York   New Rochelle

Melbourne   Sydney

Published by the Press Syndicate of the University of Cambridge
The Pitt Building, Trumpington Street, Cambridge CB2 1RP
32 East 57th Street, New York, NY 10022, USA
296 Beaconsfield Parade, Middle Park Melbourne 3206, Australia

First published 1983

Printed in the United States of America

Library of Congress Cataloging in Publication Data
Main entry under title:
The birth of particle physics.
"Based on the lectures and round-table
discussion of the International Symposium on
the History of Particle Physics, held at
Fermilab in May 1980" – Foreword
Includes index.
1. Particles (Nuclear physics) – Congresses.
2. Nuclear physics – History – Congresses.
I. Brown, Laurie M.   II. Hoddeson, Lillian.
III. International Symposium on the History
of Particle Physics (1980:Fermilab)
QC793.B57   539.7   82–1162
ISBN 0 521 24005 0   AACR2

Contents

Contributors

Carl D. Anderson
2915 Loraine Road
San Marino, California 91108

Herbert L. Anderson
Physics Division MS 434
Los Alamos Scientific Laboratory
P. O. Box 1663
Los Alamos, New Mexico 87545

Pierre V. Auger
12 Rue Emile Faguet
75014 Paris, France

Jeno M. Barnóthy*
833 Lincoln Street
Evanston, Illinois 60201

Madeline Barnóthy*
833 Lincoln Street
Evanston, Illinois 60201

Gilberto Bernardini
Scuola Normale Superiore
Istituto de Fisica
Piazza dei Cavaliere 7
I-56100 Pisa, Italy

Laurie M. Brown
Physics Department
Northwestern University
Evanston, Illinois 60201

Robert Chasson*
Department of Physics
University of Denver
Denver, Colorado 80208

Marcello Conversi
Istituto di Fisica "G. Marconi"
Universitá Degli Studi
Piazzale Delle Science 5
I-00100 Rome, Italy

Paul A. M. Dirac
Department of Physics
Florida State University
Tallahassee, Florida 32306

Satio Hayakawa
Physics Department
Nagoya University
Chikusa-ku
Nagoya, Japan 464

Lillian Hoddeson
Physics Department
University of Illinois
Urbana, Illinois 61801

Willis E. Lamb, Jr.
Department of Physics
University of Arizona
Tucson, Arizona 85721

*Audience participant.

Cesare Mansueto Giulio Lattes
Instituto de Fisica "Gleb Wataghin"
Univerdade Estadual de Campinas
Campinas, São Paulo, Brazil

Leon Lederman
Fermilab
P.O. Box 500
Batavia, Illinois 60510

Louis Leprince-Ringuet
86 Rue de Grenelle
Paris 7, France

Robert E. Marshak
202 Fincastle Drive
Blacksburg, Virginia 24060

M. G. K. Menon
Council of Scientific & Industrial
Research
Technology Bhaven
New Mehrauli Road
New Delhi 110029, India

Donald F. Moyer
2025 Sherman Avenue
Evanston, Illinois 60201

Yoichiro Nambu*
The Enrico Fermi Institute
University of Chicago
5630 S. Ellis Avenue
Chicago, Illinois 60637

H. Victor Neher
760 Calabasas Road
Watsonville, California 95076

Abraham Pais*
Department of Physics
Rockefeller University
1230 York Avenue
New York, New York 10021

Oreste Piccioni
Physics Department B-019
University of California, San Diego
La Jolla, California 92093

*Audience participant.

Bruno B. Rossi
Center for Space Research
Room 37-667
Massachusetts Institute of Technology
Cambridge, Massachusetts 02139

Silvan S. Schweber
Martin Fisher School of Physics
Brandeis University
Waltham, Massachusetts 02154

Julian Schwinger
Department of Physics
University of California, Los Angeles
Los Angeles, California 90024

Robert W. Seidel
History of Science and Technology
470 Stephens Hall
University of California
Berkeley, California 94720

Robert Serber
Department of Physics
P.O. Box 133
Columbia University
New York, New York 10027

Dudley Shapere
Department of Philosophy
University of Maryland
College Park, Maryland 20742

Dmitry Skobeltzyn
P. N. Lebedev Physical Institute
Academy of Science USSR
Leninsky Prospect, 53
II7924 GSP
Moscow B-333, USSR

Roger H. Stuewer
School of Physics and Astronomy
University of Minnesota
Minneapolis, Minnesota 55455

Takehiko Takabayasi
Department of Physics
Nagoya University
Chikusa-ku
Nagoya Japan, 464

Robert W. Thompson
5648 Dorchester
Chicago, Illinois 60637

Spencer R. Weart
Center for the History of Physics
American Institute of Physics
335 East 45th Street
New York, New York 10017

Charles Weiner
c/o STS Program
Room B-231
Massachusetts Institute of Technology
Cambridge, Massachusetts 02139

Victor F. Weisskopf
Department of Physics and Astronomy
Massachusetts Institute of Technology
Cambridge, Massachusetts 02139

Foreword

This book is based on the lectures and round-table discussions at the International Symposium on the History of Particle Physics, held at Fermilab in May 1980.

The organizers of the symposium, Laurie Brown and Lillian Hoddeson, argue that elementary particle physics evolved out of cosmic-ray and nuclear physics in the period 1930–50. In this same period, relativistic quantum field theory provided a theoretical structure that could be tested in the atom and extended into the subnuclear domain.

The idea was to explore these issues at a conference at which the participants in these events would reconstruct the happenings with their contemporaries, with following generations of particle physicists, and with historians of science. In planning this symposium, Hoddeson and Brown were guided by the experience and advice of Roger Stuewer, who had organized a successful symposium on the history of nuclear physics several years earlier.

The theoretical underpinnings were addressed by Paul Dirac, Victor Weisskopf, and Satio Hayakawa. Early cosmic-ray discoveries were described by Carl Anderson, Gilberto Bernardini, and Bruno Rossi. Quantum field theory was treated by Julian Schwinger, and the successful application to the atom was described by Willis Lamb, the field theorist who carried out the epochal atomic experiment. Robert Marshak and Robert Serber connected quantum field theory to the subnuclear phenomena observed in the 1940–50 cosmic-ray data and in the early postwar accelerator studies. These then were the principal speakers. Audience participation was very lively; some was sufficiently relevant to be included as short chapters, such as those of Oreste Piccioni and Robert Thompson.

It will be obvious to our readers that our speakers can only be representative of the heroes of the period under study. We are keenly aware that important contributors were not able to attend the symposium. The historians in our symposium called our attention to the omission of many important developments. In partial remedy, the editors have included a number of post-symposium papers. Dmitry Skobeltzyn, H. Victor Neher, Pierre V. Auger, Louis Leprince-Ringuet, Marcello Conversi, Takehiko Takabayasi, Cesare M. G. Lattes, and Julian Schwinger (on Sin-itiro Tomonaga) helped to fill some of the gaps in the symposium's program. We recognize that there are still many omissions; there is room for additional symposia and much need for detailed scholarly work in this seminal period in the history of particle physics.

For the physicists at Fermilab, who tend to be obsessed with the future, the symposium gave us an opportunity to turn away, however momentarily, and pause from the routine – to step away from our fascinations with quarks and gluons and the exotica of constituent physics, to look up from our scintillation counters, microprocessors, wire chambers, Čerenkov counters, and all that regalia – and to renew contact with our culture and listen to the giants on whose shoulders we try to stand. The giants who lectured here also wrote the books we studied and established the physics upon which we base our work. They brushed away the cobwebs that obscured the beautiful theory of quantum electrodynamics, and they *observed*. The style of their observations is our heritage.

In a more personal vein, two of the speakers were my own teachers. Willis Lamb, at Columbia University during the years 1946–51, taught me 80 percent of my graduate courses. In those times we students worked very hard, but so did our professors. In fact (you may not believe this), I remember everything Willis taught me. His chapter reminds us how much the atom taught us about the world.

The second of my teachers present was Gilberto Bernardini. As a visiting professor from Rome, he brought the students at Columbia an insight into the exciting world of cosmic-ray physics. I don't remember anything Gilberto taught me. No, that is not quite true. I do remember something he taught me, and it was an interesting thing: He taught me to be naïve. He taught me to marvel at simple things that are really not so simple. I remember once when we had finished making a counter and were looking at the pulses on an oscilloscope. Yes, in those days we had oscilloscopes! A man off the street looking at the oscilloscope would see green lines in a broken television tube. But the student of physics was much more sophisticated; when he saw these green lines pulsing up and down, he knew that it signified the passage of a particle, either an α particle or a μ meson. The passage of a particle through a counter happens today at Fermilab. We still look at oscilloscopes here, I'd like to reassure you, and we still see these pulses, and they signify the passage of particles through counters. This abstract happening is accepted very calmly. But Gilberto got hysterical when he saw these pulses. Of course, he is Italian. His excitement at the fact that you can interpret something so abstract as the passage of an ultramicroscopic particle and make that deduction from the green traces on the oscilloscope was a lesson which, generalized, is the essence of the subject with which we are concerned.

Leon M. Lederman

Editors' acknowledgments

It is a pleasure to thank all of those who participated in the symposium and contributed to the preparation of this volume. We sincerely regret that we are able to thank individually only a few of those who deserve credit here.

The institutions we want to thank are the following: Fermilab, which is supported by the U.S. Department of Energy and operated by the Universities Research Associates (URA), for generously hosting the symposium and for supplying many varieties of support ranging from office space and secretarial services to sustenance; the Sloan Foundation; the Division of Particles and Fields of the American Physical Society, for supplementary grants; and the Center for History of Physics of the American Institute of Physics, for support for tape recordings.

The committees we would like to thank include the Organizing Committee (Hans Bethe, Leon Lederman, Roger Stuewer, Spencer Weart, Robert R. Wilson, and ourselves), for help in selecting topics and speakers; Fermilab's History Committee (current members Richard Carrigan, chairman, Francis Cole, Thomas Collins, Lillian Hoddeson, Drasko Jovanovic, Lee Teng, Roger Thompson, Donald Young; past members Edwin Goldwasser, Richard Lundy, and Robert R. Wilson), for making the pivotal decision that Fermilab support a symposium on the history of particle physics and for continuing help with the program and arrangements; the Arrangements Committee (Betsy Anderson, Joanie Bjorken, Richard Carrigan, Helen Peterson, and May West), for masterminding the operation of the symposium; and the Exhibit Committee (Saundra Cox, Angela Gonzales, and Jose Poces), for producing an unusual and historically illuminating display of photographs and apparatus gathered from many individuals and institutions.

Many other committees contributed to the symposium in many ways that were not obvious, for example, in authorizing the travel support for individual symposium participants, and we are very grateful to them all.

Among the dozens of individuals whose participation in designing our undertaking was crucial, four stand out: Leon Lederman, May West, Helen Peterson, and Roger Stuewer. Had they not made their essential contributions, the symposium could not have taken place. Lederman, Fermilab's director, consistently provided enthusiastic advice and support, both moral and material. By recognizing and emphasizing publicly that the history of particle physics is part of the cultural heritage of our time, a heritage that deserves to be preserved and understood in detail, he set a precedent for scientific research leadership that has already, since our symposium, been followed at other institutions. May West, of Fermilab's library staff, generously provided necessary support services at every stage: initial invitations to speakers, preregistration and registration, handling telephone calls and symposium correspondence, processing transcripts and manuscripts. To Helen Peterson we are indebted for imaginative and effective supervision of countless essential arrangements: allocation of funds and staff; design of preregistration, registration, and symposium procedures; physical accommodations at the symposium; and the hosting of foreign guests; to name but a few. The seminal role of Roger Stuewer of the University of Minnesota included his support and advice during the early planning of the symposium, deriving from his conception and organization of the first similarly structured symposium on the history of nuclear physics at Minnesota in 1977, our model and guide.

Among the many other Fermilab employees who contributed to the symposium or to this volume, we want to thank especially the following: Richard Carrigan, for general support during the symposium in countless essential functions, including auditorium arrangements and the hosting of guests; Drasko Jovanovic, for a tour of Fermilab during the symposium, help with the exhibits, and financial assistance; Susan Grommes, for months of typing and other support services; Alfred Brenner and John Ingebretsen of the Fermilab computing department, for essential technical assistance in the preparation of this book; Joanie Bjorken, for an excellent spouse-and-friends program and a delightful May-wine garden party that all the participants at the symposium enjoyed; Chris Quigg and the Fermilab theory group for summer support for one of us (L. M. B.) while editing this volume; Roger Thompson, for audio assistance, miscellaneous support, and library facilities; Anne Burwell and Raeburn Wheeler, for computer services; Margaret Pearson and Fred Coleman, for public information services; Thomas Collins, for technical advice; Judy Ward, Jackie Coleman, and Ellen Carr Lederman, for miscellaneous general support; Robert Armstrong, for transportation and other essential site services; Herman White, for tours at the symposium; Angela Gonzales, for the symposium poster; Brad Cox and other contributors, for the exhibit booklet; Bud Stanley and Richard Skokan, for recording and other auditorium facilities; John Barry, Cynthia Sazama-Reay, and Bill Ross, for cafeteria arrangements; Ruth Ganchiff, for arranging the concert; Rick Fenner and the photography department at Fermilab; Eileen McWayne, for telexing; Sybil Krebs and her staff, for duplicating; Pam Naber and her staff, for housing arrangements.

Non-Fermilab staff we want to thank for their contributions include the following: speakers, symposium chairmen, historians, and other symposium participants; Betsy Anderson, for the program booklet, operation of auditorium microphones during the symposium, and telephone arrangements for Carl Anderson's talk; Jeanne Laberrigue, for communicating the papers of Pierre Auger and Louis Leprince-Ringuet; Robert Chasson, for securing the contribution from H. Victor Neher; Gordon Baym, for general support and technical assistance; Brigitte Brown, for general assistance and the operation of microphones during the symposium; Albert Wattenberg, for technical advice; Dana Wade, Judy Cohen, and Sharon Soper, for typing; Wenyuan Qian, for editorial assistance; Kathy Johnson, for photography; Mary K. Gaillard, for translation of the chapters by Pierre Auger and Louis Leprince-Ringuet; Michiji Konuma, Victoria Davis, and Riccardo Levi-Setti, for contributions to the exhibit; and Kyle Wallace, for many helpful conversations pertaining to the preparation of this volume.

Photographs of the symposium

At May wine buffet, left to right: Laurie Brown, Julian Schwinger, Lillian Hoddeson, Peter Galison, Satio Hayakawa, Bruno Rossi (credit: Ryuji Yamada).

Paul A. M. Dirac (credit: Fermilab Photography Department).

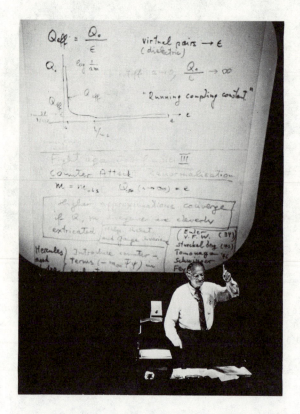

Victor Weisskopf delivering his talk at the symposium (credit: Fermilab Photography Department).

Left to right: Satio Hayakawa, Abraham Pais, Robert Marshak
(credit: Fermilab Photography Department).

Left to right: Gilberto Bernardini and his former graduate student
Leon Lederman (credit: Kathy Johnson).

Panel discussion, left to right: Herbert Anderson, Samuel Schweber, Victor Weisskopf, Paul Dirac, Gilberto Bernardini, Robert Seidel (credit: Fermilab Photography Department).

Panel discussion, left to right: Dudley Shapere, Charles Weiner, Robert Serber, M. G. K. Menon, Bruno Rossi, Satio Hayakawa (credit: Fermilab Photography Department).

Yoichiro Nambu and Takehiko Takabayasi (credit: Kathy Johnson).

Spencer Weart, Robert Wilson and M. G. K. Menon, conversing at symposium coffee break (credit: Fermilab Photography Department).

Willis Lamb conversing with participant at the symposium (credit: Fermilab Photography Department).

Part of symposium Planning Committee in the Milton G. White History of Accelerators Room at Fermilab, left to right: Richard Carrigan, Brigitte Brown, Laurie Brown, Donald Moyer, May West, Joan Bjorken, and Lillian Hoddeson (credit: Fermilab Photography Department).

Part I

Introduction

1 The birth of elementary particle physics: 1930–1950

LAURIE M. BROWN AND LILLIAN HODDESON

In the 1930s and 1940s, physicists significantly revised their views on the elementary constituents of matter, which during the 1920s had been widely assumed to be only the electron and the proton. This revision and the problems it posed gave birth to the field of modern elementary particle physics. The Fermilab symposium on which this volume is based was held to illuminate the study of this historical process through lectures by important participants and through discussions among physicists and historians. We believe that these chapters and the supplementary material that was collected after the symposium provide a useful examination of the origins of particle physics.

However, this volume is not a definitive or even comprehensive history of those origins. The lack of prior historical work in this area provided both a challenge and a rationale, but at the same time a handicap; during and after the symposium we became aware of important omissions. Some gaps were filled with postsymposium contributions, but others remain, as will be pointed out in this chapter. In addition, much of the material consisted of personal recollections, valuable to the historian as source material, but not balanced and critical historical accounts. People cannot be totally objective about the events in which they participate; we tend unconsciously to reinterpret history in terms of present-day values. Thus, expert witnesses may, in subtle ways, undercut the program of the historian to understand the past in its own terms.

Other problems stem from professional differences between physicists and historians of science. Physicists must keep their eyes on the cutting edge of current research; they aim to advance the frontier and, if possible, to do so before anyone else. This requires formulating

3

general concepts applicable to a large class of phenomena – abstract models that exclude all but the most relevant attributes of real objects and processes. But historians of science are concerned with identifying and characterizing the sequence of earlier frontiers and with explaining, within the context of their period, how each frontier led to the next. To do this, the historian must examine, in each period, and in rich complexity, the particular circumstances that influenced a given development. Such differences of focus can interfere with communication between physicists and historians.[1] We were, indeed, in inviting physicists to deliver the major historical lectures at the symposium, asking them to step out of character and view particle physics from an alien point of view. It is remarkable that they not only accepted this almost impossible assignment but did extremely well at it – a success we attribute, in part, to the presence of historians in the audience.

The professional differences between the two groups did produce tensions at the symposium (notably in the two round-table discussions, Chapters 16 and 17) that were generally constructive. They generated debate and called forth efforts to be more precise and to document their statements. It is through this tension that this volume differs from other books of recollections.[2] In retrospect, we believe that we should have gone further in this direction. We suggest that in future symposia of this kind, historians, as well as physicists, be invited to present papers to the mixed group.

Besides relating episodes in the adventure of twentieth century physics and serving as a source of data and references, this volume is a collection of viewpoints that may suggest working hypotheses for future work on the history of particle physics. One such hypothesis, used as a guide in selecting symposium speakers, was that particle physics emerged out of the turbulent confluence of three initially distinct bodies of research: nuclear physics, cosmic-ray studies, and quantum field theory. By the mid-1930s, where these fields overlapped there was conflict and apparent paradox, partly resolved by the end of the 1940s (although the resolution posed new and urgent problems). Because an earlier symposium held at the University of Minnesota in Minneapolis-St. Paul in 1977 had already considered many of the relevant nuclear physics issues,[3] we concentrated on those origins of particle physics derived from cosmic-ray studies and quantum field theory. To help the reader find a suitable vantage point from which to assess the individual contributions, we shall attempt in this introductory chapter to sketch the broad landscape of particle physics in the 1930s and 1940s: The first section will characterize the three bodies of research out of which particle physics emerged during the 1930s; the second section will deal with a problem not adequately discussed at the symposium, the application of a suspect quantum electrodynamics (QED) to the analysis of cosmic-ray events; the third section will review important experiments and theories that implied the existence of new particles; the remaining three sections will place the results in a broader framework.

The sources of particle physics

By 1930, relativity and quantum mechanics were established, and it was clear that matter is made of nuclear atoms. But the excitement of the new physics was far from over. Indeed, the next half century was characterized by startling experimental and theoretical discoveries and an assortment of puzzles wherever one looked – in particular, in the three currents that flowed together to make particle physics.

The atomic nucleus[4]

The consensus of scientists in 1930 was that matter was made of two elementary particles, electrons and protons, negative and positive electricity.[5] The neutral atom of mass number A and charge number Z was believed to contain A protons and $A - Z$ electrons in its nucleus and Z electrons in its outer shells. All forces relevant on the atomic scale were believed to be electromagnetic.

Some hoped that the tiny nucleus, containing most of the mass but only a million-millionth part of the volume of the atom, could itself be treated as a quantum mechanical system.[6] Other voices argued that just as a new dynamic theory (quantum mechanics) was needed to pass from the "human scale" to the atom, so a new generalization of the physical laws might be needed at the microcosmic distances in the nucleus.[7] Indeed, since early in the century it had been predicted (notably by Hendrik Antoon Lorentz) that the electromagnetic theory of Maxwell would fail at distances smaller than the "classical electron radius" ($r_0 = e^2/mc^2 \sim 2.8 \times 10^{-13}$ cm, where e and m are the charge and mass of the electron and c is the speed of light), which happens to be approximately the nuclear radius.

The difficulties encountered in treating the nucleus as a quantum mechanical system of protons and electrons were these: The nucleus was supposed to contain A protons and $A - Z$ electrons, but when the latter number is odd, as for ^{6}Li and ^{14}N, the spin and statistics (i.e., whether Bose or Fermi type) are incorrect. Also, unpaired electron spins in the nucleus would imply hyperfine splitting of atomic spectral lines on a scale about 1,000-fold larger than is observed. Another nuclear puzzle was that it was impossible (in the relativistic quantum theory of the electron) to confine the light electron within the small nucleus. Finally, there was the continuous spectrum of β-decay electron energies, which called into question even the conservation of energy.

Radical suggestions were seriously considered for modifying the mechanics and/or the electrodynamics and even the conservation laws. But the resolution was to hinge on the existence of new particles: the neutron, discovered by James Chadwick in 1932, and the neutrino, proposed in 1930 by Wolfgang Pauli and incorporated in a theory of β decay by Enrico Fermi in 1934.[8] These two neutral particles permitted the banishment of electrons from nuclear models. The positron, discovered in the cosmic rays in 1932 by Carl David Anderson, was used to complete the picture of nuclear β decay when Irène Curie and Frédéric Joliot produced artificially radioactive light elements that decayed by positron emission.

The cosmic rays

Cosmic rays were discovered (Table 1.1) as a result of investigations, after 1900, of fine-weather '"atmospheric electricity" (i.e., in the absence of an electrical thunderstorm). At about this time it was realized that the nonzero electrical conductivity of the atmosphere implied the presence of an ionizing agent. Indeed, minute amounts of radioactive substances, first discovered in 1896 by Henri Becquerel, were found to be present in the earth, air, and water (and in the measuring instruments themselves), and these did provide ionizing rays. However, after all known sources of ionization had been accounted for, there remained a "residual" conductivity, even in closed vessels that were heavily shielded. This phenomenon implied the existence of a "penetrating radiation" of unknown origin.[9]

The altitude dependence of the atmospheric conductivity was investigated by means of balloon flights, mainly in central Europe, and notably by Victor F. Hess in Austria. The manned balloons carried sealed electrometers whose rates of discharge first decreased with altitude, but then (above 2 km) began a marked increase. This pattern of ionization suggested, if it did not prove, the existence of an extrater-

Table 1.1 *Sequence of development of cosmic-ray physics*[a]

I. *Prehistory (to 1911, especially from 1900)*

"Atmospheric electricity" during calm weather; conductivity of air measured by electrometers; connection with radioactivity of earth and atmosphere; interest was also geophysical and meteorological.

II. *Discovery (1911-14) and exploration (1922-30)*

Balloons carrying observers with electrometers measured the altitude dependence of ionization and showed that there is an ionizing radiation that comes from above; these measurements began in 1909 and continued (at intervals) to about 1930, in the atmosphere, under water, earth, etc.; the primaries were assumed to be high-energy photons from outer space; search for diurnal and annual intensity variations; study of energy inhomogeneity.

III. *Particle physics, early (1930 – 47)*

Direct observation of the primaries was not yet possible, but "latitude effect" showed that they were charged particles; secondary charged-particle trajectories were observed with cloud chambers and counter telescope arrays, and momentum was measured by curvature of trajectory in a magnetic field; discovery of positron and of pair production; soft and penetrating components; radiation processes and electromagnetic cascades; meson theory of nuclear forces; discovery of mesotron (present-day muon); properties of the muon, including mass, lifetime, and penetrability; two-meson theory and the meson "paradox."

IV. *Particle physics, later (1947-53)*

Observation of particle tracks in photographic emulsion; discovery of pion and pion-muon-electron decay chain; nuclear capture of negative pions; observation of cosmic-ray primary protons and fast nuclei; extensive air showers; discovery of the strange particles; the strangeness quantum number.

V. *Astrophysics (1954 and later)*

Even now the highest-energy particles are in cosmic rays, but such particles are rare; studies made with rockets and earth satellites; primary energy spectrum, isotopic composition; x-ray and γ-ray astronomy; galactic and extragalactic magnetic fields.

[a]In successive periods, at least one change occurred that was so significant that it required a totally new interpretation of the previous observations and theories.

restrial source for the penetrating radiation, so that eventually (by the late 1920s) one spoke of the *cosmic rays* (Table 1.1). Until 1930, their specific ionization (ions per cubic centimeter per second) was the only property systematically observed. The vertical variation of the rate of discharge of an electrometer was interpreted as measuring an absorption length of the rays in air or water and showed them to have increasing penetrability (hardening) with depth.[10] At that time, the main interest in the rays was geophysical and astrophysical, although Robert A. Millikan proposed a theory of cosmic-ray origins that involved a new physical mechanism.[11]

The focus changed at the end of the 1920s, when two methods, the cloud chamber with magnetic field and coincidence counting, were used to study the individual behavior of charged particles produced by collisions of the primaries with air molecules. Both were adapted from techniques used to study x rays and radioactivity. They were flexible, permitting a variety of experiments to be performed, and they could be combined. Their descendants are the principal tools used today to study the interactions of elementary particles, whether the source be cosmic rays or accelerators. The pioneers in this enterprise were Walther Bothe and Werner Kolhörster in Berlin and Dmitry Skobeltzyn in Leningrad.

In 1924, Bothe and Hans Geiger had measured the degree of simultaneity of emission of the recoil electron and the scattered x-ray photon in Compton scattering, in order to test a statistical theory proposed by N. Bohr, H. A. Kramers, and J. C. Slater.[12] The Bothe-Geiger apparatus used two Geiger point counters, each connected to a separate string electrometer. Counts were registered by deflection of the string, and both electrometers were photographed by a camera with motor-driven film transport. By the time they finished the experiment, Bothe and Geiger had achieved a time resolution of 10^{-4} sec.

Kolhörster, a colleague of Bothe at the Physikalisch-Technische Reichsanstalt in Charlottenburg, outside Berlin, and an experienced cosmic-ray worker, pointed out in 1928 that by aligning two point counters in a vertical array, Bothe's coincidence technique could be used to make a γ-ray telescope for cosmic rays.[13] A few days later, Bothe and Kolhörster reported similar observations with the far more efficient Geiger-Müller tube counter. By mid-1929 they established that a 4.1-cm gold block placed between the counters reduced the coincidence rate by only 24%, and they concluded from this that the primary rays had "corpuscular nature."[14] Until then they had been thought to be high-energy photons and had been called (e.g., by Hess) "ultra γ rays."

To young Bruno Rossi, at the physics laboratory of the University of

Florence in Arcetri, Italy, these results reported from Berlin "came like a flash of light revealing the existence of an unsuspected world," and he began immediately the exploration of this world.[15] He soon found a way to improve the technique using a vacuum tube circuit to detect the coincident discharges of the tube counters, achieving greater flexibility and time resolution, and by using three out-of-line counters, he discovered that there was a great abundance of secondary radiation, later identified as "cascade showers."

Meanwhile, in Leningrad, Skobeltzyn, who had been studying radio-active γ radiation, began using the Wilson cloud chamber to observe the trajectories of cosmic-ray particles in a magnetic field, where a charged particle's track is curved, with a radius of curvature directly proportional to the particle momentum and inversely proportional to the magnetic field. He also noted that tracks appeared to be associated with each other, to a degree difficult to account for by the scattering processes known at that time.[16] Skobeltzyn's work was the first visual method for observing processes of particles with energies higher than those available from radioactive sources.

Skobeltzyn's counterpart in California was Carl David Anderson, who had been using a cloud chamber to study photoelectrons produced by x rays, and who wanted to move on to Compton collisions of nuclear γ rays; instead, at the urging of his boss, Millikan, he began tooling up (in 1930-1) a cloud chamber and a strong magnetic field to observe cosmic-ray interactions. Anderson was to discover two new particles in cosmic rays: the positron and the muon.[17]

The other major step forward was the invention and use in 1932 of the counter-controlled cloud chamber by P. M. S. Blackett and G. P. S. Occhialini.[18] In such a chamber, both the expansion and camera are activated by an electronic pulse from a counter array that selects a class of events, so that, in effect, the incident particle "takes its own picture." Soon after Anderson had discovered what he referred to as "easily deflectable positives," Blackett and Occhialini used their new instrument to observe electron pair production (suggested by P. A. M. Dirac's electron theory) and cascade showers. In the words of Hanson, "they discovered that the 'Anderson particle' and the 'Dirac particle' were the same particle."[19] By 1930, therefore, the technical framework had been established for the spectacular cosmic-ray discoveries of the next two decades (including new particles) made using counter and cloud chamber techniques.

Relativistic spinning electrons and quantum fields

The prediction of the positron, which began not as a triumph but as an embarrassment, produced a profound alteration in our notion of "particle." In his first article on the new electron theory, Dirac stated that he looked for "some incompleteness" in the Heisenberg and Schrödinger treatments of the point electron "such that when removed, the whole of the duplexity phenomena [would] follow without arbitrary assumptions."[20] He was referring to the spinning-electron model of Samuel Goudsmit and George Uhlenbeck, and Pauli's use of that model to account for details of atomic spectra by a doubling of allowed states.[21] The incompleteness was found to be the lack of relativistic invariance, which he corrected by proposing a new wave equation consistent with his transformation theory.[22] An important result was that the electron was forced to have spin of one-half and magnetic moment of one Bohr magneton. Furthermore, in the new theory, the state of the electron was described by four wave functions, rather than one as in Schrödinger's theory, or two as in Pauli's nonrelativistic theory of the spinning electron. There were thus two new degrees of freedom related to the existence of states of both signs of the energy.

The embarrassment came from Einstein's theory of relativity, in which, according to the relation $E = mc^2$, negative energy always means negative mass. This was a meaningless concept, as C. G. Darwin pointed out.[23] Dirac could not evade this issue. Much later he said:

> The problem of the negative-energy states puzzled me for quite a while. The main method of attack to begin with was to try to find some way of avoiding the transitions to negative-energy states, but then I approached the question from a different point of view. I was reconciled to the fact that the negative-energy states could not be excluded from the mathematical theory, and so I thought, let us try and find a physical explanation for them.[24]

He was thus led to consider a world in which all, or nearly all, of the infinite number of negative energy states were already occupied by an electron – "filled," if we take account of Pauli's exclusion principle, which prevents more than a single electron from occupying a given state. Transitions from a positive to a negative energy state would practically not occur. There remained the problem of the physical interpretation of the occasional unoccupied state, or "hole," in the sea of negative energy electrons. After a period of doubt, the hole was recognized to be the positive electron; and it became fully legitimate in 1932 when it was experimentally observed.[25]

The theory of holes was based on "a very close analogy provided by the chemical theory of valency," in which an unfilled hole in an otherwise closed atomic shell behaves like a positively charged valence electron.[24] In the case of the negative-energy states, there is no positive background charge to neutralize the vacuum; also, one must deal not with a finite number of electrons but with an infinite number of electrons. The last entails an infinite number of quantum mechanical degrees of freedom and thus, in turn, quantum fields.

Relativistic electron theory and the quantum theory of fields were both on the agenda of theoretical physics after the invention of quantum mechanics by Heisenberg and Schrödinger in 1925 and 1926. Both emerged from the fertile brain of Paul Dirac, once he stopped being "strongly fascinated by Bohr orbits," as he says in Chapter 2, and formulated his general quantum Hamiltonian dynamics. In his pioneering work of February 1927 on quantum electrodynamics (QED), Dirac proposed a solution to the problem of the wave-particle duality, which had puzzled physicists since Einstein hypothesized the light quantum in 1905.[26] At the end of this paper, Dirac summarized its contents as follows:

> The problem is treated of an assembly of similar systems satisfying the Einstein-Bose statistical mechanics, which interact with another different system, a Hamiltonian function being obtained to describe the motion. The theory is applied to the interaction of an assembly of light-quanta with an ordinary atom, and it is shown that it gives Einstein's laws for the emission and absorption of radiation.
>
> The interaction of an atom with electromagnetic waves is then considered, and it is shown that if one takes the energies and phases of the waves to be q-numbers satisfying the proper quantum conditions instead of c-numbers, the Hamiltonian function takes the same form as in the light-quantum treatment. The theory leads to the correct expressions for Einstein's A's and B's.

(The As and Bs are light-quantum emission and absorption probability amplitudes.) From this we can see that Dirac treated the electromagnetic field as a Bose-Einstein gas of light quanta. The analogous treatment for a Fermi-Dirac gas, applicable to electrons, was given the following year by Pascual Jordan and Eugene Wigner.[27] The Jordan-Wigner type of quantization, designed to forbid more than one electron to occupy a given state, was just what Dirac needed to formulate his theory of holes.

The growing theoretical realization that the holes could not be protons, that they must have electronic mass, that they would constitute a new type of matter (antimatter), and that they were realizable in nature is discussed by Dirac (Chapter 2) and by Weisskopf (Chapter 3).[28] Here we want to emphasize how the electron theory reinforced the need, already present, for a quantum theory of fields.

In his 1927 papers on the quantum theory of the electromagnetic field, Dirac quantized only the radiation part of the field, consisting of transverse waves. The Coulomb interaction was considered a part of the energy of the "matter" system (i.e., the charged particles). This separation is convenient and often is a calculational necessity; however, as Gregor Wentzel has remarked, it "not only appears contrary to the spirit of Maxwell's theory, but also raises questions from the view-point of relativity theory . . . the splitting is not [relativistically] invariant."[29]

Thus, in 1929, Heisenberg and Pauli took up a task whose completion would require the best theoretical efforts of the next two decades:

> to connect, in a contradiction-free manner, mechanical and electrodynamic quantities, electro-magneto-static interaction, on the one hand, and radiation-induced interactions on the other, and to treat them from a unified viewpoint. Especially [to take] into account in a correct manner the finite propagation velocity of electromagnetic forces.[30]

In the course of this work they discovered that the self-mass of the point electron, just as in the classical theory, was infinite. The sequel to this discovery and those to the other "infinities" of QED are described by Weisskopf in Chapter 3. The resolutions of these problems are also discussed by Schwinger in Chapters 21 and 22 (see also Table 1.2). Although it was not until the postwar period that a more self-consistent QED was achieved, nevertheless, as we shall show in the next section, the admittedly imperfect QED could still be fashioned into an effective tool for analyzing the high-energy cosmic rays.

Table 1.2. *Sequence of development of quantum field theory*

I. *Prehistory*

Classical (nineteenth century): electromagnetism (Faraday, Maxwell, Hertz, Lorentz).

Quantum (1900-27): blackbody radiation (Planck, 1900); photon hypothesis (Einstein, 1905); stationary states of atom (Bohr, 1913); atomic emission and absorption coefficients (Einstein, 1916); Bose and Fermi statistics (1924); electron waves (de Broglie, 1924); exclusion principle and spin (Pauli, Goudsmit, and Uhlenbeck, 1925); quantum mechanics of atoms and molecules (Heisenberg, Schrödinger, Dirac, Born, Jordan, 1925-6); general transformation theory (Dirac, 1927).

II. *Birth and early development (1927-9)*

Quantum electrodynamics (QED) (Dirac, 1927); relativistic electron theory (Dirac, 1928); relativistic QED (Heisenberg and Pauli, 1929); second quantization (Jordan and Klein, 1927; Jordan and Wigner, 1928); theory of holes (Dirac, 1929).

III. *Developments, difficulties, and doubts (1929-34)*

Applications of QED and Dirac theory (Klein and Nishina, 1929; Oppenheimer et al.; Bethe and Heitler, 1934); experimental tests (Meitner-Hupfeld, Tarrant, Gray, Chao, 1930); specter of infinite energy shifts (Oppenheimer, Waller, 1930); specter of infinite vacuum polarization (Dirac, 1932).

IV. *New fields (1934-46)*

Meson theory of nuclear forces (Yukawa, 1935); scalar field theory (Pauli and Weisskopf, 1934); beta decay theory (Fermi, 1934); relativistic spin-one theory (Proca, 1936); developments of meson theory (Fröhlich, Heitler, Kemmer, Yukawa, Sakata, Taketani, Kobayasi, 1938); "infrared" radiation (Bloch and Nordsieck, 1937); S matrix (Wheeler, 1937; Heisenberg, 1943).

V. *Renormalization (1947 and later)*

Lamb shift (Lamb and Retherford, 1947); electron magnetic moment (Foley and Kusch, 1948); calculation of Lamb shift (Bethe, 1947); renormalized relativistic QED (Tomonaga, Schwinger, Feynman, Dyson, 1948-9).

Quantum electrodynamics and the cosmic rays

Several chapters in this volume discuss the doubts that had arisen by the early 1930s about the validity of QED. Weisskopf (Chapter 3) describes the dramatic struggles against internal contradictions (the "infinities"); in his view, it is inner logic or perhaps a heuristic that determines the outcome, rather than confrontation with experiment. Anderson (Chapter 7) mentions the early doubts about relativistic quantum theory that arose from the scattering of γ rays at moderate energies. Serber (Chapter 12) relates Oppenheimer's pessimism about the validity of QED, based on the same experiments.

What was missing, however, was the view (inescapable for those of us who have heard Hans Bethe lecture on the subject or have read Walter Heitler's book[31]) that the "bad" high-energy behavior of QED had to be met and tamed, or sidestepped if necessary, because cosmic-ray workers in the 1930s were already observing particles of an energy for which QED might have been failing. Without being able to do justice to this issue in the available space, a few remarks may be helpful.

The disturbing experiments at moderate energies, somewhat larger than 1 MeV, corresponding to twice the rest energy of the electron, showed a much greater scattering of high-energy γ rays than was predicted by the Compton-effect calculation based on Dirac's relativistic electron theory.[32] Anderson relates how C. Y. Chao's absorption measurements with an electroscope at Caltech encouraged him to explore the same phenomenon in a more detailed way with the cloud chamber. The excess absorption was found, by 1933, to be due to electron-positron pair production, and the excess "scattering" was due to photons produced by pair annihilation. This resolved the "doubts at 2 mc^2," which, however, moved then to 137 mc^2, as Serber says.

According to the ideas of relativity and quantum mechanics, a particular energy implies a characteristic length.[33] There is considerable evidence, not recalled here, to show that a breakdown in normal physics (classical or quantum, electrodynamical or mechanical, relativistic or not – anything considered normal in the first three decades of this century) was anticipated within a distance comparable to the nuclear radius, a length that happens to agree with the so-called classical electron radius.[34] This particular concern could be put to rest only after the relatively successful nuclear theory of Heisenberg based on the neutron (1932), the establishment of Dirac's hole theory through the discovery of pair production and annihilation (1933), and the explanation of nuclear β decay by Pauli's neutrino and Fermi's weak interaction theory (1934).

The doubts about the validity of QED at energies of the order of 137 mc^2 are indirectly corroborated by Anderson (Chapter 7), who says that in 1934, among themselves, the Caltech group spoke of " 'green' electrons and 'red' electrons – the green electrons being the penetrating type, and the red the absorbable type." But the green electrons did not behave exactly like electrons. Although the ionization energy loss formulas for very fast charged particles were considered to be accurate, there seemed to be a problem with the radiation formulas, even in 1934.[35] Referring to Anderson's analysis of his cloud chamber photographs, Bethe and Heitler said that *"the theoretical energy loss by radiation is far too large to be in any way reconcilable with the experiments of Anderson."* For the particles of energy 300 MeV (the assumed green electrons), Anderson found an energy loss per centimeter of lead of 35 MeV, whereas Bethe and Heitler concluded that "it seems impossible that the theoretical energy loss can be smaller than about 150 million volts per centimetre lead for Anderson's electrons."[36]

Instead of suggesting that these strangely behaving electrons might be some other particle, Bethe and Heitler proposed a possible "explanation" that reveals the spirit of the time:

> This can perhaps be understood for electrons of so high an energy. The de Broglie wave-length of an electron having an energy greater than $137 \, mc^2$ is smaller than the classical radius of the electron, $r_0 = e^2/mc^2$. One should not expect that ordinary quantum mechanics which treats the electron as a point-charge could hold under these conditions. It is very interesting that the energy loss of the fast electrons really proves this view and thus provides the *first instance in which quantum mechanics apparently breaks down for a phenomenon outside the nucleus.* We believe that the *radiation of fast electrons will be one of the most direct tests for any quantum-electrodynamics to be constructed.*[36]

The problem was not with QED but with the assumption that Anderson's penetrating high-energy "green" electrons were electrons. They were, in fact, mesotrons (now called muons) about 200 times as massive. Three years were to pass before anyone had the courage (or the faith in QED) to ascribe the observations to new particles.

Although it was tempting to explain away discrepancies between observed high energy phenomena and theoretical expectations by appealing to a breakdown of QED at small distances, at a "fundamental length," or at the corresponding large momentum, this became impossible by 1937; by that time, through a complex series of steps, QED showed itself to be not only useful after all, in spite of its menacing "infinities," but also the indispensable means for understanding the nature of the cosmic rays.[37] A major step in this process was solution of the problem of the energy loss of fast charged particles by ionization, treated earlier by Bohr (1913) and Fermi (1924) and now calculated by quantum mechanics and extended to relativistic energies.[35] Although the electrodynamics of such energetic particles were questioned, E. J. Williams showed in 1933 that the important momentum transfers involved were small and that in a suitably chosen reference frame the collisions were gentle ones, not involving high energies or small distances.[38]

In another step, taken in 1934, Bethe and Heitler calculated the relativistic formulas for bremsstrahlung (x-ray production) and electron pair creation.[36] Although, as noted earlier, they found significant disagreement with Anderson's results (assuming that he was looking at electrons), Williams and C. F. von Weizsäcker showed that no disagreement with theory was to be expected, *even if QED were to break down at 137* mc^2. Again, the argument was based on looking at the collisions in a suitable rest frame.[39] As Williams said in his 1935 article, "We find that the quantum mechanics which enter into the existing treatments really concerns energies of the order of mc^2 however big the energy of the electron or photon."[40] (Nevertheless, as Serber notes, Oppenheimer "struggled to maintain his lack of faith." Arguing that "the origin of the critique of the theoretical formulae lies in classical electron theory," he asserted that "the application of electron theory becomes dubious" whenever appreciable high frequency components are present, "in distinction to v. Weizsäcker and Williams.")[41]

By 1937, the usefulness of QED had also been demonstrated by the explanation it provided for the behavior of the "soft component" of the cosmic rays, the cascade showers. An energetic photon collides with an atom of air or other material to produce an electron-positron pair. The charged particles produce photons by bremsstrahlung and by the annihilation processes; the process cascades until the energy is spread among a number of lower-energy charged particles, which then lose their energy by ionization or other collision processes. Many physicists contributed to the solution of this problem; the first successes were due to Homi J. Bhabha and Heitler and to J. F. Carlson and Oppenheimer.[42]

However, the infinities of QED remained; to obtain useful results, they had to be ignored or thought of as corrections that would be "small," were they calculable in finite terms.

Particles envisioned and particles seen

The particle discoveries of the early 1930s (if we can call the neutrino proposal a "discovery") permitted the banishing of electrons from the nucleus. On the heels of the neutron discovery, Heisenberg made a model of the nucleus as a nonrelativistic quantum mechanical system of neutrons and protons in which the neutron was to some extent treated as an elementary particle, the neutral counterpart of the proton.[43] However, within this model, Heisenberg tried to incorporate a neutron pictured as a tightly bound compound of proton and electron, in which the electron loses most of its properties (notably its spin, magnetic moment, and fermion character). The dominant nuclear force was to consist of the exchange of this abused electron.

After Enrico Fermi's successful β-decay theory gave the neutrino a more legitimate status than it had previously enjoyed, attempts were made (although not by Fermi) to incorporate electron-neutrino pair exchange into the Heisenberg nuclear picture, the so-called Fermi-field model.[44] However, it was shown to be impossible to fit simultaneously the range and strength of nuclear forces together with nuclear β decay. Hideki Yukawa, in Japan, made a bold imaginative stroke in an attempt to resolve this conflict, introducing a new theory of nuclear forces that required the existence of a new type of particle, a fundamental massive boson.[45] The particle was to carry the electronic unit charge (either positive or negative), and its exchange was to be the agent of Heisenberg's charge-exchange nuclear force. Its mass was determined, from the range of nuclear forces, to be about 200 electron masses. Furthermore, Yukawa's meson (as it later became known) was to be capable of decaying into an electron and neutrino, Yukawa's proposed mechanism for nuclear β decay. Finally, it was predicted to be a part of the cosmic-ray flux.

The discovery in the cosmic rays in 1937, by Anderson and others, of charged particles of both signs, about 200 times as massive as the electron, was greeted by some as a fulfillment of Yukawa's prediction.[46] Some properties of the particles (mass, charge, lifetime) were determined before or during World War II; others (e.g., spin, parity, interactions) were not unambiguously determined until the large accelerators came into use at the turn of the 1950s. The fact that such properties as were known, other than the charge, did not provide a satisfactory match between the (observed) cosmic-ray meson and Yukawa's (postulated) nuclear force meson stimulated the flowering of new field theories that went beyond QED.

Because these new field theories had even worse divergence difficulties than QED, and because their strong interactions made perturbation methods far more questionable, there again arose practical, as well as esthetic, demands for curing or circumventing "the infinities" of field theory. The theoretical struggle was double-pronged: one effort, described here mainly by Hayakawa (Chapter 4) and Marshak (Chapter 23), was to find a version of meson theory that agreed with the cosmic-ray meson's behavior; another, discussed here by Serber (Chapter 12) and Schwinger (Chapter 21), was to find a meson theory to fit the nuclear forces, whose complications came to be better known.[47] An important success of the second approach was the symmetric meson theory of nuclear forces, by Nicholas Kemmer, which established the utility of the concept of isospin and also called for the existence of a charged triplet of positive, negative, and neutral mesons;[48] the neutral meson, whose two-photon decay initiates the majority of cascade showers in the cosmic rays, was not observed until it was artificially produced in 1950.[49]

Numerous other important field theoretical developments in the 1930s included the creation of field theories of particles of spin zero (1934), the Pauli-Weisskopf theory (which Pauli liked to call anti-Dirac), and the theory of massive particles of spin one (QED is a theory of *massless* spin one particles) by the Romanian physicist Alexandru Proca (1936).[50] The two-meson theory, proposed in Japan during World War II, about 4 years before the observation of the pion in Bristol, England (and proposed in the United States just before the actual discovery), is discussed by Hayakawa (Chapter 4) and Marshak (Chapter 23).

The great developments leading to the formulation of renormalized QED by Richard P. Feynman, Julian Schwinger, Sin-itiro Tomonaga, and Freeman Dyson are described by Weisskopf (Chapter 3), Lamb (Chapter 20), and Schwinger (Chapters 21 and 22), and the important experimental work that helped to stimulate it (as well as Bethe's first successful calculation of a higher order effect) is described by Lamb (Chapter 20).[51] The end of our period was characterized by a confidence that QED was the model for a correct theory of the elementary particles.

We now summarize the outstanding early experimental evidence concerning the new particles. Anderson observed the positron in 1932; whether or not positrons were "seen" in cloud chamber pictures before they were "recognized" as such by Anderson is debatable.[52] The existence of showers, containing approximately equal numbers of electrons and positrons arising from the pair production process, was demonstrated by Blackett and Occhialini soon after Anderson's discovery;[53] these showers form the bulk of the soft (absorbable) component of cosmic rays, responsible for much of the ionization that the earliest cosmic-ray workers had measured with their electroscopes.

The cosmic-ray meson (muon), so much discussed at this symposium, is the main part of the hard (penetrating) component of the sea-level cosmic rays. This component was seen as early as 1929 in the first absorption measurements made on individual cosmic-ray particles and was perhaps suggested by even earlier measurement.[54] But only in 1937 were these cosmic-ray mesons claimed, by Neddermeyer and Anderson, to be new charged particles (i.e., neither electrons nor protons) on the basis of their ability to penetrate a 1-cm thickness of platinum.[55] While similar observations were being made elsewhere (e.g., in England and France), the preferred interpretation was that there was a breakdown of QED at high energies. That was the view of Blackett, who called the particles electrons and considered that a modification of the radiation formulas was in order.[56] Two French cloud chamber groups emphasized that there were "two species of corpuscular rays" (like Anderson's red and green electrons) differing in their penetrating power; however, they did not insist on any new particles.[57] Two observations of mesons stopping in the gas of a cloud chamber permitted a determination of their masses sufficient to show them to be roughly 200 times the electron mass (or about one-tenth of the proton mass).[58]

The next step was the false identification of the muon with Yukawa's nuclear meson, posing this problem: If the cosmic-ray meson were Yukawa's strongly interacting particle, why did it not seem to interact at all? The remaining story of the muon, the determination of its mass, lifetime, and interaction properties, and the growing sense of bewilderment and paradox in the confrontation of experiment with the supposed theory, is developed here by Bernardini (Chapter 8), Rossi (Chapter 11), Piccioni (Chapter 13), and Conversi (Chapter 14).

The grand finale was the revelation by the Bristol group, by means of a new nuclear photographic emulsion technique, of the pion, Yukawa's nuclear meson, and its decay into the muon, the cosmic-ray meson.[59] The remarkable spirit and great achievements of the group at Bristol, just after World War II, are dealt with by Lattes in Chapter 19.

While one outstanding puzzle was being solved, new ones were being born: The first cloud chamber photographs of V particles, later to be known as examples of the *strange particles*, were being taken by G. D. Rochester and C. C. Butler at the University of Manchester, England.[60] This subject is addressed by Thompson (Chapter 15) and also is mentioned by Marshak (Chapter 23). Cloud chamber and emulsion techniques were to reveal many new phenomena of elementary particles during the 1950s, first in the cosmic rays and then at accelerators, until the bubble chamber became the dominant visual technique. The theta-tau puzzle, involving the decay of the strange K meson, caused the first questioning of parity conservation. The muon puzzle, discussed by Bernardini in Chapter 8, arose out of the solution of the π-μ paradox. The observation of the complete decay chain (pion $\rightarrow$ muon $\rightarrow$ electron), together with the long muon lifetime, strongly suggested that the muon was a heavy version of the electron, in modern terms, a second-generation lepton (Chapter 8). (The problem of the muon, strangeness, charm, etc., is known today as the puzzle of the "generations" of quarks and leptons.)

Unification and diversification

Many physicists today believe that we are approaching a new synthesis in our view of matter, in which the world will be seen as made up of a few types of elementary particles, interacting by means of a small number of forces, both particles and fields being aspects of a few or perhaps even a single quantum field.[61] An important reason for this confidence is the apparent success of the theory of the unified electroweak field, which was partly confirmed in 1973 by the discovery of neutral weak currents predicted by that theory.[62] The 1979 Nobel lectures in physics deal with this subject, but also with speculative theories of more advanced type, having names such as "electronuclear grand unification" and "extended supergravity."[63]

The mood of the three lectures is one of barely qualified optimism. Glashow, for example, while cautioning against the adoption of a "premature orthodoxy," favorably contrasts the present with 1956, when he began theoretical physics and "the study of elementary particles was like a patchwork quilt." He continues:

Things have changed. Today we have what has been called a "standard theory" of elementary particle physics in which strong, weak, and electromagnetic interactions all arise from a local symmetry principle. It is, in a sense, a complete and apparently correct theory, offering a qualitative description of all particle phenomena and precise quantitative predictions in many instances. There are no experimental data that contradict the theory. In principle, if not yet in practice, all experimental data can be expressed in terms of a small number of "fundamental" masses and coupling constants. The theory we now have is an integral work of art: the patchwork quilt has become a tapestry.[63]

These remarks are reminiscent of other far-reaching syntheses: not only the "mechanical philosophy" of the eighteenth century and the "electromagnetic synthesis" at the end of the nineteenth century but also physics as it appeared about 50 years ago, at the beginning of the period dealt with in this volume. Then it was believed that there were only two fundamental material particles (electron and proton), only two fundamental forces (gravitation and electromagnetism), and that the fundamental laws were known (relativity and quantum mechanics). Accordingly, as Hawking reports, shortly after Paul Dirac published his relativistic wave equation for the electron, Max Born said that "physics, as we know it, will be over in six months."[64]

Although the positron discovery of August 1932 was a validation of Dirac's theory, that particle (and the neutron, neutrino, and meson) totally destroyed the synthesis that appeared to be at hand in 1930. As Millikan said: "Prior to the night of 2 August, 1932, the fundamental building-stones of the physical world had been universally supposed to be simply protons and negative-electrons."[65] Progress in the 1930s, and the next few decades, would lie not in unification of forces and a reduction in the number of elements but rather in diversification – in the discovery of new particles, in the enlargement of the particle concept, and in the recognition of new nuclear forces, both strong and weak. During the 1930s and 1940s there were discovered, besides the first antiparticle (the positron), the second baryon (the neutron), the second lepton (the muon), a neutral massless lepton (the neutrino, although actually first detected in 1953), the first massive field quanta, both charged and neutral (the pions), and the strange particles. By 1950, the modern idea of families of particles and the distinction between hadrons and leptons had already emerged. As Marshak suggests in Chapter 23, the idea of the universal weak interaction was also in the air. For hadrons there was the beginnning of what Weisskopf called the "third spectroscopy" (i.e., after those of atoms and nuclei), although all three cases involve not only spectroscopy but also structures. The path toward possible unification, which looked attainable for a few years after resolution of the π-μ paradox, thus led through a minefield of the most diverse phenomena.

The elementary particle structure of the world

One way to look at the development of fundamental physics in the twentieth century is as a war over the concept of the Newtonian mass point – a struggle to maintain, in the face of evidence to the contrary, the notion of an elegantly simple world, built up out of tiny rigid bodies, carrying mass and occurring in only two species, one positively charged and the other negatively charged. But observations have not upheld this ideal; they have pointed toward far greater complexity in nature. Modern elementary particles are no longer very Newtonian: They are described by intricate philosophical concepts, including indeterminacy, complementarity, strangeness, and charm. In place of orbits they have wave functions or probability "clouds." They may be charged or neutral; they may have mass or be massless; they may be extended or pointlike (and yet "spin"); they may be long-lived or may decay within a fraction of a second; they may be created or annihilated! There are now hundreds of "elementary particles" (although the present trend is toward reducing this scheme to a much smaller set of subparticles).

This enlargement of the concept of the elementary particles was initially met by much skepticism, disbelief, and fierce resistance. Some of the greatest battles occurred in the time span considered in this volume. Two of them, the battles over the neutron and the neutrino, belong also to nuclear physics and were discussed at the Minnesota symposium.[2] One of the speakers there reflected: "I remember being quite shocked when it dawned on me [in 1934] that the neutron, an 'elementary particle' as I had by that time already learned to speak of it, might decay by β-emission with a half-life that I could roughly estimate . . . to be about half an hour or shorter."[66]

The battles over the positron and over the two mesons, referred to in several chapters in this volume, illustrate the psychological resistance of physicists to admitting new particles to their cherished scheme. Dirac, in his first paper on the positron, and Anderson, in tune with what he called the "spirit of conservatism," both initially identified this new particle as a proton. Dirac even tried to make an argument for increasing the positron mass to the size of the proton mass, realizing that the new particles could not be protons only after Hermann Weyl proved mathematically that the holes had to have the same mass as electrons.

Yukawa had virtually no support for his proposed nuclear meson outside Japan until the μ meson was observed; Bohr's response to the proposal by the Kyoto group that there is a neutral meson in addition to the charged one was "Why do you want to create such a particle?"[67] And the tantalizing π-μ paradox during 1937-47 arose out of the reluctance to admit that there could be a second particle having a mass similar to that of the Yukawa particle but in other respects behaving differently.

Hanson attempted to explain "the universal reluctance to accept a new particle" circa 1932 by means of a historical argument. From Greek times, and especially with the work of William Gilbert in the late sixteenth century, Benjamin Franklin in the mid-1700s, James Clerk Maxwell in 1860, Michael Faraday in 1833, Wilhelm Weber in 1871, and J. Johnstone Stoney during 1874-91 (who coined the name electron for the "natural unit of electricity"), all evidence indicated the existence of only two types of electrical particles: positive resinous particles and negative vitreous particles. After 1913, J. J. Thomson, Lord Rutherford, N. R. Campbell, and O. W. Richardson were speaking of positive as well as negative electrons, but by 1919 the positive charge in their scheme had become associated with the nucleus of the hydrogen atom (i.e., it had become the proton). Hanson claimed that with this long historical tradition for believing in only one positive particle and one negative particle, "the very conception of a third particle beyond the proton and the electron seemed unsupportable."[68] (In fact, the idea that there might be other particles had been around since the 1880s, but it was only in the background and did not influence the progress of fundamental physics.[69])

Sociological influences also affected the conception of elementary particles. These influences are reflected by Serber (Chapter 12), who presents "a view from Berkeley," and by Hayakawa (Chapter 4), who describes "meson physics in Japan," emphasizing that physics is studied not only by solitary individuals but also by communities – the international community of physicists, and also local and national groups, with distinctive orientations and esprit. In the 1930s, most physics communities opposed postulating new particles. Yukawa's innovative leap and that of the authors of the two-meson theory probably were aided as much by their isolation as by the encouragement of their small peer groups. Hayakawa's description of Yukawa's theory and the two-meson hypothesis (Chapter 4), Takabayasi's account of the Japanese style in particle physics (Chapter 18), and Schwinger's lecture on Tomonaga (Chapter 22) give some idea of the social relationships existing among Japanese physicists.[70]

We have already contrasted the psychological attitudes toward QED, for example, of Dirac and Weisskopf on the one hand and of Bethe and Heitler on the other. The approach of the former could be called conceptual or mathematical, that of the latter pragmatic or physical. Another characterization might be abstract or idealistic as opposed to concrete or materialistic. One may ask whether or not there was a local or even national tradition bearing on this distinction and also whether or not it correlated with the degree of communication existing between the theorists and experimentalists of a particular subculture.

Consider the idealistic tendencies of the northern European group of Bohr, Pauli, Schrödinger, and Heisenberg. The members of this tradition did not propose new particles (with the exception of Pauli, in postulating the neutrino); in general, they believed that if there were to be a breakdown of physics theory, the cause would not be the wrong choice of building blocks but rather wrong theories. Thus Heisenberg believed that quantum mechanics would break down at small distances (as Newtonian mechanics had done earlier in the century), that there was probably something wrong with our concept of space-time, that it should be quantized, and that there should be a fundamental length. Bohr thought that there might be a breakdown of energy conservation and the laws of electrodynamics. However, proposing new particles turned out to be the correct approach to solving the problems at hand.

New particles were more often proposed in the 1930s by the members of another tradition, which seems to contrast with the more abstract northern European one: an "island tradition," more concrete, more oriented toward model building, and having a closer relationship between theory and experiment. In England, for example, this tradition was exemplified by the work of Maxwell, J. J. Thomson, and Rutherford. And it is perhaps notable that Bethe and Heitler were originally more abstract in their approach to physics when they lived on the Continent, but like Rudolf Peierls and Kemmer, they became markedly more pragmatic in their research in the 1930s when they worked in England. Other examples of this more concrete orientation are provided by Japan and Italy – both of them, like England, geographically peripheral, and all three, perhaps coincidentally, famous for their love of nature and gardening, for the things of the earth.

The willingness of the Japanese physicists of the Yukawa school to postulate new particles and fields during the 1930s may be related to the philosophy advocated by Taketani. Whereas Western philosophy emphasizes only phenomenological versus fundamental theory, this Japanese philosophy includes three stages; in an additional "*substantialistic*" stage, between the phenomenological and the dynamical, one identifies the substances or materials of which a system is made.

To admit new particles at a time when such admission was opposed on psychological, historical, social, and philosophical grounds required taking a step that was not only bold but also nonrational: One had to believe in the new particles in the face of opposition, before sufficient evidence had been accumulated to fully support this belief. Yukawa has emphasized this component in elementary particle physics and has suggested that the answer to "What are the basic constituents of matter?" is in the same class of knowing as is the answer to "Is a fish happy swimming in the water?"[71] (See also Takabayasi's Chapter 18 in this volume.) Neither question is answerable on entirely rational grounds, although acceptable answers must comply with everything else known or believed. At the present time, Yukawa might argue, given all that we know, is there any reason to doubt that the fish is happy? Or is there reason to doubt that matter is composed of quarks, leptons, and other subatomic particles? The accepting, though nonrational, stance usually is more fruitful than the skeptical one, because it leads the researcher to new confrontations with nature.[72]

Dirac's remarks in Chapter 2 about the resistance of human beings to adopting new and unfamiliar ideas are in the same vein. People, as he points out, have a tendency to be conservative; the resistance to admitting to new particles in the 1930s is one example, and he lists others. For Dirac, following the clues provided by the mathematics, defines a program for overcoming conservatism in physics – a program for advancement that he described in his second major paper on the positron:

> The most powerful method of advance that can be suggested
> at present is to employ all the resources of pure mathematics
> in attempts to perfect and generalize the mathematical for-
> malism that forms the existing basis of theoretical physics,

and *after* each success in this direction, to try to interpret the new mathematical features in terms of physical entities.

The prediction of the positron, made in this way, was for Dirac but a "small step according to this general scheme of advance."[73]

Although Dirac's personal solution to the problem of resistance to new ideas would not be suitable for the majority of physicists, his comments about the need to overcome this resistance seem to be widely applicable. They relate to our earlier discussion of synthesis and diversification in the progress of physics: The pull through paradoxes and inconsistencies toward diversification, and away from a given familiar synthesis, is resisted by the conservative human tendency; only when such pulls exceed the psychological resistance to them can physics advance.

The happy thirties and forties[74]

Researchers in the 1930s and 1940s were strongly affected by the overwhelming economic, social, and political upheavals of that period. To list but a few: the economic depression, which took away many jobs and financial security; the rise of fascism in Europe, which displaced many physicists (including Weisskopf, Bethe, Peierls, Fermi, and Rossi) from their homes in Germany and Italy, at the same time dissolving much of the research establishments in those countries; the political controls on philosophy (including physics) in certain countries; the brutal war, with its diversion from scientific research to military technology, its bombings and destruction; the death camps; the economic shortages; the breakdown of communications between countries; the occupations. These developments have been dealt with by other authors, but their impacts on physics have not yet been fully examined.[75]

These impacts are mentioned, although not adequately discussed, in this volume. For example, Weisskopf (Chapter 3), one of the German refugees, refers to the hard times during the Depression and to his refugee experience, which included the immediate task of bringing relatives out of danger. Similarly, Rossi (Chapter 11) mentions the shock of losing his position in Italy, stopping off at Manchester, where he was able to carry out some experimental work, and then arriving in the United States, where he immediately went to work on the problem of the decay of the μ meson. He comments on differences in the styles of experimental physics in the United States and Italy. Conversi

(Chapter 14), who remained in Italy during the war, recalls dictating his thesis while hearing Mussolini declare war on France and Great Britain. And Piccioni (Chapter 13), who also worked in Italy during the war, tells how, when Rome was bombed in July 1943 by the Americans, he and Conversi took refuge in a classroom in the basement of Liceo Virgilio, near Vatican City, and continued to work there on the preparations for their famous experiment on the capture and noncapture of μ mesons. This work continued even during the Nazi occupation of Rome, when having to hide from the Germans was an everyday chore.

Hayakawa (Chapter 4) and Takabayasi (Chapter 18) discuss how Japanese physicists worked before and during the war; under very difficult circumstances they sustained a rich intellectual life in Tokyo and Kyoto and made remarkable progress in physics. Marshak (Chapter 23), Weiner (chapter 17), and others refer to the communications breakdown between Japan and the United States and to some of its consequences, in particular, preventing the news of the two-meson theory and of Tomonaga's work on renormalization from arriving in the United States in time to influence subsequent or parallel developments there by Marshak and Bethe and by Schwinger and Feynman. Weiner also mentions the difficulties Japan faced in building accelerators at the start of the war and some of the means Japanese physicists developed to overcome their intellectual isolation. Schwinger (Chapter 22) describes the peculiar circumstance that he and Tomonaga (two physicists whose names mean "oscillator" or "shaker"), who carried out parallel work on renormalization theory, also worked on similar problems during the war – on the behavior of microwaves in waveguides and cavity resonators in connection with the magnetron, both making use in this work of Heisenberg's paper on the scattering matrix, and thus familiarizing themselves with a viewpoint that, Schwinger reflects, "would help both work out the concept of self-consistent subtraction, or renormalization." Lamb (Chapter 20) tells how his war work on the microwave magnetron influenced his subsequent microwave measurement of the fine structure of hydrogen, which provided the basis for the Schwinger-Tomonaga-Feynman renormalized field theory.

Many vital social issues were not raised at the symposium. For example, consider the impact of the occupations on physics. In Japan, the American occupation during 1945-51 produced a slowdown in nuclear physics research by explicitly prohibiting experimental nuclear physics. But at the same time the occupation helped to establish the institutional basis for Japan's rapid progress in nuclear physics during the 1950s and 1960s.[76] Other issues were touched on far too lightly. For example, why did knowledge about the human potential for mass murder and torture on a large scale not curtail physics more than it did? Weisskopf suggested that research was one of the few psychological escapes physicists had during those difficult times – "to have these wonderful things with which we were occupied" helped them to preserve their sanity.

In the postwar period, particle physics (and other physics subfields) grew very rapidly. Among the factors influencing this postwar boom were the greater internationalism of science resulting from the war, new experimental techniques developed as part of the weapons programs, new funding mechanisms that grew out of wartime support for research (and resulted, for example, in the United States, in the National Science Foundation and the Atomic Energy Commission), the new widespread appreciation of the value of science for national security, the sudden reentry into physics of graduate students and other researchers who after approximately four years away were anxious to make up for lost time, and the closer relationship between theory and experiment resulting from the experience of the large wartime projects such as building the bomb and developing radar for defense. All of these larger issues need to be illuminated in detailed scholarly studies, for they are inseparable from the intellectual development of physics; scholars will need to probe them deeply in order to understand the birth of particle physics.

Notes

1 Martin J. Klein, "The Use and Abuse of Historical Teaching in Physics," in *History in the Teaching of Physics*, ed. by Stephen G. Brush and Allen L. King (Hanover, N.H.: University Press of New England, 1972), pp. 12-18.

2 Some exceptions: Roger H. Stuewer (ed.), *Nuclear Physics in Retrospect; Proceedings of a Symposium on the 1930s* (Minneapolis: University of Minnesota Press, 1979); Charles Weiner (ed.), *History of Twentieth Century Physics* (New York: Academic Press, 1977).

3 Stuewer (ed.), *Nuclear Physics in Retrospect* (Note 2).

4 The early 1930s picture of nuclear physics portrayed in this section is discussed in detail by E. Rutherford, J. Chadwick, and D. C. Ellis, *Radiations from Radioactive Substances* (Cambridge University Press, 1930) and G. Gamow, *Constitution of Atomic Nuclei and Radioactivity* (London: Oxford University Press, 1931). See also the works cited in Note 2.

5 See, for example, Robert A. Millikan, "The Electron," in *Encyclopaedia Britannica*, 14th ed. (1929), Vol. 8, pp. 336-40.

6 See Gamow, *Constitution of Atomic Nuclei* (Note 4).

7 Joan Bromberg, "The Impact of the Neutron: Bohr and Heisenberg," *Historical Studies in the Physical Sciences*, Vol. 3, ed. by R. McCormmach (Princeton: Princeton University Press, 1971), pp. 307-41.

8 To use the terminology of Mituo Taketani's "three stage methodology" mentioned in Chapter 18, the problem of the atomic nucleus in 1930 was believed to be "essentialistic" (i.e., finding a correct dynamics), but it was in fact "substantialistic" (i.e., identifying the constituents correctly).

9 Victor F. Hess, *The Electrical Conductivity of the Atmosphere and Its Causes* (New York: Van Nostrand, 1928), translated from the German original by L. W. Codd.

10 To give but a single example: R. A. Millikan and G. Harvey Cameron, "High Frequency Rays of Cosmic Origin. III. Measurements in Snow-Fed Lakes at High Altitudes," *Phys. Rev. 28* (1926), 851-68.

11 Theories of cosmic-ray origins lie outside the scope of this symposium. However, for information on Millikan's theory and his dispute of 1932 with Arthur H. Compton, see Chapters 5, 6, 7, and 16; see also Daniel J. Kevles, *The Physicists* (New York: Knopf, 1978), especially pp. 179-80 and 240-42.

12 Also outside our scope; see the excellent account in Roger H. Stuewer, *The Compton Effect* (New York: Science History Publications, 1975).

13 Werner Kolhörster, "Eine neue Methode zur Richtungsbestimmung von Gamma-Strahlen," *Naturwiss. 16* (1928), 1044-5;

14 W. Bothe and W. Kolhörster, "Eine neue Methode für Absorptionmessungen an sekundaren β-Strahlen," *Naturwiss. 16* (1928), 1045; W. Bothe and W. Kolhörster, "Das Wesen der Höhenstrahlung," *Z. Physik 56* (1929), 751-77. The "tube counter" was far more sensitive than the "point counter." It was first described by H. Geiger and W. Müller in "Elekronenzählrohr zur Messung Schwächster Aktivitäten," *Naturwiss. 16* (1928), 617-18. A fuller account is H. Geiger and W. Müller, "Das Elektronenzähler," *Physik Zeit. 29* (1938), 839-41. See Chapter 5 in this volume for a discussion of the timing of this work.

15 For a lucid discussion of the Bothe-Kolhörster work and for an excellent (though personal) semipopular treatment of cosmic-ray history, see Bruno Rossi, *Cosmic Rays* (New York: McGraw-Hill, 1964); the quotation is from page 43.

16 See Chapter 5 in this volume for details; also, see Norwood Russell Hanson, *The Concept of the Positron* (Cambridge University Press, 1963) for Skobeltzyn's work (and that of C. D. Anderson) and reactions thereto.

17 See Chapter 7 in this volume. The classic papers on the positron and muon are the following: C. D. Anderson, "The Apparent Existence of Easily Deflectable Positives," *Science 76* (1932), 238-9; Seth H. Neddermeyer and Carl D. Anderson, "Note on the Nature of Cosmic-Ray Particles," *Phys. Rev. 51* (1937), 884-6.

18 P. M. S. Blackett and G. P. S. Occhialini, "Some Photographs of the Tracks of Penetrating Radiation," *Proc. Roy. Soc. (London) A139* (1933), 699-727.

19 Hanson, *Concept of the Positron* (Note 16), p. 159.

20 The problematic relationship between the positron's prediction and discovery has been the subject of philosophical and historical investigation. See, for example, Hanson, *Concept of the Positron* (Note 16) and Donald Franklin Moyer's three-part series "Origins of Dirac's Electron, 1925-1928," *Am. J. Phys. 49* (1981), 944-9, "Evaluations of Dirac's Electron, 1928-1932," *Am. J. Phys. 49* (1981), 1055-62, "Vindications of Dirac's Electron, 1932-1934," *Am. J. Phys. 49* (1981), 1120-5.

21 P. A. M. Dirac, "The Quantum Theory of the Electron," *Proc. Roy. Soc. (London) A117* (1928), 610-24 and *A118* (1928), 351-61. "Duplexity" is Dirac's translation of Heisenberg's "Zweideutigkeit," without its primary meaning of "ambiguity." However, compare this somewhat contradictory statement of 1972: "I was not interested

in bringing the spin of the electron into the wave equation, did not consider the question at all and did not make any use of Pauli's work. The reason for this is that my dominating interest was to get a relativistic theory agreeing with my general physical interpretation and transformation theory." P. A. M. Dirac, "Recollections of an Exciting Era," in Weiner (ed.), *History of Twentieth Century Physics* (Note 2), p. 139.

22 "Transformation theory" is Dirac's generalization of Heisenberg's algebra of non-commuting operators related to atomic transitions. P. A. M. Dirac, "The Fundamental Equations of Quantum Mechanics," *Proc. Roy. Soc. (London) A109* (1925), 642-53; P. A. M. Dirac, "The Physical Interpretation of the Quantum Dynamics," *Proc. Roy. Soc. (London) A113* (1927), 621-41. A bibliography of Dirac's extant writings until 1971, compiled by Jagdish Mehra, is given in *Aspects of Quantum Theory*, edited by Abdus Salam and E. P. Wigner (Cambridge University Press, 1972).

23 C. G. Darwin, "Wave Equations of the Electron," *Proc. Roy. Soc. (London) A118* (1928), 654-80.

24 See Dirac, "Recollections" (Note 21), p. 144.

25 For a more extended treatment of the history of the concept of the "hole," and antiparticles in general, see Hanson, *Concept of the Positron* (Note 16), Moyer, "Evaluations" (Note 20), and Dirac, "Recollections" (Note 21). For more historical detail, see P. A. M. Dirac's "The Prediction of Antimatter," the first H. R. Crane Lecture, April 17, 1978, at the University of Michigan, Ann Arbor. It is of interest that the analogous concept of holes in solids appeared at about the same time that the positron was conceived as a hole in the Fermi sea of the vacuum. But these two ideas seem to have developed quite independently of each other. The concept of positive electronic states in solids was introduced (although not explained or interpreted) in a paper by Rudolf Peierls on the anomalous Hall effect: "Zur Theorie der galvanomagnetischen Effekte," *Z. Physik 53* (1929), 255-66, submitted late December 1928, one year before Dirac's paper, "A Theory of Electrons and Protons," *Proc. Roy. Soc. (London) A126* (1930), 360-65, was received on December 6, 1929 by the Royal Society of London. Like Dirac, Peierls worked out the effect he was interested in, using quantum mechanics, and from the mathematics came to his conclusion that the behavior is "als wäre ein Elektron von dem Zustand mit dem grössten negativen k in dem mit dem grössten positiven k übergangen." The concept of the solid state hole as a vacancy in a nearly filled band, acting as if occupied by a positively charged electron, was not, however, made explicit until the paper by Werner Heisenberg in 1931: "Zum Paulischen Ausschliessungprincip," *Ann. Physik 7* (1931), 888-904. Although the research literature of 1929-31 gives no indication of any mutual influence in the development of the two "hole" concepts, there might have been some communication over the issue, and this possibility merits investigation. We would like to thank Dr. Egon Loebner for a helpful discussion of this point.

26 J. Bromberg, "Dirac's Quantum Electrodynamics and the Wave-Particle Equivalence," in Weiner (ed.), *History of Twentieth Century Physics* (Note 2). P. A. M. Dirac, "The Quantum Theory of the Emission and Absorption of Radiation," *Proc. Roy. Soc. (London) A114* (1927), 243-65; this is paper 1 in a useful volume of reprints: *Quantum Electrodynamics*, ed. by Julian Schwinger (New York: Dover, 1958), referred to hereafter as *QE*.

27 P. Jordan and E. Wigner, "Ueber das Paulische Äquivalenzverbot," *Z. Physik 47* (1928), 631-51. The Fermi-Dirac statistics were published by Fermi at the beginning of 1926 and by Dirac at the end of 1926, as Dirac candidly describes in Chapter 2.

For historical accounts, see L. H. Hoddeson and G. Baym, "The Development of the Quantum Mechanical Electron Theory of Metals: 1900-28," *Proc. Roy. Soc. (London) A371* (1980), 3-23, and Lanfranco Belloni, "A Note on Fermi's Route to Fermi-Dirac Statistics," *Scientia 113* (1978), 421-9.

28 Articles dealing with the theories discussed in this section are to be found in *Aspects of Quantum Theory*, edited by Abdus Salam and E. P. Wigner (Cambridge University Press, 1972), *The Physicist's Conception of Nature*, edited by Jagdish Mehra (Dordrecht, Holland: D. Reidel, 1973), *Theoretical Physics in the Twentieth Century*, edited by M. Fierz and V. F. Weisskopf (New York: Interscience, 1960).

29 Gregor Wentzel, "Quantum Theory of Fields (until 1947)," in Fierz and Weisskopf (eds.), *Theoretical Physics* (Note 28). In the same article, Wentzel also praises "the novelty and boldness of Dirac's approach." The simplicity of Dirac's treatment was reviewed pedagogically by Enrico Fermi in "Quantum Theory of Radiation," *Rev. Mod. Phys. 4* (1932), 87-132.

30 W. Heisenberg and W. Pauli, "Zur Quantendynamik der Wellenfelder," *Z. Physik 56* (1929), 1-61, and *59* (1930), 168-90. For the infinite self-mass of Heisenberg and Pauli, see J. R. Oppenheimer, "Two Notes on the Probability of Radiative Transitions," *Phys. Rev. 35* (1930), 939-47, and I. Waller, "Bemerkungen über die Rolle die Eigen-energie des Elektrons in der Quantentheorie der Strahlung," *Z. Physik 62* (1930), 673-6.

31 Walter Heitler, *The Quantum Theory of Radiation*, 2nd ed. (Oxford: Oxford University Press, 1945).

32 O. Klein and Y. Nishina, "Ueber die Streuung von Strahlung durch freie Elektronen nach der neuen relativistischen Quantenelektrodynamik von Dirac," *Z. Physik 52* (1929), 853-68. For a contemporary account of anomalous absorption, see Gamow, *Constitution of Atomic Nuclei* (Note 4).

33 Energy (E) and length (L), as well as momentum (P) and time (T), are all simply related according to the ideas of relativity and quantum mechanics. For a particle of mass M, the characteristic quantities are $E = Mc^2$, $P = Mc$, $T = \hbar/Mc^2$, and $L = \hbar/Mc$, where c is the velocity of light and $\hbar$ is Planck's constant divided by 2π. The length corresponding to the electron rest energy is the electron Compton wavelength (divided by 2π), that is, about 4×10^{-11} cm. This length is about 137 times smaller than a typical atom and 137 times larger than a typical nucleus. (The multiplier 137 is the reciprocal of the well-known fine-structure constant.)

34 Joan Bromberg, "The Impact of the Neutron: Bohr and Heisenberg" (Note 7); Laurie M. Brown, "The Idea of the Neutrino," *Phys. Today 31* (Sept. 1978), 23-8; David C. Cassidy, "Cosmic ray showers, high energy physics, and quantum field theories: Programmatic Interactions in the 1930s," *Historical Studies in the Physical Sciences 12*:1 (1981), 1-39.

35 Christian Møller, "Zur Theorie des Durchgangs schneller Elektronen durch Materie," *Ann. Physik 14* (1932), 531-85; H. Bethe, "Bremsformel für Elektronen relativistischer Geschwindigkeit," *Z. Physik 76* (1932), 293-9; F. Bloch, "Bremsvermögen von Atomen mit mehreren Elektronen," *Z. Phys. 81* (1933), 363-76.

36 H. Bethe and W. Heitler, "On the Stopping of Fast Particles and on the Creation of Positive Electrons," *Proc. Roy. Soc. (London) A146* (1934), 83-112 (the italics are those of Bethe and Heitler).

37 Heisenberg persisted in his belief in a "fundamental length," although a smaller one, in connection with Fermi's β-decay theory and the so-called burst phenomena (i.e., apparent multiple-particle production in an elementary high-energy process). See Cassidy, "Cosmic ray showers" (Note 34).

38 E. J. Williams, "Applications of the Method of Impact Parameter in Collisions," *Proc. Roy. Soc. (London) A139* (1933), 163-86.

39 E. J. Williams, "Nature of the High Energy Particles of Penetrating Radiation Formulae," *Phys. Rev. 45* (1934), 729-30; E. J. Williams, "Correlation of Certain Collision Problems with Radiation Theory," *Kgl. Danske Videnskab. Selskab, Mat.-Fys. Medd. 13* (1935, No. 4), 1-50; C. F. von Weizsäcker, "Ausstrahlung bei Stössen sehr schneller Elektronen," *Z. Physik 88* (1934), 612-25.

40 E. J. Williams, "Correlation of Certain Collision Problems" (Note 39), p. 4.

41 J. R. Oppenheimer, "Are the Formulae for the Absorption of High Energy Radiations Valid?" *Phys. Rev. 47* (1935), 44-52.

42 H. J. Bhabha and W. Heitler, "The Passage of Fast Electrons and the Theory of Cosmic Showers," *Proc. Roy. Soc. (London) A159* (1937), 432-45; J. F. Carlson and J. R. Oppenheimer, "On Multiplicative Showers," *Phys. Rev. 51* (1937), 220-31. The shower phenomena themselves had already been discussed in 1934 by various workers (e.g., Blackett and Rossi) at the London conference: *International Conference on Physics, London, 1934, Vol. 1, Nuclear Physics* (Cambridge University Press, 1935).

43 W. Heisenberg, "Ueber den Bau der Atomkerne," *Z. Physik 77* (1932), 1-11, *78* (1932), 156-64, and *80* (1933), 587-96.

44 E. Fermi, "Versuch einer Theorie der β-Strahlen. I," *Z. Physik 88* (1934), 161-71.

45 Hideki Yukawa, "On the Interaction of Elementary Particles. I," *Proc. Phys.-Math. Soc. Japan 17* (1935), 48-57. See also, for other references to Yukawa's papers on meson theory, Chapter 4 in this volume and Laurie M. Brown, "Yukawa's Prediction of the Meson," *Centaurus 25* (1981), 71-132. For meson theory in general, see V. Mukherji, "A History of the Meson Theory of Nuclear Forces from 1935 to 1952," *Archive for History of Exact Sciences 13* (1974), 27-102.

46 J. R. Oppenheimer and R. Serber, "Note on the Nature of the Cosmic-Ray Particles," *Phys. Rev. 51* (1937), 1113; E. C. G. Stueckelberg, "On the Existence of Heavy Electrons," *Phys. Rev. 52* (1937), 41-2; Hideki Yukawa, "On a Possible Interpretation of the Penetrating Component of the Cosmic Rays," *Proc. Phys.-Math. Soc. Japan 19* (1937), 712-13.

47 Important contributions to both types of meson physics (unfortunately underemphasized in this symposium) were made by the German group, under Heisenberg's influence, both before and during World War II. See W. Heisenberg (ed.), *Cosmic Radiation*, translated from German by T. H. Johnson (New York: Dover, 1946); Walter Bothe and Siegfried Flügge (senior authors), *Nuclear Physics and Cosmic Rays*, 2 volumes (*Fiat Review of German Science, 1939-1946*, published by the Office of Military Government for Germany).

48 N. Kemmer, "The Charge-Dependence of Nuclear Forces," *Proc. Cambridge Phil. Soc. 34* (1938), 354-64. This is the first introduction of an "internal" symmetry group into particle and nuclear physics. Among its consequences is the equality (up to small electromagnetic corrections) of the three pion masses and of the neutron and proton masses.

49 On the neutral meson: Its two-photon decay was proposed and estimated by Shoichi Sakata and Yasutaka Tanikawa, "The Spontaneous Disintegration of the Neutral Mesotron (Neutretto)," *Phys. Rev. 57* (1940), 548. It was proposed as the origin of cascade (Auger) showers by H. W. Lewis, J. R. Oppenheimer, and S. A. Wouthuysen, "The Multiple Production of Mesons," *Phys. Rev. 73* (1948), 127-40. The observation at the Berkeley synchrotron was made by J. Steinberger, W. K. H. Panofsky, and J. Steller, "Evidence for the Production of Neutral Mesons by Photons," *Phys. Rev. 78* (1950), 802-5.

50 W. Pauli and V. Weisskopf, "Ueber die Quantisierung der skalaren relativistischen Wellengleichung," *Helv. Phys. Acta 7* (1934), 709-31; A. Proca, "Sur la Théorie ondulatoire des électrons positifs et négatifs," *J. Phys. Radium 7* (1936), 347-53; A. Proca, "Théorie non relativiste des particules a spin entier," *J. Phys. Radium 9* (1938), 61-6. See also T. Toró, "Alexandru Proca and the Vector Mesons in Hadron Physics," *Noesis* (Romania) *4* (1978), 87-90.

51 See also H. M. Foley and P. Kusch, "On the Intrinsic Moment of the Electron," *Phys. Rev. 52* (1948), 412.

52 In *Concept of the Positron* (Note 16), pp. 135-9, Hanson claims that Dirac, Blackett, and Bethe all expressed to him "their conviction that tracks were encountered, but not identified, long before Anderson's discovery." However, Skobeltzyn, in a letter to Hanson (24 October 1960, pp. 181-3 of *Concept of the Positron*), makes this objection: "It is true that as early as 1931 (but not earlier as you suggested) I, prior to others, observed electron-positron pairs, not being able to identify them, however." See also Skobeltzyn's remarks on Hanson in Chapter 5 of this volume.

53 See Notes 18 and 42.

54 See Notes 14 and 15. A series of absorption measurements showing "hardening" of the cosmic rays was made in the 1920s by Robert A. Millikan and G. Harvey Cameron (see Note 10).

55 Neddermeyer and Anderson (Note 17).

56 P. M. S. Blackett and J. G. Wilson, *Proc. Roy. Soc. (London) A160* (1937), 304-23.

57 Jean Crussard and Louis Leprince-Ringuet, "Étude dans le grand électro-aimant de Bellevue de traversées d'écrans par des particules du rayonnement cosmique," *Compt. Rend. 204* (1937), 240-2; Pierre Auger and Paul Ehrenfest, Jr., "Clichés de rayons cosmiques obtenus avec une chambre de Wilson-Blackett dans des condition spéciales," *Journ. de Phys. 6* (1935), 255-6.

58 J. C. Street and E. C. Stevenson, "Penetrating Corpuscular Component of the Cosmic Radiation," *Phys. Rev. 51* (1937), 1005 (abstract); Y. Nishina, M. Takeuchi, and T. Ichimiya, "On the Nature of Cosmic-Ray Particles," *Phys. Rev. 52* (1937), 1198-9.

59 C. M. G. Lattes, H. Muirhead, G. P. S. Occhialini, and C. F. Powell, "Processes Involving Charged Mesons," *Nature 159* (1947), 694-7. For a complete account of the nuclear emulsion technique and its many spectacular results, see C. F. Powell, P. H. Fowler, and D. H. Perkins, *The Study of Elementary Particles by the Photographic Method* (New York: Pergamon, 1959).

60 G. D. Rochester and C. C. Butler, "Evidence for the Existence of New Unstable Particles," *Nature 160* (1947), 855-7. See also, for a general survey, G. D. Rochester and J. G. Wilson, *Cloud Chamber Photographs of the Cosmic Radiation* (New York: Academic Press, 1952), and W. Gentner, H. Maier-Leibnitz, and W. Bothe, *An Atlas of Typical Expansion Chamber Photographs* (New York: Interscience, 1954).

61 See, for example, Stephen Hawking, *Is the End in Sight for Theoretical Physics*, lecture on assuming the Lucasian Chair (Cambridge University Press, 1980).

62 F. J. Hasert et al., "Observation of Neutrino-like Interactions without Muon or Electron in the Gargamelle Neutrino Experiment," *Phys. Letters B46* (1973), 138-40.

63 Steven Weinberg, "Conceptual Foundations of the Unified Theory of Weak and Electromagnetic Interactions," *Rev. Mod. Phys. 52* (1980), 515-23; Abdus Salam, "Gauge Unification of Fundamental Forces," *Rev. Mod. Phys. 52* (1980), 525-38; Sheldon Lee Glashow, "Toward a Unified Theory: Threads in a Tapestry," *Rev. Mod. Phys. 52* (1980), 539-43.

64 Quoted by Hawking (Note 61).

65 R. A. Millikan, *Electrons* (Cambridge University Press, 1935), p. 320; quoted by Hanson (Note 16).

66 Maurice Goldhaber, "The Nuclear Photoelectric Effect and Remarks on Higher Multipole Transitions: A Personal History," in Stuewer (ed.), *Nuclear Physics in Retrospect* (Note 2), pp. 83-110 (the quotation is from p. 88).

67 According to Mituo Taketani.

68 Hanson, *Concept of the Positron* (Note 16), pp. 152-9.

69 Joan Bromberg, "The Concept of Particle Creation before and after Quantum Mechanics," *Historical Studies in the Physical Sciences 7* (1976), 161-91; J. L. Heilbron in Weiner (ed.), *History of Twentieth Century Physics* (Note 2), p. 45.

70 Besides these chapters, other useful orientations are available: *Science and Society in Modern Japan, Selected Historical Sources*, ed. by Shigeru Nakayama, David L. Swain, and Eri Yagi (Cambridge, Mass.: M.I.T. Press, 1974); *Particle Physics in Japan, 1930-1950*, 2 volumes ed. by L. M. Brown, M. Konuma, and Z. Maki (Kyoto: Research Institute for Fundamental Physics, 1980).

71 The issue is raised in an essay written in 1966: Hideki Yukawa, *Creativity and Intuition*, translated from the Japanese by John Bester (Tokyo: Kodansha International, 1973), pp. 69-72. The illustration is a Taoist fable, taken from the seventeenth chapter, "The Autumn Flood," of the Chinese *Book of Chuangtse*: "One day, Chuangtse was strolling beside the river with Huitse. Huitse, a man of erudition, was fond of arguing. They were just crossing a bridge when Chuangtse said, 'The fish have come up to the surface and are swimming about at their leisure. That is how fish enjoy themselves.' Immediately Huitse countered this with: 'You are not a fish. How can you tell what a fish enjoys?' 'You are not me,' said Chuangtse. 'How do you know that I can't tell what a fish enjoys?' 'I am not you,' said Huitse triumphantly. 'So of course I cannot tell about you. In the same way, you are not a fish. So you cannot tell a fish's feelings.' " Chuangtse asserted, nevertheless, that he *knew* that the fish were happy! Yukawa regarded this interchange as "an indirect comment on the question of rationalism and empiricism in science" and said that he was more in sympathy with Chuangtse than with Huitse in the argument.

72 Richard P. Feynman has expressed a similar view: "A very great deal more truth can become known than can be proven." (Nobel lecture Stockholm, 1965); reprinted in *Physics Today*, August 1966, pp. 31-44 (quotation on p. 43).

73 P. A. M. Dirac, "Quantised Singularities in the Electromagnetic Field," *Proc. Roy. Soc. (London) A133* (1931), 60-72.

74 A reference to the talk by Hans Bethe, "The Happy Thirties," given at the Minnesota symposium (Note 2), pp. 11-31, in which he mentioned that "politically the thirties were anything but happy" generally, but still a happy period for nuclear physics.

75 The following bibliography offers a small sampling of the rich literature available on the social context of physics in the 1930s and 1940s (a more detailed bibliography is available from John Heilbron and Bruce Wheaton, Office of Science and Technology, University of California, Berkeley): Alan D. Beyerchen, *Scientists under Hitler: Politics and the Physics Community in the Third Reich* (New Haven: Yale University Press, 1977); Max Born, *My Life, Recollections of a Nobel Laureate* (New York: Charles Scribner's Sons, 1975); Arthur Compton, *Atomic Quest: A Personal Narrative* (New York: Oxford University Press, 1956); Freeman Dyson, *Disturbing the Universe* (New York: Harper & Row, 1979); Laura Fermi, *Illustrious Immigrants. The Intellectual Migration From Europe 1930-41* (Chicago: University of Chicago Press, 1971); Donald Fleming and Bernard Bailyn (eds.), *The Intellectual Migration* (Cambridge, Mass.: Harvard University Press, 1969); Otto Frisch, *What Little I*

Remember (Cambridge University Press, 1979); Margaret Gowing, *Britain and Atomic Energy: 1939-1945* (New York: St. Martin's Press, 1964); Daniel S. Greenberg, *The Politics of Pure Science* (New York: New American Library, 1967); Morton Grodzins and Eugene Rabinowitch (eds.), *The Atomic Age, Scientists in National and World Affairs*, articles from the *Bulletin of the Atomic Scientists 1945-1962* (New York: Basic Books, 1963); Werner Heisenberg, *Physics and Beyond. Encounters and Conversations* (New York: Harper & Row, 1971); Richard G. Hewlett and Oscar E. Anderson, *The New World, 1939/1946 (A History of the United States Atomic Energy Commission), Vol. I* (University Park: Pennsylvania State University Press, 1962); Richard G. Hewlett and Francis Duncan, *Atomic Shield, 1947/1952 (A History of the United States Atomic Energy Commission), Vol. II* (University Park: Pennsylvania State University Press, 1969); David Irving, *The Virus House* (London: Kimber, 1964); Daniel J. Kevles, *The Physicists, The History of a Scientific Community in Modern America* (New York: Alfred A. Knopf, 1978). Many articles in the *Bulletin of the Atomic Scientists* deal with the social context for physics in the 1930s and 1940s: R. E. Marshak, "The Rochester Conferences," *Bulletin of the Atomic Scientists* (June 1970); Philip M. Morse, *In at the Beginnings: A Physicist's Life* (Cambridge, Mass.: M.I.T. Press, 1977). The oral history interview collection of the Center for the History of Physics of the American Institute of Physics is a mine of information about the social context of physics in the 1930s and 1940s. This collection includes interviews carried out as part of the center's program in the recent history of physics and astronomy, interviews carried out by Thomas Kuhn and associates as part of the Quantum History Project, and additional interviews (such as the series by Lillian Hoddeson with solid state physicists). *Proceedings of the Tenth International Congress on the History of Science, Ithaca, N.Y., 1962* (Paris: Hermann, 1964); Emilio Segrè, *Enrico Fermi, Physicist* (Chicago: University of Chicago Press, 1970); John C. Slater, *Solid-State and Molecular Theory; A Scientific Biography* (New York: Wiley, 1975); Alice Kimball Smith, *A Peril and a Hope: The Scientists' Movement in America, 1945-1947* (Cambridge, Mass.: M.I.T. Press, 1971); Alice Kimball Smith and Charles Weiner (eds.), *Robert Oppenheimer: Letters and Recollections* (Cambridge, Mass.: Harvard University Press, 1980); Roger H. Stuewer (ed.), *Nuclear Physics in Retrospect. Proceedings of a Symposium on the 1930s* (Minneapolis: University of Minnesota Press, 1979); M. Taketani, *Atomic Energy and Scientists* (Tokyo: Keiso-shobo, 1968, in Japanese); Spencer R. Weart, *Scientists in Power* (Cambridge, Mass.: Harvard University Press, 1979); Spencer R. Weart, "The Physics Business in America: A Statistical Reconnaissance," in N. Reingold, *Science in the American Context* (Washington: Smithsonian Press, 1978), pp. 295-358; Spencer R. Weart, "Scientists with a Secret," *Physics Today* 29 (1976), 23-30; Spencer R. Weart, "Secrecy, Simultaneous Discovery, and the Theory of Nuclear Reactors," *American Journal of Physics* 45 (1977), 1049-60; Charles Weiner, "Physics in the Great Depression," *Physics Today* 23 (1970), 31-8; Charles Weiner, "1932 – Moving into the New Physics," *Physics Today* 25 (1972), 40-9; Charles Weiner, "Physics Today and the Spirit of the Forties," *Physics Today* 26 (1973), 23-8; Charles Weiner (ed.), *History of Twentieth Century Physics. Proceedings of the International School of Physics "Enrico Fermi," course 57, Varenna, Italy* (New York: Academic Press, 1977); Charles Weiner, *Exploring the History of Nuclear Physics* (New York: American Institute of Physics, 1972); Charles Weiner, "Cyclotrons and Internationalism: Japan, Denmark and the United States, 1935-1945," in *Proceedings of the Fourteenth International Congress of the History of Science, 1974, No. 2* (Tokyo: Science Council of Japan, 1975), pp. 353-65 and references therein; Charles Weiner, "Institutional Settings for Scientific Change:

Episodes from the History of Nuclear Physics," in A. Thackray and E. Mendelsohn (eds.), *Science and Values* (New York: Humanities Press); Jane Wilson (ed.), *All in Our Time. The Reminiscences of Twelve Nuclear Pioneers* (Chicago: Bulletin of the Atomic Scientists, 1975). See also Notes 70 and 76.

76 Lillian Hoddeson, "Establishing KEK in Japan and Fermilab in the U.S.: Internationalism, Nationalism, and High Energy Accelerators," *Social Studies of Science 13* (1983).

Part II

Theoretical underpinnings

2　The origin of quantum field theory

PAUL A. M. DIRAC

Born 1902, Bristol, England; B.Sc., Bristol University, 1921;
Ph.D., Cambridge University, 1926; Nobel Prize, 1933, for relati-
vistic electron theory and the prediction of antimatter; Florida
State University.

I would like to discuss the early days in the development of
quantum mechanics and quantum field theory. I would like you to
understand the excitement and the frustrations as they appeared to a
young physicist who was living at that time.

Bohr orbits

I came to Cambridge as a research student in 1923, and I was
appointed to work with R. H. Fowler as my supervisor. I then heard
about Bohr orbits for the first time in my life. It was quite a surprise to
me that one had such a detailed knowledge of the inside of an atom,
that one could actually make use of Newton's equations of motion in
studying how electrons behaved in an atom. I had previously been an
engineering student at Bristol University and had heard nothing of all
this work, and I was fascinated by the Bohr orbits.

There were a good many serious problems in those days. First of all,
the successes of the theory occurred only in problems where one was
concerned essentially with just a single electron, applications in which
just one electron was doing the work. Even then there were some
grave difficulties in understanding this one electron. In the first place,
the number of states for the electron appeared to be twice as great as
the theory would lead one to expect, and that completely upset all the
arrangements of putting electrons together into an atom and building
up the periodic table.

How could one understand this doubling of the number of states for an electron? Electron spin was unknown at that time, and people thought that the doubling must come from the interaction of that electron with the other electrons. Werner Heisenberg took this question very seriously, and he introduced a special name for this doubling. He called it *Zweideutigkeit*. One might translate that as "duplexity." An electron in an atom, interacting with other electrons, had this duplexity appearing that meant a doubling in the number of states.

There were some people thinking about electron spin in those days, but there was a lot of basic opposition to such an idea. One of the first was Ralph de Laer Kronig. He got the idea that the electron should have a spin in addition to its orbital motion. He was working with Wolfgang Pauli at the time, and he told his idea to Pauli. Pauli said, "No, it's quite impossible." Pauli completely crushed Kronig.

Then the idea occurred quite independently to two young Dutch physicists, George Uhlenbeck and Samuel Goudsmit. They were working in Leiden with Professor Paul Ehrenfest, and they wrote up a little paper about it and took it to Ehrenfest. Ehrenfest liked the idea very much. He suggested to Uhlenbeck and Goudsmit that they should go and talk it over with Hendrik Lorentz, who lived close by in Haarlem. They did go and talk it over with Lorentz. Lorentz said, "No, it's quite impossible for the electron to have a spin. I have thought of that myself, and if the electron did have a spin, the speed of the surface of the electron would be greater than the velocity of light. So, it's quite impossible." Uhlenbeck and Goudsmit went back to Ehrenfest and said they would like to withdraw the paper that they had given to him. Ehrenfest said, "No, it's too late; I have already sent it in for publication."

That is how the idea of electron spin got publicized to the world. We really owe it to Ehrenfest's impetuosity and to his not allowing the younger people to be put off by the older ones. The idea of the electron having two states of spin provided a perfect answer to the duplexity.

However, the great problem at that time was how Bohr orbits interact with each other. Consider the helium atom, or any other atom with several electrons. It contains two or more electrons, each moving in its own Bohr orbit. These Bohr orbits must interact with each other. How can one deal with this interaction? That was the main problem that I was concerned with as a young research student at Cambridge during 1923-5. The only clue seemed to be that one should use Hamiltonian methods. Hamiltonian methods were very successful in dealing with the Bohr orbits, and it seemed to me that some development in the use of these methods and Hamiltonian interaction theory would be required.

I ought to mention that there was some work going on at the time quite independent of Bohr orbits. Here the prime instigator was Albert Einstein. Already in 1917 Einstein had thought about the interaction of an atom with the electromagnetic field around it, considering the atom to be absorbing and emitting photons, and he introduced some coefficients to describe this interaction. An atom in a higher state could emit and drop to a lower state with the emission of a photon, and that emission process was described by Einstein's coefficient A. Then he introduced a coefficient B to describe an absorption process, with an atom in a lower state absorbing a photon, and jumping to a higher state. He found that in order that there may be equilibrium, with the radiation satisfying Planck's law, it was necessary to have a process of stimulated emission. If the atom is capable of emitting radiation and if there is radiation falling on it of the right frequency, the emission will be stimulated in excess of the spontaneous emission, by an amount which is also given by the coefficient B. Einstein worked out the relationship between the coefficients A and B just from statistical considerations, the need to be able to have equilibrium with Planck's law of radiation.

These coefficients of Einstein were each associated with two states of an atom. They had nothing to do with Bohr orbits. The people who were working with Bohr orbits were working all the time with coefficients which depended only on a single state. Einstein, by introducing these coefficients depending on two states, had a theory independent of the Bohr orbit theory.

Max Born and Heisenberg developed this idea of the Einstein coefficients, and they set up a more general theory for an atom interacting with radiation, a theory of dispersion. This theory was based entirely on the Einstein coefficients and had nothing to do with Bohr orbits.

I read about this work, but I was not strongly impressed by it, and that was really my mistake. I was so fascinated by the Bohr orbits that I believed that substantial progress in atomic theory could be made only with an understanding of the interaction between Bohr orbits. There I was quite wrong.

Werner Heisenberg and Niels Bohr at Bohr Institute conference, 1934 (credit: Paul Ehrenfest, Jr.).

Heisenberg's mechanics

The final breakthrough came from Heisenberg, in 1925, when he set up his new theory of matrix mechanics, a theory based just on quantities, matrix elements as they became later, each associated with two atomic states and not just one. That was a theory quite independent of Bohr orbits.

Heisenberg broke away from the idea of Bohr orbits and was led to the formulation of a new dynamics. This new dynamics was based essentially on a new algebra, an algebra in which the product $\mathbf{u} \cdot \mathbf{v}$ for two dynamical variables is not the same as the product $\mathbf{v} \cdot \mathbf{u}$. It was a big surprise to me when I heard about this work and suddenly realized that it was the key to making progress in atomic theory. It showed me how completely wrong I was, sticking to Bohr orbits, when what one really wanted was a new mathematics which should be quite independent of the previous mathematics and should lead to some quite new equations.

When one looks over the present situation in quantum field theory, I think one should realize that it is very similar to the pre-Heisenberg situation. People at the present time are making the mistake of continually trying to develop the physical ideas which they have gotten used to: physical ideas which usually are expressed in terms of Feynman diagrams; and I believe that this essential reliance on Feynman diagrams and trying to introduce artificial processes of renormalization to get over the difficulties is the same kind of mistake that I was making in 1924 and 1925 in insisting on the Bohr orbits. What one really needs is a new kind of mathematics. One needs new equations to express the interaction between basic quantities in physics; one should not stick to the standard ideas and just try to push on with them.

That is how quantum mechanics started. It was a wonderful revelation to me to see the great power which physicists then acquired from using the noncommutative algebra which Heisenberg had been led to, or rather forced to adopt. (I think he was rather reluctant in accepting the idea that noncommutative algebra should really be a basic feature of our dynamics, but still his ideas forced him in that direction.) It was found to be possible to develop this new mechanics in a way which was formally very close to the old classical mechanics expressed in Hamiltonian form. In this development, it was the mathematics which really dominated the progress that was made. The correct understanding of the physical relationships brought in by the new mathematics did not become clear until a few years later. The equations themselves came first.

It was found that one could set up equations with the new mechanics closely analogous to all the important equations of classical mechanics. One merely had to express those equations in Hamiltonian form and then use a general relationship connecting the quantity $\mathbf{u} \cdot \mathbf{v} - \mathbf{v} \cdot \mathbf{u}$ in the new dynamics with the Poisson bracket of the two corresponding variables in the classical theory.

Schrödinger's waves

I was very excited by this and was completely engrossed in these relationships. It was some months later that Erwin Schrödinger brought out his wave equation. I was not very interested in that at first, because I believed that Heisenberg's ideas provided all that was really necessary. They provided a basis that one could go on extending as much as one liked; one did not need a new set of basic ideas such as Schrödinger had introduced. Schrödinger had introduced the idea of waves connected with particles.

I should mention that it was, first of all, Louis de Broglie who got the idea that particles in general should be associated with waves. Einstein had already made this association in the connection between photons and waves of light. He had done this in 1905, long before, and de Broglie extended it to particles of all kinds. De Broglie's idea came entirely from the beauty of the mathematics that one got by setting up the equations in relativistic form.

De Broglie's ideas were for a free electron by itself, and Schrödinger extended them to apply to an electron moving in an electromagnetic field. As soon as he got his general equation, he applied it to the hydrogen atom. The result that he obtained was not in agreement with experiment, because Schrödinger did not know at that time about the spin of the electron. He was extremely disappointed by this failure. He told me about it many years later. He believed that the whole idea of his wave equation was wrong. He was terribly dejected, and he abandoned it altogether. Then it was some months later that he recovered from his depression sufficiently to go back to this work, look over it again, and to see that if he did it in a nonrelativistic approximation, so far as a nonrelativistic system was concerned, his theory was in agreement with observation. He published his equation then as a nonrelativistic equation.

You may wonder how it appears that Schrödinger's early papers were all nonrelativistic, although they were inspired by de Broglie waves and the de Broglie waves were built up from relativistic ideas. It was in this indirect way that it came about. Schrödinger lacked courage to publish an equation that gave results in disagreement with observation. He should have had that courage; he would then have published a second-order equation in $\partial/\partial t$, an equation that was later to be known as the Klein and Gordon equation, although it was discovered by Schrödinger before Oskar Klein and Walter Gordon and was the first wave equation that he worked with. But Schrödinger would only publish something that was not in direct disagreement with observation. People were rather timid in those days, I suppose, and it was left to Klein and Gordon to publish an equation which is now accepted as the correct equation for a charged particle without spin.

When this theory of Schrödinger's appeared, I was a bit annoyed by it because I did not want to be disturbed at all from the development of Heisenberg's ideas and from following up the analogy between Heisenberg's mechanics and Newtonian mechanics. It was wrong of me to have this hostility, because Schrödinger's theory was shown by Schrödinger himself and others to be equivalent in its mathematical consequences to the Heisenberg theory, and Schrödinger's theory did provide new insights. It provided new directions for development which one would not have thought of just keeping to the Heisenberg theory.

The main new development which is suggested is what one gets if one applies the theory to two or more particles of a similar nature. It then becomes possible to think of a Schrödinger wave function which is symmetrical between the two particles, or, alternatively, a wave function which is antisymmetrical. These symmetry ideas I doubt that one would ever have thought of if one had just kept to the Heisenberg picture. But they do suggest themselves fairly naturally as soon as one brings in the Schrödinger picture and joins it on to one's ideas of the Heisenberg picture.

I soon saw the value of this development of symmetry/antisymmetry relations and studied them. I found that if one allows only symmetrical states (which one can do, because if one starts off with a symmetrical state, it always remains symmetrical), one is led to a theory of a number of similar particles which satisfy a statistics different from the classical statistics. They satisfy a statistics that had already been proposed in 1924 by Satyandra Nath Bose. This was before the big discovery of Heisenberg.

Bose had been concerned only with the question of how to account for Planck's law of radiation. Einstein also had been much concerned with this question. The big problem was a logical one, namely, that classical mechanics is evidently wrong for studying Planck's law of radiation, and one ought to be able to manage without using a wrong theory. How can one get Planck's law without using classical mechanics? Bose thought of a way of doing it by just arranging to have the total energy distributed among the various possible states for the oscillators which compose the electromagnetic field and allowing equal probabilities for any number of degrees of excitation of each of the oscillators. Working that out, one gets Planck's law without making use of classical ideas. Bose sent his work to Einstein. Einstein was enthusiastic about it and sent it in for publication. That started the new statistics of Bose. It turned out that quantum mechanics, when one restricts it with the requirement that the wave functions are to be symmetrical between the similar particles, just leads to the Bose statistics for those particles.

Let us consider the other kind of symmetry, which arises when we restrict the wave function to be antisymmetrical between the various particles. Under these conditions you cannot have two particles in the same state. That is just what you need for electrons in order to get Pauli's exclusion principle. So this kind of symmetry requirement is needed for electrons.

So we have the two kinds of statistics, the Bose statistics, which is applicable to photons, and the other kind of statistics which goes with antisymmetrical wave functions and is applicable to electrons. Enrico Fermi had written about this other kind of statistics, but I had forgotten about Fermi's paper when I wrote my own work on the subject, and I made no reference to Fermi in it. Fermi wrote to me, pointing out that he had been the first to propose this kind of statistics, and I had to agree with him and to apologize to him for forgetting about his paper. The reason for my forgetfulness was that I was interested only in papers which had a relationship to the fundamental problems of atomic theory. I had read Fermi's work at the time it was published, but it did not stick in my mind at all. I did not see that it was of any importance, because it did not have any bearing on the main problem at that time, which was how the Bohr orbits interact.

In that way, one got started on the idea of two kinds of particles, bosons and fermions. It appears that all the particles known in nature belong to one of these two classes. I do not know if there is any obvious reason why that should be so, but it does happen to be so. With this development of the theory, one sees how in passing from the Heisenberg to the Schrödinger formulation one is led to ideas which one probably would never have had if one had just stuck to the Heisenberg formulation and built up the analogy between the new quantum mechanics and the classical mechanics.

That illustrates what I think is a rather general principle in the development of theoretical physics; namely, one should allow oneself to be led in the direction which the mathematics suggests. The mathematics really push one into thinking of symmetrical and antisymmetrical states, and one must follow up this mathematical idea and see what its consequences are, even though one gets led to a domain which is completely foreign to what one started with.

I might mention some other examples where I think it is advisable to allow oneself to be led forward by the mathematical ideas. One of those was some mathematics which led to the possible existence of the monopole. It's a very beautiful mathematics, but up to the present it has not proved to be of any value in physics. It might still do so in the future. This is a good example of how mathematics can lead us in a direction we would not take if we only followed up physical ideas by themselves.

Again, in connection with the symmetrical and antisymmetrical states, one could consider other kinds of symmetry which the mathematics allows. If one handles this question generally, one gets the idea of permutation operators. You have a wave function applying to an assembly of particles of the same nature, and you can apply a permutation operator to it, in which you make some permutation of the similar particles. Now all the particles known in nature are such that permutation operators applied to them lead to plus or minus one, because the particles are all bosons or fermions. But still, you might apply the permutation operators to some of the variables describing the particle, and not all of them. If the particles have spin, for instance, you might apply a permutation operator to the position variables of the particle, and not to the spin variables. That is something that the mathematics allows. In that way one can get some new operators that appear as dynamical variables in quantum mechanics. They appear quite independently of anything which classical mechanics would suggest. The mathematics leads you into a new direction. This particular development does turn out to be of some value in applications to atoms containing several electrons. It helps one to understand the multiplet structure in the spectra of such atoms.

Then again, we can introduce other dynamical variables quite independently of anything which classical theory suggests. These other dynamical variables must be such that we can multiply them together and get new variables of the same nature; that is to say, these variables must form a group. This process is used very much nowadays for describing the internal structure of the modern particles. One introduces new variables forming some group. Determining the correct group to use presents a problem; there is no general theory about it. One has to try various possibilities and compare the results with experiment. All the modern theories of the particles which appear in high-energy physics involve these dynamical variables referring to the elements of some group. We have there mathematical developments which are quite independent of anything that is suggested in classical mechanics.

Radiation theory

There we have the beginning of the ideas of quantum mechanics, and it became necessary to make further developments. I worked out a theory for the interaction of an atomic system with electromagnetic radiation, taking the electromagnetic radiation to be an external perturbation acting on the atomic system, and I found that this external perturbation could lead to transitions of the atomic system in which it could absorb a quantum of energy, or, alternatively, emit a quantum of energy, with a jump from one state to another. That led to a theory that reproduced the Einstein coefficients B – the coefficients that govern the absorption of radiation and the stimulated emission. These are the first coefficients that were deduced on the basis of the new mechanics. But that method was inadequate for explaining the Einstein coefficients A for spontaneous emission.

I should mention that the Einstein coefficient for stimulated emission referred to an effect that was extremely small, quite hopeless to observe at the time it was proposed, but people have found ways of enhancing its effect, and it now forms the basis of the theory of lasers and is most important. We owe the essential idea of lasers to Einstein's work of 1917.

In further development of the quantum mechanics of Heisenberg and Schrödinger, it occurred to me to apply a process that has been called second quantization. One simply takes the wave function of Schrödinger's theory, the wave function $\psi(q)$ (q is a variable specifying a point in the domain of the wave function), and supposes that all the $\psi(q)$'s for different values of q are made into operators (Heisenberg operators) instead of just being numbers. Then one takes the conjugate quantities $\bar{\psi}(q)$ and makes them also into operators. The $\psi(q)$'s are supposed to commute with each other, and likewise the $\bar{\psi}(q)$'s, but the $\bar{\psi}(q)$'s are supposed not to commute with the $\psi(q)$'s. It seemed to me that this was a very interesting idea, and I wondered where it would lead. I worked it out and found that it just led to an assembly of similar systems, all satisfying the Bose statistics.

It was a bit of a disappointment to find that nothing really new came out of the idea. I thought at first it was a wonderful idea and was very much looking forward to getting something really new out of it, but it turned out to be just a new way of going back to the idea of an assembly satisfying Bose statistics.

But still, it did give a new way of looking at Bose statistics, and in particular it focused attention on the operators $\psi(q)$ and $\bar{\psi}(q)$. They turned out to be the operators that increase the excitation of one of the oscillators by one quantum, or else reduce the excitation by one quantum. These operators of emission and absorption then become the most useful basic variables to use in the description of an assembly of bosons.

I proceeded from these ideas to consider an assembly of photons interacting with some outside atomic system, and the result of following through this calculation, treating it by a standard perturbation method, was to give a theory of the interaction of these photons with the outside system, in which one of them can be emitted spontaneously. It gave again the Einstein coefficients B; but it gave, in addition, the Einstein coefficient A for spontaneous emission. We then had a complete theory of radiation. The whole of the Einstein theory then followed from quantum mechanics. One only has to apply quantum mechanics to an assembly of photons interacting with some outside system, an atom, or something equivalent.

Instead of working with a picture of the photons as particles, one can use instead the components of the electromagnetic field. One thus gets a complete harmonizing of the wave and corpuscular theories of light. One can treat light as composed of electromagnetic waves, each wave to be treated like an oscillator; alternatively, one can treat light as composed of photons, the photons being bosons and each photon state corresponding to one of the oscillators of the electromagnetic field. One then has the reconciliation of the wave and corpuscular theories of light. They are just two mathematical descriptions of the same physical reality.

An assembly of fermions

This work about the emission and absorption of radiation follows from applying a method of second quantization that is appropriate to Bose statistics. Soon after that, Pascual Jordan and Eugene Wigner proposed another method of second quantization that was appropriate to a set of fermions. This new kind of second quantization was again based on absorption and emission operators, but they were absorption and emission operators for fermions.

Now, there can never be two fermions in the same state, so that if we take an operator of emission of a fermion, applied twice over, we must get zero. Calling these emission operators η, η^2 is equal to zero. Similarly, the absorption operator, $\bar{\eta}$ let us say, is again an operator that we cannot apply twice over without getting zero; so $\bar{\eta}^2 = 0$. It is a bit strange to have these operators with zero squares, but there is nothing wrong with it mathematically.

We have relationships connecting the η's with $\bar{\eta}$'s that are formally very similar to the corresponding relationships between boson emission and absorption operators. The only difference is in the sign of some of the terms in the equations that express the basic commutation relations.

When I first heard about this work of Jordan and Wigner, I did not like it. The reason was that in the case of the bosons we had our operators that were closely connected with the dynamical variables that describe oscillators. We had operators that had classical analogues. In the case of the Jordan-Wigner operators, they had no classical analogues at all and were very strange from the classical point of view. The square of each of them was zero. I did not like that situation. But it was wrong of me not to like it, because, actually, the formalism for fermions was just as good as the formalism I had worked out for bosons.

I had to adapt myself to a rather different way of thinking. It was not so important always to have classical analogues for everything. What we really needed was to have dynamical variables satisfying commutation relations or anticommutation relations that would be consistent with one another. In terms of such dynamical variables, one can proceed to build up a reasonable quantum theory quite independently of whether or not there is a classical analogue. If there is a classical analogue, so much the better. One can picture the relationships more easily. But if there is no classical analogue, one can still proceed quite definitely with the mathematics. There were several times when I went seriously wrong in my ideas in the development of quantum mechanics, and I had to adjust them.

Negative energies

Another question that concerned me very much was the negative energies that appear as possible in any relativistic theory. Einstein's formula for the energy of a system with given momentum involves a square root, and the result is that the value for the energy, mathematically, can be either positive or negative. That has been known for a very long time, since 1905. It did not bother people to begin with, because one had only to assume that the world had started off with all particles in positive energy states. They would then always stay in positive energy states. They could not jump the barrier from a positive energy state to a negative energy state. So one could forget all about the existence of negative energy states.

With the development of quantum theory, that was no longer possible, because jumps can take place discontinuously from one energy level to another. If we start off a particle in a positive energy state, it may jump into a negative energy state.

For a long time that problem did not disturb people very much, simply because there were other more serious difficulties. There were the fundamental difficulties of understanding the stability of atoms, understanding the way the electrons interact in an atom, understanding the spin – and these dominated the problems that people were concerned with. But these other difficulties gradually got cleared away, one by one, until the main surviving fundamental difficulty was that provided by the negative energy states for the electron.

That bothered me very much. It bothered me especially because I had worked out a general transformation theory of quantum mechanics, a theory that enabled one to transform wave functions from one set of commuting variables to another set and to apply the result to determine the probability of any set of commuting variables having specified values. This theory I liked very much, and it seemed to me essential that one should retain it. With this theory it was necessary that one should have a wave equation for the electron which was linear in the operator of time differentiation $\partial/\partial t$. Such a linear equation in $\partial/\partial t$ would go against Einstein's principle of treating time essentially on the same basis as the x, y, and z coordinates.

People had been working quite a bit with a spinless electron, an electron described by the Klein-Gordon equation, which is a relativistic equation, but which could not be interpreted in terms of my general transformation theory of quantum mechanics and was therefore unacceptable to me. Other physicists with whom I talked at that time were not so obsessed by the need for having quantum theory agreeing with this general transformation theory, and they were rather inclined to let it go as it was. But I just stuck to this problem. That led me eventually to think of a new wave equation in which we have a wave function involving four components, and this wave equation satisfied very well the requirement that the electron should have a spin. It gave correctly the magnetic moment, as well as the angular momentum of the spin. But it brought me into the problem of the negative energy states. With four components for the wave function, one gets a doubling that is required for the two components of spin and then an extra doubling beyond that corresponding to the negative energy states.

And then I got the idea that because the negative energy states cannot be avoided, one must accommodate them in the theory. One can do that by setting up a new picture of the vacuum. Suppose that in the vacuum all the negative energy states are filled up. The possibility of doing that arises because the exclusion principle of Pauli prevents more than one electron being in any state. We then have a sea of negative energy electrons, one electron in each of these states. It is a bottomless sea, but we do not have to worry about that. The picture of a bottomless sea is not so disturbing, really. We just have to think of the situation near the surface, and there we have some electrons lying above the sea that cannot fall into it because there is no room for them.

There is, then, the possibility that holes may appear in the sea. Such holes would be places where there is an extra energy, because one would need a negative energy to make such a hole disappear. Also, such a hole would move as though it had a positive charge. It has an absence of negative charge; so in that respect, also, it appears as a positive charge. Thus the holes appear as particles with positive energy and positive charge.

When I first got this idea, it seemed to me that there ought to be symmetry between the holes and the ordinary electrons, but the only positively charged particles known at that time were the protons; so it seemed to me that these holes had to be protons. I lacked the courage to propose a new kind of particle. I should say that there were good grounds for belief at that time that there were only two particles, two basic charged particles – electrons and protons. There were just two kinds of electricity, positive and negative, and one needed one particle for each kind of electricity. In those days the climate of opinion was very much against the idea of proposing new particles. I certainly did not dare to do it; so I published my idea as a theory of electrons and protons, and I believed that maybe the difference in mass between the electrons and protons would come about in some way from the interaction between the electrons. But I realized the difficulties were enormous because the difference in mass was so great.

I was soon assailed by other physicists on the grounds that there could not be this difference between the mass of the new particles, the holes, and the mass of ordinary electrons. The person who most definitely came out against it was Hermann Weyl; he was essentially a mathematician and was not so much disturbed by physical realities but was very much dominated by mathematical symmetries. He said quite categorically that the new particles formed by these holes would have to have the same mass as the electrons, and I came around to that point of view.

We all know the consequences of that. The new particles were given the name of antielectrons and were afterward discovered by the experimenters. The first was Carl Anderson, one of the chief physicists whom we have to thank for that. Thus this question has now been resolved.

The infinities

A good many problems were solved, but there were some serious difficulties remaining. The most important one was that when you try to set up an accurate theory for electrons interacting with the electromagnetic field, you are led to a Schrödinger equation which you are unable to solve. You try to solve it by standard perturbation methods, and when you get to the second-order terms, you are led to infinities. The only conclusion you can come to is that the equation does not have solutions. That is a very fundamental difficulty which appeared quite early in the theory of quantum electrodynamics. It is still a fundamental difficulty, and has not been resolved.

A great deal of work has been done on how to handle this difficulty, and we shall be hearing more about it in other chapters. I do not want to go into the details of it, but I do want to emphasize my point of view. This chapter is intended to build up to the climax that we ought to accept that there is something basically wrong with our theory of the interaction of the electromagnetic field with electrons. By basically wrong I mean that the mechanics is wrong, or the interacting force is wrong. What is wrong with the theory is just about as serious as what was wrong with the Bohr orbit theory.

When one is studying the problems of present-day high-energy physics, if one is as old as I am, one cannot help being reminded of the sense of frustration that we had in the pre-Heisenberg era. We had ideas that worked pretty well up to a certain point, but, basically, nothing was right. It would seem that something has to be changed – just as fundamental as the change that we had to make in departing from Bohr orbits. We need some new mathematics just as surprising, just as foreign to what we are used to, as was Heisenberg's noncommutative algebra at the time when people were just working with Bohr orbits.

I have been studying this question for many years. I believe that the only true answer will be obtained when someone is able to think of this new mathematics. There are serious limitations in the kind of mathematics that people are using nowadays. One of these limitations that I have been much concerned with recently (I have been studying it now for a good many years) involves the nature of the basic equations which we have.

The basic equations

There are two forms of these equations: the equations of motion of the Heisenberg theory and the Schrödinger wave equation. People usually think that these two forms of equations are equivalent and that one can pass freely from one to the other and can use whichever one wants for handling the particular problem that one is concerned with. These two formulations are equivalent in the case of a system with a finite number of degrees of freedom, but they are not equivalent when one is dealing with quantum electrodynamics, where the number of degrees of freedom is infinite.

Here we are still dealing with Heisenberg dynamical variables, and these dynamical variables can still be expressed as operators in a Hilbert space. Hilbert space is just like Euclidean space in n dimensions, where n goes up to infinity, all the way through the natural numbers. (We can formulate it also in terms of continuous variables, or make lots of transformations, but that does not matter for this discussion.) It is a space that has an infinite number of dimensions, but the infinity is what mathematicians call denumerable. We can count them all – one, two, three, four – going on to infinity in that way, including them all.

Now, the Schrödinger wave function is a vector in Hilbert space when we are dealing with a dynamical system involving a finite number of degrees of freedom. But not in quantum electrodynamics. The Schrödinger wave function is then a vector in a space with a much larger number of dimensions than there are in Hilbert space. In fact, the number of dimensions needed is roughly 2^n, where n is a denumerable infinity.

You can see how this very large number of dimension arises. You take all the possible states for one of the electrons and use a system for counting them – one, two, three, and so on. But for each of these states there may be either an electron present or no electron present. There are thus two possibilities for each of these states. The total number of possibilities is 2 to the power of the number of states. That means 2 to the power of a denumerable infinity. And with that number of dimensions in your space, you have a space that is much larger than Hilbert space.

The Schrödinger wave function in quantum electrodynamics is not a vector in Hilbert space; it is something much more complicated. If we are dealing with the Heisenberg picture, we can stick to Hilbert space, use operators in Hilbert space, and do all our mathematics in terms of Hilbert space. If we want to go over to the Schrödinger wave functions, we need a space of a much larger number of dimensions. I think that this space that we then need is the space provided by spinors in Hilbert space. I have done quite a bit of work on them and have written a little book about them. They are probably the basic mathematical quantities needed to describe the states of quantum electrodynamics.

In any case, we have this complicated new kind of quantity that shows itself up in the standard treatment of these questions through what we call the vacuum-to-vacuum transitions, with the Feynman diagrams for expressing transitions. The vacuum-to-vacuum transitions have to be taken into account in the Schrödinger picture, but they are of no importance physically, and they add enormously to the complexity of the calculations. So I believe that we ought to try to keep to the Heisenberg picture, in which we do not have these vacuum-to-vacuum transitions appearing anywhere at all, and we have the possibility of working with mathematical equations in Hilbert space. I spent a lot of time trying to determine the correct equations to use. I am fairly certain that the correct basic equations have not yet been discovered.

Some new relativistic equations are needed; new kinds of interactions must be brought into play. When these new equations and new interactions are thought out, the problems that are now bewildering to us will get automatically explained, and we should no longer have to make use of such illogical processes as infinite renormalization. This is quite nonsense physically, and I have always been opposed to it. It is just a rule of thumb that gives results. In spite of its successes, one should be prepared to abandon it completely and look on all the successes that have been obtained by using the usual forms of quantum electrodynamics with the infinities removed by artificial processes as just accidents when they give the right answers, in the same way as the successes of the Bohr theory are considered merely as accidents when they turn out to be correct.

3 Growing up with field theory: the development of quantum electrodynamics[1]

VICTOR F. WEISSKOPF

Born 1908, Vienna, Austria; Ph.D., Göttingen, 1931; research associate to Schrödinger, Bohr, Pauli; director-general of CERN, 1961-5; quantum mechanics, electron theory, nuclear physics; Massachusetts Institute of Technology (emeritus).

In 1928 I came to the University of Göttingen as a graduate student in order to work toward a Ph.D. degree in theoretical physics. Before that, I had attended some introductory courses in physics and mathematics for two years at the University of Vienna. I remember in particular a course in general classical theoretical physics by Professor Hans Thirring (father of the theorist currently active in Vienna). His teaching, as well as that of Paul Ehrenfest when he was guest professor at Göttingen, had a decisive influence on my attitude toward physics because of its clear and simple presentations and its emphasis on the essential physical insights rather than on mathematical formalisms. I recall Ehrenfest's remark: "Physics is simple but subtle."

When I arrived in Göttingen in 1928, nonrelativistic quantum mechanics was in full development. During the few years since its inception, many new ways had been opened up for an understanding of the structure of atoms, of the formation of molecules, of the physics of solids, and in particular of the electric and magnetic properties of metals. "Never have so few done so much in such a short time."[2]

This chapter is devoted to the development of quantum electrodynamics,[3] which was born in 1927 when P. A. M. Dirac published his famous paper on this subject.[4] It was communicated by Niels Bohr himself. He wrote in the opening paragraph that "hardly anything has been done up to the present on quantum electrodynamics." That was

56

an understatement. Nothing had been done up to that time on quantum electrodynamics.

The pre-Dirac time

Classical electrodynamics started in 1862 when James Clerk Maxwell created his equations connecting the electric field **E** and the magnetic field **B** with the charge density ρ and the current density **j**:

$$\operatorname{curl} \mathbf{B} - \frac{1}{c}\dot{\mathbf{E}} = \frac{4\pi}{c}\mathbf{j},$$

$$\operatorname{curl} \mathbf{E} + \frac{1}{c}\dot{\mathbf{B}} = 0,$$

$$\operatorname{div} \mathbf{E} = 4\pi\rho,$$

$$\operatorname{div} \mathbf{B} = 0. \tag{1}$$

These equations, together with the expression for the Lorentz force acting on a system carrying charge and current,

$$\mathbf{F} = \int d^3x \left(\rho \mathbf{E} + \frac{1}{c}\mathbf{j} \times \mathbf{B} \right), \tag{2}$$

led to an understanding of light as an electromagnetic wave, of the radiation emitted by moving charges and of the effects of radiation on charged bodies. The results were splendidly verified by Heinrich Hertz in 1885 for radiations emitted and absorbed by antennas.

The physicists tried to apply equations (1) and (2) to atomic radiation. They were stymied by two facts: First, ρ and **j** in atoms were unknown to them; second, they faced a fundamental difficulty when the statistical theory of heat was applied to the radiation field. The number of degrees of freedom of a radiation field in a volume V per frequency interval $d\omega$ is $V(\omega^2 d\omega)/\pi^2 c^3$, and if each degree is supposed to get an energy $kT/2$ according to the equipartition theorem, the total energy density becomes infinite; the empty space will be an infinite sink of radiation energy. Furthermore, apart from this distressing result, the classical theory of light had no explanation of the daily experience that incandescent matter changes its color with rising temperature from red to yellow and then to white. The physicists must have felt before 1900 as the neurophysiologists of today feel without any explanation of what memory is.

Then came quantum theory. It developed with increasing speed within a quarter century beginning with Max Planck's insight into the nature of blackbody radiation in 1900, followed by Albert Einstein's revolutionary idea of the existence of the photon in 1905, by Bohr's atomic model in 1913, and by Louis de Broglie's daring hypothesis of the wave-particle duality of particles in 1924. It reached its peak with the formulation of quantum mechanics by Werner Heisenberg, Erwin Schrödinger, Dirac, Wolfgang Pauli, and Bohr in 1925.

The ideas were so new and unaccustomed that even the most experienced physicists had difficulties accepting them. It was about 1914 when Max von Laue and Otto Stern went for a walk in the neighborhood of Zurich on a hill called Uetliberg. This name resembles the name of a hill called Ruetli on which the representatives of the original cantons of Switzerland came together in 1307, swearing allegiance to their newly won independence from Austria and to defend it with all their power. This event is referred to as Ruetlischwur. Von Laue and Otto Stern made a vow that they would give up physics if and when those newfangled ideas by Niels Bohr about the hydrogen atom should turn out to be correct. They dubbed it the Uetlischwur. In contrast to the valiant Swiss, they did not keep their vow.

The difficulties of the classical theory disappeared at one stroke – not without bringing about other difficulties, about which much more will be said soon. Of course, the problem of heat radiation was immediately solved, and the reasons for the sharp characteristic spectral lines of each atomic species became evident. Atomic stabilities, sizes, and excitation energies could be derived from first principles: The chemical forces turned out to be a direct consequence of quantum mechanics; chemistry became part of physics.

We graduate students at Göttingen in 1928 were introduced in a rather unsystematic way to all these new and exciting developments. Regular courses in quantum mechanics were not yet given. Our sources of information were informal discussions, collective readings of the new papers, most of which were rather opaque for beginning graduate students of those days. But two lecture series remain in my memory, one by Gerhard Herzberg on atomic physics and the other by Walter Heitler on applications of quantum mechanics. The lecturers, hardly much older than the students, gave us exciting accounts of the new physics. The content of Herzberg's course is still available in book form and remains – 50 years later – one of the best introductions to atomic physics.[5]

Back to the problem of the radiation of atoms. Did quantum mechanics furnish the expressions for ρ and **j** within the atoms in order to calculate the interaction of atoms with light? Not really. We were told at that time that the matrix elements $\langle a|\rho|b \rangle$ and $\langle a|\mathbf{j}|b \rangle$ between two stationary states a, b of the atom play the roles of charge and current responsible for the radiation connected with the quantum transition from a to b or vice versa. The atom was considered as an "orchestra of oscillators," and the matrix elements determined the strengths of those oscillators ascribed to each pair of states.

Actually, the Schrödinger equation allowed the calculation of transitions under the influence of an external radiation field, that is, the absorption of light and the forced emission of an additional photon in the presence of an incident radiation. The field of an incident light wave could be considered as a perturbation on the atom in the initial state; it was possible by means of the Schrödinger equation to calculate the probability of a transition that turned out to be proportional to the intensity of the incident light wave. However, the emission by a transition from a higher state to a lower state in a field-free vacuum could not be treated. One had either to use the oscillator model, taking the emission to be the classical radiation of the oscillators, or to use the Einstein relations. From the latter it follows that the probability of spontaneous emission from b to a is equal to the absorption probability from a to b when the light intensity per frequency interval $d\omega$ is put equal to a certain value I_0:

$$I_0 = \frac{\hbar\omega^3}{\pi^2 c^2}. \tag{3}$$

This happens to be the light intensity when each degree of freedom of the radiation field contains one photon. According to this rule, the probability of spontaneous emission is equal to the probability of a forced emission by the fictitious radiation field of equation (3).

But why? According to the Schrödinger equation, any stationary state should have an infinite lifetime when there is no radiation present.

Quantization of the radiation field and its coupling to nonrelativistic systems

Dirac's fundamental paper in 1927, entitled "The Quantum Theory of the Emission and Absorption of Radiation," changed all that. Quantum mechanics must be applied not only to the atom via the Schrödinger equation but also to the radiation field. Dirac made use of an old idea of Ehrenfest (1906) and Peter Debye (1910) to describe the electromagnetic field in empty space as a system of quantized oscillators. In the presence of atoms or of other systems of charged particles, the coupling between the charged particles and the field is expressed by an interaction energy

$$H' = e\int \mathbf{j}\cdot\mathbf{A}d^3x, \tag{4}$$

where $\mathbf{j}$ is the current density of the particles. The value e of the particle charge is inserted here as an explicit factor, and $\mathbf{A}$ is the vector potential. Both $\mathbf{j}$ and $\mathbf{A}$ are operators in the quantized system of the atom and the field oscillators. Expression (4) is a direct consequence of Maxwell's equations. The Hamiltonian of the combined system then has the form

$$
\begin{aligned}
H &= H_0 + H', \\
H_0 &= H_{\text{field}} + H_{\text{atom}},
\end{aligned}
\tag{5}
$$

where H_{field} is the Hamiltonian of the isolated field oscillators and H_{atom} is the Schrödinger Hamiltonian of the atom isolated from the electromagnetic fields.

The Hamiltonian H_0 describes field and atom without interaction. The effects of H' are treated as a perturbation on the system H_0. The stationary states of H_0 are characterized by

$$(\ldots,n_i,\ldots; a), \tag{6}$$

where the n_i are the occupation numbers of the radiation oscillators (the numbers of photons present in each oscillator i), and a indicates the stationary state of the atom.

The states (6) are no longer stationary when the perturbation energy H' is taken into account. The theory yields simply and directly the laws of emission and absorption of light. Indeed, the state $(\ldots 0,0,\ldots; a)$ of an atom, the excited state a without any radiation present, is not stationary according to the Hamiltonian (5). A first-order perturbation calculation gives a probability $P_{ab}d\Omega$ per unit time for a transition from a to a state b, of lower energy, accompanied by the emission of a photon of frequency $\omega = (\epsilon_a - \epsilon_b)/\hbar$ into the solid angle $d\Omega$ and with a polarization vector $\mathbf{s}$:

$$P_{ab}d\Omega = \frac{e^2}{\hbar c}\frac{\pi}{2\hbar\omega^2}I_0|\mathbf{s}\cdot\mathbf{j}_{ab}|^2 d\Omega; \tag{7}$$

I_0 is given by the expression (3). The matrix element is determined (for a one-electron system) by

$$\mathbf{j}_{ab} = \int \psi_a^* \mathbf{j} e^{i\mathbf{k}_{ab}\cdot\mathbf{x}}\psi_b d^3x, \tag{8}$$

where **j** is the operator of the current and $\mathbf{k}_{ab}$ is the wave vector of the emitted quantum. The effect of the size of the system compared to the wavelength is taken into account by the exponential; it was neglected in the oscillator picture (dipole approximation). According to (7), spontaneous emission appears as a forced emission caused by the zero-point oscillations of the electromagnetic field, which are always present, even in a space without any photons.

This was the start of an interesting development in theoretical physics. After Einstein had put an end to the concept of aether, the field-free and matter-free vacuum was considered to be truly "empty space." The introduction of quantum mechanics changed this situation, and the vacuum gradually became "populated." In quantum mechanics an oscillator cannot be exactly at its rest position except at the expense of an infinite momentum, according to Heisenberg's uncertainty relation. The oscillatory nature of the radiation field therefore requires zero-point oscillations of the electromagnetic fields in the vacuum state, which is the state of lowest energy. The spontaneous emission process can be interpreted as a consequence of these oscillations.

An important contribution to the physical understanding of the quantized radiation field was a paper by Bohr and Leon Rosenfeld in which a number of "Gedanken experiments" were described for measuring electromagnetic field strengths.[6] Their considerations demonstrated that there exist uncertainty relations like Heisenberg's between different field strengths, in full accord with the quantization of the fields as oscillators. For example, the x component of the electric field and the y or z component of the magnetic field cannot be simultaneously well defined.

Dirac's theory produced all the results regarding the absorption and emission of light by atoms that had previously been obtained by unreliable arguments. The results followed from the Hamiltonian (5) when the interaction energy (4) was treated as a first order perturbation. Some other radiation phenomena such as photon scattering processes, resonance fluorescence, and nonrelativistic Compton scattering of photons by electrons appear only in the second order of the perturbation treatment, but the theory gave an excellent account of all radiation phenomena in the order in which they first appear. The higher approximations give rise to difficulties that will be discussed later.

Greatly impressed by the success and simplicity of Dirac's treatment of the interactions between the radiation field and atomic systems, Eugene P. Wigner and I tried to understand the phenomenon of the natural width of spectral lines on the basis of his theory.[7] As perturbation theory is insufficient for the treatment of this special effect, because it considers only transition probabilities per unit time and does not include the exponential decay of an excited state, we applied a method that assumed exponential time dependence from the outset. This could be done only by restricting the equations to a few atomic levels.[8] The results were considered surprising: The natural width Γ_{ab} of a spectral line emitted by a transition from a to b turned out to be the sum of the widths of the levels: $\Gamma_{ab} = \Gamma_a + \Gamma_b$, where Γ_a and Γ_b are the reciprocal lifetimes of the levels. A weak line (small Γ_a) will be broad if Γ_b is large. Today this result seems almost obvious, but at that time the thinking about emission was so closely tied to the picture of the atom as an orchestra of oscillators that the line width was expected to be given by the transition probability from a to b, that is, by Γ_a.[9] Later on, I applied the same methods to the treatment of resonance fluorescence.[10]

The quantized radiation field and its coupling to relativistic systems

In 1928 Dirac published two papers on a new relativistic wave equation of the electron. It was his third great contribution to the foundations of physics; the first was the reformulation of quantum mechanics, the "transformation theory"; the second was the theory of radiation. The Dirac equation was supposed to replace Schrödinger's equation for cases where electron energies and momenta were too high for a nonrelativistic treatment. It immediately gave rise to three great triumphs:[11]

1. The spin $\hbar/2$ of the electron appeared to be a natural consequence of the relativistic wave equation.

2. The g factor of the electron necessarily had the value 2, and the value of the magnetic moment of the electron followed directly from the equation.

3. When applied to the hydrogen atom, the equation yielded the correct Sommerfeld formula for the fine structure of the hydrogen spectrum.

The coupling of the quantized radiation field with the Dirac equation made it possible to calculate the interaction of light with relativistic electrons. The most important results were the derivation of the Klein-Nishina formula for the scattering of light by electrons,[12] the Møller formula for the scattering of two relativistic electrons,[13] and the emission of photons when electrons are scattered by the Coulomb field of nuclei (bremsstrahlung).[14]

In spite of these amazing successes, a number of serious difficulties turned up immediately, and it took a long time to solve them. It is interesting to note that the shakedown of nonrelativistic quantum mechanics took only two to three years (1925-7), although it was based on some of the most revolutionary ideas of physics, whereas an understanding of the consequences of the Dirac equation took a much longer time.

All of this took place during my graduate studies. For us, the Dirac equation was a great puzzle, and we had difficulty grasping the significance of the successes mentioned earlier. It was much more mysterious to us than Dirac's transformation theory and his theory of radiation. Imagine a student who had just gone through the conceptual problems of ordinary quantum mechanics and who begins to feel not at ease but barely capable of dealing with Schrödinger wave functions and Heisenberg noncommuting matrices suddenly facing wave functions with four components and with strange transformation properties of which he has never heard before. It was somewhat discouraging.

A great help for all of us was an article by Enrico Fermi, "The Quantum Theory of Radiation," which appeared early in 1932.[15] It contained a lucid presentation of Dirac's radiation theory, his relativistic wave equation, and the foundations of quantum electrodynamics. It used what one calls today the "Coulomb gauge" and thus avoided the difficulties of longitudinal quanta that caused so much trouble to Heisenberg and Pauli when they tried to develop a consistent theory.[16] Hans Bethe spoke for many of us when he wrote: "Many of you, probably, like myself, have learned their first field theory from Fermi's wonderful article."[17]

Now to the serious difficulties of the theory. They came from the existence of states of negative kinetic energy or negative mass. There was no way to get rid of them. If one tried to exclude them from the Hilbert space of the electron, the space became incomplete; furthermore, the Klein-Nishina formula could not be derived without them. Taken at face value, the existence of these states would imply that the hydrogen atom is not stable because of radiative transitions from the ordinary states to the states of negative energy.

The properties of these impossible states were constantly in the center of discussion during those years. George Gamow referred to electrons in these states as "donkey electrons" because they tend to move in the opposite direction from the applied force.

The triumph and curse of the filled vacuum

It was again Dirac who proposed a way out of the difficulty.[18] As it happens with the ideas of great men, it was not only a way out of a difficulty but also a seminal idea that led to the recognition of the existence of antimatter and ultimately to the development of field theory with all its concomitant insights into the nature of matter. He made use of the Pauli principle and assumed that, in the vacuum, all states of negative kinetic energy are occupied. This was the second step in the development of "populating" the vacuum. Later on, this step was somewhat mitigated by eliminating the notion of the actual presence of these electrons, but the fluctuations of matter density in the vacuum remained as an additional property of the vacuum besides the electromagnetic vacuum fluctuations.

Dirac's daring assumption had most disturbing consequences, such as an infinite charge density and infinite (negative) energy density of the vacuum. Some of these impossible consequences were circumvented later, as reported in the next section. However, the assumption not only solved most of the problems of the negative energy states but led to an impressive and unexpected broadening of our views about matter.

First of all, the transitions from positive to negative energy states were excluded, and the stability of the atoms was assured. Furthermore, Dirac's assumption was eventually seen to require the existence of processes in which one particle from the "sea" of filled negative states would be lifted to a state of positive energy, if the necessary energy were supplied by absorption of photons or by other means. A hole in the sea and a normal particle would be created. The hole would have all the properties of a particle of opposite charge. Moreover, a particle could fall back into a hole with the emission of a photon with the right amount of energy and momentum. This, of course, would be a process of particle-antiparticle annihilation. Thus Dirac's assumption led to the recognition of the existence of antiparticles and the existence of two new fundamental processes: pair creation and pair annihilation.

In the beginning, these ideas seemed incredible and unnatural to everybody. No positive electron had been seen at that time; the asymmetry of charges (positive for the heavy nuclei, negative for the light electrons) seemed to be a basic property of matter. Even Dirac shrank from the concept of antimatter and tried to interpret the positive "holes" in the sea of the vacuum electrons as protons. It was soon recognized, however, that this interpretation would again lead to an unstable H atom and that the holes must have the same mass as the particles.[19] Antimatter ought to exist. Indeed, the positron was found by Carl D. Anderson in 1932; the antiproton was discovered 25 years later, because its production needed energy concentrations several thousand times higher than were available before the invention of the synchrocyclotron.

We should realize that these theoretical predictions of new fundamental processes and new properties of matter were made before even the slightest experimental evidence was known. On the contrary, all previous evidence contradicted the symmetry between positive and negative charges. These predictions rank among the greatest intellectual achievements in natural science.[20]*

Once the idea of the filled vacuum took hold, it was relatively easy to calculate the cross section for the annihilation of an electron and a positron into two photons and the cross section for pair creation by photons in the Coulomb field of atomic nuclei.[14,21] It is astonishing that it took more than three years after the identification of the holes with positrons before the pair creation in a Coulomb field was calculated. All that was needed was a very simple determination of a transition probability. It illustrates the wonder and incredulity that those ideas encountered during the first years.

Today it is hard to realize the excitement, the skepticism, and the enthusiasm aroused in the early years by the development of all the new insights that emerged from the Dirac equation. A great deal more was hidden in the Dirac equation than the author had expected when he wrote it down in 1928. Dirac himself remarked in one of his talks that his equation was more intelligent than its author. It should be added, however, that it was Dirac who found most of the additional insights.

The formulas derived for the creation of pairs and for radiative scattering (bremsstrahlung) also gave an excellent account of the development of cosmic-ray cascade showers in matter, once the incoming

---

* *Ed. note*: See also Joan Bromberg, "The Concept of Particle Creation before and after Quantum Mechanics," in *Historical Studies in the Physical Sciences, Vol. 7*, ed. by R. McCormmach (Princeton: Princeton University Press 1976), pp. 161-91. Before and around 1900, positive and negative electrons making a neutral doublet were sometimes called "neutrons." See N. Feather, "A History of Neutrons and Nuclei: Part 1," *Contemporary Physics 1* (1960), 191-203; J. L. Heilbron, "Lectures on the History of Atomic Physics, 1900-1922," in *History of Twentieth Century Physics*, ed. by C. Weiner (New York: Academic Press, 1977), pp. 40-108 (especially p. 45).

energy is transformed into electrons and photons. It is interesting to observe how this success was interpreted. First it was considered as proof that radiation theory and pair creation are valid even at very high energy. Then, when it turned out that a part of the cosmic rays do not form showers (we know now that this is the part consisting of muons), doubts were expressed as to the validity of radiation theory at high energies (see Chapter 12). But, as shown by Fermi and later by C. F. von Weizsäcker and E. J. Williams, the effect of a Coulomb field on a fast-moving electron can be expressed as the effect of light quanta when a suitable system of reference is used (the system in which the electron is at rest).[22] Such an analysis of the production of cascade showers shows clearly by this method that only energies and momenta of the order mc^2 and mc are exchanged in the relevant processes. Hence the shower production does not test the theory at high energies, nor can any deviation from the expected showers be explained by a breakdown of the theory at high energies. Indeed, electron accelerators of many GeV were needed to test the theory at large energies. Recent measurements with electron-positron colliders have shown radiation theory to be valid at least up to energy exchanges of 100 GeV.

How unreasonable the idea of antimatter seemed at that time may be illustrated by the fact that many of us did not believe in the existence of an antiparticle to the proton because of its anomalous magnetic moment. The latter was measured by Otto Stern in 1933 and could be interpreted as an indication that the proton does not obey the Dirac equation. The fundamental character of the matter-antimatter symmetry and its independence of the special wave equations were recognized only very slowly by most physicists.

The following conclusions must be drawn from the new interpretation of the negative energy states in the Dirac equation. There are no real one-particle systems in nature, not even few-particle systems. Only in nonrelativistic quantum mechanics are we justified to consider the hydrogen atom as a two-particle system, but not so in the relativistic case, because we must include the presence of an infinite number of vacuum electrons. Even if we consider the filled vacuum as a clumsy description of reality, the existence of virtual pairs and of pair fluctuations shows that the days of fixed particle numbers are over.

Particles must be considered as the quanta of a field, just as photons are the quanta of the electromagnetic field; such quanta are created or destroyed. The theory of the interaction of charged particles with the radiation field has become a field theory, a theory in which two (or more) quantized fields interact: the matter field(s) and the radiation field.

The formulation of that aspect of field theory makes use of a method called second quantization. It is the inappropriate name of a formalism for the treatment of many particle problems that was devised by Pascual Jordan and Oskar Klein and by Jordan and Wigner.[23] There is no additional quantization involved. It is a formalism appropriate to field theory in which creation and destruction operators are introduced that increase or decrease the number of particles in certain quantum states. The field amplitudes are expressed as linear combinations of these operators. It is a direct generalization of the quantization of the electromagnetic field as decomposed into oscillator amplitudes. The operator of an oscillator amplitude contains matrix elements only between states that differ by one unit of excitation. The corresponding operator either adds (creates) or subtracts (destroys) a quantum of the oscillator.

There are essential differences between a field of particles with spin 1/2 and the radiation field. The former describes the behavior of fermions, whereas the latter is an example of a boson field. In the classical limit, the boson field is a classical field whose field strength is a well-defined function of space and time. The fermion fields cannot have a classical limit, because no more than one fermion can be put into one wave; its classical limit is a particle with a well-defined momentum and position. It is interesting to note that, so far, the constituents of matter have all turned out to be fermions interacting by means of boson fields.

Furthermore, the interaction between fermion and boson fields in its simplest form necessarily is bilinear in the fermion fields and linear in the boson fields. This is indicated by the form (4), because the currrent density is a bilinear expression of the particle wave functions. One cannot construct a Lorentz invariant expression that is linear or cubic in the spinor wave functions. Boson fields (vector or scalar), however, may appear linearly in the interaction.

When the fields are expressed in terms of creation and annihilation operators, the form of the interaction can be interpreted in the following way: The fundamental interaction between fermions and bosons consists of the product of two fermion creation and/or destruction operators b^+ and b, and one boson operator a or a^+: b^+ba or b^+ba^+. It is interpreted as a change of state of a fermion, destroyed in one state and created in another accompanied with either the emission or absorption of a boson. But there are also terms of the form b^+b^+a or bba^+ that give rise to the creation or annihilation of a fermion pair with a corresponding boson emission or absorption.

The fight against infinities I: elimination of the vacuum electrons

In spite of all successes of the new hole theory of the positron, the infinite charge density and the infinite negative energy density of the vacuum made it very difficult to accept the theory at its face value. A war against infinities started at that time.[24] It was waged with increasing fervor by the developers of quantum electrodynamics when more intricate infinities appeared besides those mentioned earlier, as will be described in the subsequent sections.

When I had the privilege of working with Pauli in Zurich, I was indirectly involved in these fights. A lively correspondence developed between Heisenberg and Pauli; every letter from Heisenberg was intensely discussed. More often than not I was asked to write the first draft of Pauli's critical answers to some of Heisenberg's suggestions.

There is a rather primitive way to take care of the infinite charge density by a slight change in the definition of charge and current. It amounts to the following argument: Because the theory is completely symmetric in regard to electrons and positrons, it would be equally valid to construct a theory in which the positrons are the particles and the electrons are the holes in a sea of positrons that occupy negative energy states. The actual theory then could be considered as a superposition of these two theories, one with an infinite negative charge density and the other with an infinite positive one. This combination also serves to emphasize the symmetry between matter and antimatter. The vacuum charge densities cancel; the corresponding expressions for charge and current indeed give a more satisfactory description of the phenomena.

It was recognized in 1934 by J. Robert Oppenheimer and Wendell Furry that the creation and destruction operators are most suitable for turning the liability of the negative energy states into an asset, by interchanging the role of creation and destruction of those operators that act on the negative states.[25] This interchange can be done in a consistent way without any fundamental change of the equations. The consequences are identical to those of the filled-vacuum assumption, but it is not necessary to introduce that disagreeable assumption explicitly. Particles and antiparticles enter symmetrically into the formalism, and the infinite charge density of the vacuum disappears. One can even get rid of the infinite negative energy density by a suitable rearrangement of the bilinear terms of the creation and destruction operators in the Hamiltonian. After all, in a relativistic theory the vacuum must have vanishing energy and momentum. There remains, however, the unpleasant fact of the existence of vacuum fluctuations without any energy.

The fundamental interaction between charged fermions and photons now contains three basic processes: the scattering of a fermion, with the emission or absorption of a photon, and the creation or annihilation of a fermion-antifermion pair, with the emission or absorption of a photon. All electrodynamic interaction processes are combinations of these fundamental steps.

Surprisingly enough, it took many years before the physicists realized the great advantages of this new formalism. One still read about the "hole theory" of positrons in papers written in the late 1940s, when renormalization was the topic of the day.

An interesting episode in the fight for the elimination of vacuum electrons was the quantization of the Klein-Gordon relativistic wave equation for scalar particles. I was appointed assistant to Pauli in Zurich in 1934 – one of those strokes of luck in my life as a physicist. During that time I studied the properties of the Klein-Gordon equation for charged scalar particles. It seemed to be a rather academic activity, because no scalar particle was known at that time. In that theory, the charge density is $(\dot{\phi}^*\phi - \dot{\phi}\phi^*)$ and is not identical to the wave intensity, which is $|\phi|^2$. Therefore, it seemed possible that, under the influence of external electromagnetic fields, the total intensity $\int|\phi|^2 d^3x$ might change in time, although the total charge remains conserved. It smelled of a creation or annihilation process of oppositely charged particles. I was unable to develop the problem further because I was not accustomed to use creation and destruction operators. I went to Pauli for help, but it was not too easy to draw his attention to the problem. (I have described some of my struggles in an essay.[26]) When he finally got caught up with the problem, it attracted him because he immediately saw that the quantized Klein-Gordon equation gives rise to particles and antiparticles and to pair creation and annihilation processes without introducing a vacuum full of particles.

In the course of this work, Pauli asked me to calculate the cross section for pair creation of scalar particles by photons. It was only a short time after Bethe and Heitler had solved the same problem for electrons and positrons. I met Bethe in Copenhagen at a conference and asked him to tell me how he did the calculations. I also inquired how long it would take to perform this task; he answered, "It would take me three days, but you will need about three weeks." He was right, as usual; furthermore, the published cross sections (not checked by Pauli) were wrong by a factor of four.

Note that at the time the method of exchanging the creation and destruction operators (for negative energy states) was not yet in fashion; the hole theory of the filled vacuum was still the accepted way of dealing with positrons. Pauli called our work the "anti-Dirac paper." He considered it a weapon in the fight against the filled vacuum that he never liked. We thought that this theory only served the purpose of a nonrealistic example of a theory that contained all the advantages of the hole theory without the necessity of filling the vacuum. We had no idea that the world of particles would abound with spin-zero entities a quarter of a century later. That was the reason we published it in the venerable but not widely read *Helvetica Physica Acta*.[27]

The work on the quantization of the Klein-Gordon equation led Pauli to the famous relation between spin and statistics. Pauli demonstrated in 1936 the impossibility of quantizing equations of scalar or vector fields that obey anticommutation rules. He showed that such relations would have the consequence that physical operators do not commute at two points that differ by a spacelike interval. This would be in contradiction to causality because it would require that measurements interfere with each other when no signal can pass from one to the other.[28] Thus Pauli concluded that particles with integer spin could not obey Fermi statistics. They must be bosons. During the days of the hole theory it was obvious that particles with spin 1/2 could not obey Bose statistics because it would be impossible to "fill" the vacuum. Four years later Pauli proved the necessity of Fermi statistics for half-integer spins, also on the basis of the same causality arguments.[29]

The fight against infinities II: infinities on the attack; the infinite self mass

The infinities of the filled vacuum and of the zero-point energy of the vacuum turned out to be relatively harmless compared with other infinities that appeared in quantum electrodynamics when the coupling between the charged particles and the radiation field was considered in detail. No difficulties appeared as long as only the first terms of the perturbation treatment were taken into account; that is, those terms in which the phenomenon under consideration appears in the lowest order. It soon turned out that the higher terms always contain infinities, as Oppenheimer was the first to point out.[30*]

In 1934, Pauli asked me to calculate the self energy of an electron according to the positron theory. It was a modern repetition of an old problem of electrodynamics. In classical theory the energy contained in the field of an electron of radius a (neglecting the inside) is

$$\epsilon = \frac{1}{2} \int_a^\infty \frac{e^2}{r^4} r^2 dr = \frac{e^2}{2a}, \tag{9}$$

and it will diverge linearly if the radius a goes to zero. The corresponding calculation in the positron theory was much more complicated. One had to calculate the difference between two infinite amounts: the energy of the vacuum and the energy of the vacuum plus one electron. It could be done, and the result was equivalent to the statement that the electric field inside one Compton wavelength $\lambda_c = h/mc$ from the electron is not e/r^2 but $(e/r^2)(r/\lambda_c)^{1/2}$. It increases only as $r^{-3/2}$ when r goes to zero. The self-energy ϵ then becomes

$$\epsilon = m_0 c^2 + \frac{3}{2\pi} m_0 c^2 \frac{e^2}{hc} \log \frac{\lambda_c}{a}, \tag{10}$$

where m_0 is the intrinsic or "mechanical" mass of the electron that appears in the Hamiltonian of the electron when it is decoupled from the electromagnetic field.[31] It diverges only logarithmically.

This brings back one of the dark moments of my professional career. I made a mistake in the first publication that resulted in a quadratic divergence of the self-energy. Then I received a letter from Furry, who kindly pointed out my rather silly mistake and the fact that actually the divergence is logarithmic. Instead of publishing the result himself, he allowed me to publish a correction quoting his intervention. Since then, the discovery of the logarithmic divergence of the electron self-energy has been wrongly ascribed to me rather than to Furry.

A consistent relativistic theory requires a point electron; that is, $a \rightarrow 0$. It is worth noting, however, that the value of a for which the second term of (10) becomes half of the first is as small as 10^{-72} cm! Even the Schwarzschild radius of the electron is only 10^{-45} cm. This means that the deformation of the space around the electron is large enough to

---

* *Ed. note*: See also Note 16 and I. Waller, "Bemerküngen über die rolle der Eigenenergie des Electrons in der Quantentheorie der Strahlung," *Z. Physik* 62 (1930), 673-6.

prevent the electron from interacting with photons of that wavelength. This would provide a natural cutoff long before the electromagnetic self-energy becomes important. Unfortunately, no consistent calculation of this effect has ever succeeded.

Later on, I tried to investigate the physical reasons for this weak divergence.[32] It turned out that it comes from the interplay between the electron and the fluctuations of the vacuum. Among other effects, it is the Pauli principle that reduces the self-energy, because it forbids certain fluctuations of electron-positron pairs to take place, when the electron of the pair comes too near to the electron under consideration. This effect induces a change in charge density near the electron and reduces the self-energy.

Another somewhat more benign type of infinity appears in quantum electrodynamics when emission of photons of very low frequency is considered. Such emission takes place, for example, when an electron is scattered by a static electric field. Classical theory predicts that the emitted energy does not vanish in the limit of zero frequency. The quantum result ought to be identical with the classical one at that limit; it would indicate that the number of emitted quanta goes to infinity. This trouble, called "infrared catastrophe," can be avoided by describing this limit with the help of classical fields, as Felix Bloch and Arnold Nordsieck showed in their important paper of 1937.[33] It put an end to any worries about this kind of infinity.

The fight against infinities III: infinities on the attack; the infinite vacuum polarization

The virtual pairs endow the vacuum with properties similar to those of a dielectric medium. We may ascribe a dielectric coefficient ϵ to the vacuum. A direct calculation of this effect leads to a dielectric coefficient that consists of a constant part ϵ_0 and an additional part that depends on the electromagnetic fields and their derivatives in time and space:

$$\epsilon = \epsilon_0 + \epsilon_{field}. \tag{11}$$

The constant part ϵ_0 cannot have any physical significance because it serves only to redefine the unit of charge. Any charge Q_0 would appear as $Q = Q_0/\epsilon$. The actual value of ϵ_0 turns out to be logarithmically divergent; it goes as $\log(\Lambda/m)$, where Λ is the highest momentum considered in the calculation. The additional field-dependent term, however, turns out to be finite and therefore should have physical significance. Heisenberg, H. Euler, and I found an exact expression for static slowly varying fields, valid to all orders of $e^2/\hbar c$.[34]

Let us now consider what happens to a charge Q_0 when placed in a vacuum with a dielectric coefficient of the form (11). At large distances r, the effective charge will be Q_0/ϵ_0. When r becomes of the order $\lambda_c = h/(mc)$ or less, the second term becomes important. Calculations of this term for a Coulomb field (this is not a slowly varying field) were carried out in perturbation theory by Serber and by Edwin A. Uehling,[35] who used a method suggested by Dirac.[24] They found that $\epsilon(r)$ decreases with r when r becomes smaller than the Compton wavelength λ_c. That is because, for smaller r, only those virtual pairs contribute whose energy is larger than hc/r. The decrease is finite and calculable. The infinite value of ϵ_0 was interpreted as an indication that the intrinsic "true" charge Q_0 is infinite, so that the observed charge becomes finite and equal to $e = Q_0/\epsilon_0$ for $r \to \infty$. The decrease of ϵ with decreasing r when $r < \lambda_c$ would then amount to an increase of the effective charge Q_{eff} at those small distances.

This increase of Q_{eff} for $r < \lambda_c$ over the value e is rather small, of the order of $e/137$. A strong increase occurs only at the very small distance $r \sim \lambda_c \exp(-\hbar c/e^2)$, the same distance we discussed in connection with the self-energy, where the theory is most likely inapplicable. We then get a dependence of Q_{eff} on the distance, as shown in Figure 3.1. This must be regarded as an interesting result of quantum electrodynamics in spite of the unnatural assumption of an infinite "true" charge Q_0.

The fight against infinities IV: counterattack; renormalization

The appearance of infinite magnitudes in quantum electrodynamics was noticed in 1930. Because they occurred only when a certain phenomenon was calculated to a higher order of accuracy than the lowest one in which it appeared, it was possible to ignore the infinities and stick to the lowest-order results that were good enough for the experimental accuracy at that period. However, the infinities at higher order indicated that the formalism contained unrealistic contributions from the interaction with high-momentum photons.

Already in 1936 the conjecture had been expressed that the infinite contributions of the high-momentum photons were all connected with the infinite self-mass, with the infinite intrinsic charge Q_0, and with nonmeasurable vacuum quantities such as a constant dielectric coefficient of the vacuum.[36] Thus it seemed that a systematic theory could be developed in which these infinities were circumvented. At that time,

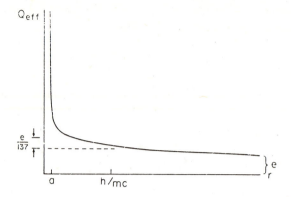

Figure 3.1. Running coupling constant in quantum electrodynamics. The effective charge Q_{eff} as a function of the distance r. The distance a is very much smaller than indicated in this drawing. From V. F. Weisskopf in *Physics Today 34* (1981), 77.

nobody attempted to formulate such a theory, although it would have been possible then to develop what is now known as the method of renormalization.

There was one tragic exception, and that was Ernst C. G. Stueckelberg.[37] He wrote several papers in which a manifestly invariant formulation of field theory was put forward. This could have been a perfect basis for developing the ideas of renormalization. Later on, he actually carried out a complete renormalization procedure in papers with D. Rivier, independent of the efforts of other authors.[38] Unfortunately, his writings and his talks were rather obscure, and it was very difficult to understand them or to make use of his methods. He came frequently to Zurich in the years 1934-6, when I was working with Pauli, but we could not follow his way of presentation. Had Pauli and I myself been capable of grasping his ideas, we might well have calculated the Lamb shift and the correction to the magnetic moment of the electron at that time.

A new impetus to such attempts came from an experimental result. Willis E. Lamb and Robert Retherford were able to measure reliably the difference in energy between the $2S_{1/2}$ and $2P_{1/2}$ states of hydrogen (Lamb shift).[39] The two states should have been exactly degenerate according to the Dirac equation applied to the hydrogen problem. Already in the 1930s the degeneracy of these two levels was in doubt from spectroscopic measurements (the so-called Pasternack effect);

William Houston and R. C. Williams found strong indications of a difference, but Lamb and Retherford, using newly developed microwave methods, definitely established the splitting and measured it with great accuracy.[40]*

It had long ago been conjectured that such a splitting should be caused by the coupling of the radiation field with the atom, but early attempts to calculate it ran into difficulties because the infinite mass and vacuum polarization appeared in the same approximation. It was Hans A. Kramers who pointed out that one ought to be able to calculate the effect by carefully subtracting the infinite energy of the bound electron from that of the free one and thereby separating the parts that contribute to the mass and charge from those of real significance.[41] Infinities are always difficult to subtract in an unambiguous way. After the Lamb shift had been measured, Bethe made an attempt to estimate the effect of the radiation coupling, simply by omitting the coupling with photons of an energy larger than mc^2.[42] This attempt was successful because most of the effect comes from the coupling with photons of lower energy that can be treated nonrelativistically. J. B. French and I calculated that difference carefully and got a well-defined result in agreement with the experiment. We believe that we were the first to arrive at that result. Then followed a tragicomical episode. We showed our method and our result to Julian Schwinger and to Richard P. Feynman. They independently tried to repeat our calculations but found a result differing by a small additive numerical constant. The trouble was that both of them got the same result. Having both Feynman and Schwinger against us shook our confidence, and we tried to find a mistake in our calculation, without success. Only seven months later Feynman informed us that it was he and Schwinger who had made a mistake! We published our paper, but in the meantime, a similar calculation was made by Norman M. Kroll and Lamb, which appeared a few months earlier than ours.[43] Self-confidence is an important ingredient that makes for a successful physicist.

This episode shows that our primitive methods of subtracting two infinities were clumsy and unreliable. Therefore, a formidable group of physicists, including Schwinger, Feynman, Freeman J. Dyson, and Sinitiro Tomonaga, developed a reliable way to deal with the infinities.[44] A method of renormalization was introduced in which the initial pa-

* *Ed. note:* Pasternack attempted a theoretical explanation: S. Pasternack, "Note on the Fine Structure of H_α and D_α," *Phys. Rev. 54* (1938), 1113.

rameters were eliminated in favor of those with immediate physical significance. In any computation of an electrodynamic result, the effects of the mass and charge redefinitions had to be incorporated. Infinite "counter-terms" were introduced into the Hamiltonian in such a manner that they compensated for the infinite mass and charge. In order to make this procedure unambiguous, it was necessary to keep the expressions in a manifestly relativistic and gauge-invariant form throughout the calculations.

The results were most encouraging. Schwinger found that the magnetic moment of the electron should be slightly larger than the Bohr magneton, a result that was observed shortly beforehand by John Nafe, Edward Nelson, and Isodor Rabi and more accurately afterward by Henry Foley and Polykarp Kusch.[45] The Lamb-shift results were recalculated in a much simpler way, radiative corrections of higher order in $e^2/\hbar c$ to scattering processes were unambiguously determined, and the vacuum polarization effects were worked out in detail; the latter found an impressive experimental confirmation in measurements of the spectrum of muonic atoms (the electron replaced by a muon); the muon moves in the region $r < (\hbar/m_e c)$ where the vacuum polarization is a 1% effect.

The war against infinities was ended. There was no longer any reason to fear the higher approximations. The renormalization took care of all infinities and provided an unambiguous way to calculate with any desired accuracy any phenomenon resulting from the coupling of electrons with the electromagnetic field. It was not a complete victory, because infinite counter-terms had to be introduced to remove the infinities. Furthermore, the procedure of eliminating infinities could be carried out only by renormalizing successively at each step of the perturbation expansion in powers of the coupling parameter. It is still not clear whether or not this method leads to a convergent series.[46] It is like Hercules's fight against Hydra, the many-headed sea monster that grows a new head for every one cut off. But Hercules won the fight, and so did the physicists. Sidney Drell characterized the situation most aptly as "a peaceful coexistence with the infinities."

Here are the signs of victory in the war against infinities:

1. The Lamb shift in hydrogen (about 2.5% is due to vacuum polarization; most of the rest is the interaction with the zero-point oscillations of the electromagnetic field):

$$\Delta\nu(2S_{1/2} - 2P_{1/2}) \quad \begin{aligned} &= 1057.862 \ (20) \ \text{MHz (experimental)} \\ &= 1057.864 \ (14) \ \text{MHz (theoretical)} \end{aligned}$$

(the units are megahertz).

2. The g factor of the electron $[a = (g - 2)/2] \times 10^3$

$$a \quad \begin{aligned} &= 1.15965241 \ (20) \ \text{(experimental)} \\ &= 1.159652379 \ (261) \ \text{(theoretical)}. \end{aligned}$$

3. Vacuum polarization: 90% of the Lamb shift in muonic helium (α particle + muon) is caused by vacuum polarization:

$$\Delta E(2S_{1/2} - 2P_{3/2}) \quad \begin{aligned} &= 1.5274 \ (0.9) \ \text{eV (experimental)} \ . \\ &= 1.5251 \ (9) \ \text{eV (theoretical)}. \end{aligned}$$

In spite of these victories, there remain nagging problems in quantum electrodynamics. There are definite indications that we understand only a partial aspect of what is going on. As mentioned earlier, the elimination of infinities is possible only in a perturbation approach; it is contingent on the smallness of $e^2/\hbar c$. But the effective coupling constant at very small (indeed, incredibly small) distances becomes larger than unity. Will there be a theory that avoids renormalization by using nonperturbative methods? Or will a future unification of electrodynamics and general relativity heal the disease of divergencies because of the fact that the dangerous distances are smaller than the Schwarzschild radius of the electron? Or will a unification of electrodynamics with strong interactions bring a solution of the problems? The unification with weak interactions by M. J. G. Veltman, Sheldon Glashow, Steven Weinberg, Abdus Salam, G. 't Hooft, and G. Ward has not served that purpose.[47]

Moreover, there is no way to understand and derive the mass of the electron within today's electrodynamics. This problem has become even more acute since heavier electrons such as the muon and the τ lepton have been discovered. There is not the slightest indication why electrons with different masses should exist. In present-day field theories, the masses are arbitrary parameters that may assume any value. The proliferation of fundamental particles with different masses is not restricted to electrons. We also face a series of quark masses that defy explanation on a field-theoretical basis.

The importance of the mass problem may be illustrated as follows. We have no explanation for the mass of the electron, that is, for the smallness of the ratio $(1836)^{-1}$ between the electron mass and the proton mass. One of the two masses may be considered as a natural unit. In our description of nature we expect three intrinsic magnitudes to appear that determine the units of our measuring system. Their values do not require any explanation. These units may well be h, c, and one mass. The small value of the mass ratio of the electron and the proton determines the properties of everything we see around us. It is the precondition of molecular architecture, of the fact that the positions of atomic nuclei are well defined within the surrounding electron clouds. Without it there would be no materials as we know them and no life. We have no idea about the deeper reasons for the smallness of that important ratio. Finally, last but certainly not least, there is no explanation in sight within the present framework of electrodynamics for the specific value of the electric charge, that is, for the ratio $e^2/\hbar c = (137)^{-1}$. We have not the slightest idea why the masses of electrons and quarks vary all over the scale, but the electric charges are all identical or differ by simple factors such as 1/3 or 2/3.

Some time ago I was told about the teachings of the Kabbalah by Gershom Scholem, the great scholar of Jewish mysticism. He explained to me that every Hebrew word is associated with a number that carries some symbolic meaning. He then asked me to tell him a few of the unsolved riddles of modern physics. When I told him about $(e^2/\hbar c)$, his eyes lit up in surprise and astonishment: "Do you know that 137 is the number associated with the word Kabbalah?" So far, this is still the best explanation for the value of the charge of the electron.

Notes

1 This chapter is based in part on the 1979 Bernard Gregory Lecture (Yellow Book 80-03 Geneva: CERN, 1980).
2 A variation on a statement by Winston Churchill about the Royal Air Force.
3 There are two studies of some interest: A. Pais, "The Early History of the Electron: 1897-1947," in *Aspects of Quantum Theory*, ed. by Abdus Salam and E. P. Wigner (Cambridge University Press, 1972), pp. 79-93; Steven Weinberg, "The Search for Unity: Notes for a History of Quantum Field Theory," *Daedalus 106* (1977), 17-35.
4 P. A. M. Dirac, "The Quantum Theory of the Emission and Absorption of Radiation," *Proc. Roy. Soc. (London) A114* (1927), 243-65; reprinted in *Selected Papers on Quantum Electrodynamics*, ed. by Julian Schwinger (New York: Dover, 1958).
5 Gerhard Herzberg, *Atomic Spectra and Atomic Structure*, 2nd ed., translated from German by J. W. T. Spinks (New York: Dover, 1944).
6 N. Bohr and L. Rosenfeld, "Für Frage der Messbarkeit der elektromagnetischen Feldgrössen," *Kgl. Danske Videnskab. Selskab, Mat.-Fys. Medd. 12* (1933), 1-65.
7 V. Weisskopf and E. P. Wigner, "Berechnung der natürlichen Linienbreite auf Grund der Diracschen Lichttheorie," *Z. Physik 63* (1930), 54-73.

8 We also had to disregard certain infinite integrals whose existence was not completely unknown at that time and that were harbingers of future troubles. See J. R. Oppenheimer, "Note on the Interaction of Field and Matter," *Phys. Rev. 35* (1930), 461-77.

9 I remember vividly that, by his uncanny intuition, James Franck predicted the correct result.

10 V. F. Weisskopf, "Zur Theorie der Resonanzfluorescenz," *Ann. Physik 9* (1931), 23-66.

11 P. A. M. Dirac, "The Fundamental Equations of Quantum Mechanics," *Proc. Roy. Soc. (London) A109* (1925), 642-53; P. A. M. Dirac, "The Physical Interpretation of the Quantum Mechanics," *Proc. Roy. Soc. (London) A113* (1927), 621-41. It turned out later that there exist relativistic wave equations for particles with different spin. Dirac's equation for a spin $\hbar/2$ is distinguished in that the energy operator appears linearly.

12 O. Klein and Y. Nishina, "Ueber die Streuung von Strahlung durch freie Elektronen nach der neuen relativistischen Quantendynamik von Dirac," *Z. Physik 52* (1929), 853-68. See also I. Tamm, "Ueber die Wechselwirkung der freien Elektronen mit der Strahlung nach der Diracschen Theorie des Elektrons und nach der Quantenelektrodynamik," *Z. Physik 62* (1930), 545-68.

13 C. Møller, "Zur Theorie des Durchgangs schneller Elektronen durch Materie," *Ann. Physik 14* (1932), 531-85.

14 W. Heitler and F. Sauter, "Stopping of Fast Particles with Emission of Radiation and the Birth of Positive Electrons," *Nature 132* (1933), 892; H. Bethe and W. Heitler, "On the Stopping of Fast Particles and on the Creation of Positive Electrons," *Proc. Roy. Soc. (London) A146* (1934), 83-112.

15 Enrico Fermi, "Quantum Theory of Radiation," *Rev. Mod. Phys. 4* (1932), 87-132.

16 W. Heisenberg and W. Pauli, "Zur Quantentheorie der Wellenfelder," *Z. Physik 56* (1929), 1-61 and *59* (1930), 168-90.

17 H. A. Bethe, at the Fermi memorial session of the American Physical Society in 1955, *Rev. Mod. Phys. 27* (1955), 253.

18 P. A. M. Dirac, "A Theory of Electrons and Protons," *Proc. Roy. Soc. (London) A126* (1930), 360-5.

19 Hermann Weyl, *The Theory of Groups and Quantum Mechanics* , translated from 2nd (revised) German edition by H. P. Robertson (New York: Dutton, 1931; reprinted New York: Dover, 1950), p. 225. German original, *Gruppentheorie and Quantenmechanik* (1928). J. R. Oppenheimer, "On the Theory of Electrons and Protons," *Phys. Rev. 35* (1930), 562-3; J. R. Oppenheimer, "Two Notes on the Probability of Radiative Transitions," *Phys. Rev. 35* (1930), 939-47; P. A. M. Dirac, "Quantized Singularities in the Electromagnetic Field," *Proc. Roy. Soc. (London) A133* (1931), 60-72.

20 The possibility of antiparticles had already been mentioned by Pauli and by Einstein. See, for example, A. Pais, "Einstein and the Quantum Theory," *Rev. Mod. Phys. 51* (1979), 863-914 (especially p. 909).

21 P. A. M. Dirac, "On the Annihilation of Electrons and Protons," *Proc. Cambridge Phil. Soc. 26* (1930), 361-75; J. R. Oppenheimer and M. S. Plesset, "On the Production of the Positive Electron," *Phys. Rev. 44* (1933), 53-5; Yoshio Nishina, Shinichiro Tomonaga, and Shoichi Sakata, "On the Photo-electric Creation of Positive and Negative Electrons," *Scientific Papers of the Institute of Physics and Chemical Research 17 (Suppl.)* (1934), 1-5.

22 E. Fermi, "Ueber die Theorie des Stosses zwischen Atomen und elektrisch geladenen Teilchen," *Z. Physik 29* (1924), 315-27; C. F. von Weizsäcker, "Ausstrahlung

bei Stössen sehr Schneller Elektronen," *Z. Physik 88* (1934), 612-25; E. J. Williams, "Correlation of Certain Collision Problems with Radiation Theory," *Kgl. Danske Videnskab. Selskab, Mat.-Fys. Medd. 13* (1935), 1-50.

23 P. Jordan and O. Klein, "Zum Mehrkörperproblem der Quantentheorie," *Z. Physik 45* (1927), 751-65; P. Jordan and E. Wigner, "Ueber das Paulische Aequivalenzverbot," *Z. Physik 47* (1928), 631-51, reprinted in Schwinger (ed.), *Quantum Electrodynamics* (Note 4).

24 W. Heisenberg, "Bemerkung zur Diracschen Theorie des Positrons," *Z. Physik 90* (1934), 209-31; P. A. M. Dirac, "Discussion of the Infinite Distribution of Electrons in the Theory of the Positron," *Proc. Cambridge Phil. Soc. 30* (1934), 150-63.

25 W. H. Furry and J. R. Oppenheimer, "On the Theory of the Electron and Positron," *Phys. Rev. 45* (1934), 245-62.

26 V. F. Weisskopf, "My Life as a Physicist," in *Physics in the Twentieth Century – Selected Essays of V. F. Weisskopf* (Cambridge, Mass.: M.I.T. Press, 1972), pp. 1-21 (especially p. 11).

27 W. Pauli and V. Weisskopf, "Ueber die Quantizierung der skalaren relativistischen Wellengleichung," *Helv. Phys. Acta 7* (1934), 709-31.

28 W. Pauli, "Théorie quantique relativiste des particules obéissant à la statistique de Einstein-Bose," *Ann. Inst. Henri Poincaré 6* (1936), 137-52.

29 W. Pauli, "The Connection between Spin and Statistics," *Phys. Rev. 58* (1940), 716-22, reprinted in Schwinger (ed.), *Quantum Electrodynamics* (Note 4).

30 See Oppenheimer (Note 8).

31 V. Weisskopf, "Ueber die Selbstenergie des Elektrons," *Z. Physik 89* (1934), 27-39; Erratum, *Z. Physik 90* (1934), 53-4.

32 V. F. Weisskopf, "On the Self-Energy and the Electromagnetic Field of the Electron," *Phys. Rev. 56* (1939), 72-85.

33 F. Bloch and A. Nordsieck, "Note on the Radiation Field of the Electron," *Phys. Rev. 52* (1937), 54-9, reprinted in Schwinger (ed.), *Quantum Electrodynamics* (Note 4).

34 W. Heisenberg and H. Euler, "Folgerungen aus der Diracschen Theorie des Positrons," *Z. Physik 98* (1936), 714-32; V. Weisskopf, "Ueber die Elektrodynamic des Vakuums auf Grund der Quantentheorie des Elektrons," *Kgl. Danske Videnskab. Selskab, Mat.-Fys. Medd. 14* (1936), 1-39, reprinted in Schwinger (ed.), *Quantum Electrodynamics* (Note 4).

35 Robert Serber, "Linear Modifications in the Maxwell Field Equations," *Phys. Rev. 48* (1935), 49-54; E. A. Uehling, "Polarization Effects in the Positron Theory," *Phys. Rev. 48* (1935), 55-63.

36 See Note 32 and Hans Euler, "Ueber die Streuung von Licht an Licht nach Diracschen Theorie," *Ann. Physik 26* (1936), 398-448.

37 E. C. G. Stueckelberg, "Relativistisch invariante Störungstheorie des Diracshen Elektrons," *Ann. Physik 21* (1934), 367-89; E. C. G. Stueckelberg, "Die Wechselwirkungskräfte in der Elektrodynamik und in der Feldtheorie der Kernkräfte (Teil I)," *Helv. Phys. Acta* (1938), 225-44.

38 D. Rivier and E. C. G. Stueckelberg, "A Convergent Expression for the Magnetic Moment of the Neutron," *Phys. Rev. 74* (1948), 218; D. Rivier and E. C. G. Stueckelberg, "Causalité et structure de la Matrice S," *Helv. Phys. Acta 23* (1950), 215-22.

39 Willis E. Lamb, Jr. and Robert C. Retherford, "Fine Structure of the Hydrogen Atom by a Microwave Method," *Phys. Rev. 72* (1947), 241-3. See also Chapter 20 in this volume.

40 W. V. Houston, "A New Method of Analysis of the Structure of H_α and D_α," *Phys.*

Rev. 51 (1937), 446-9; Robley C. Williams, "The Fine Structures of H_α and D_α Under Varying Discharge Conditions," *Phys. Rev. 54* (1938), 558-67.

41   H. A. Kramers, "Die Wechselwirkung zwischen geladenen Teilchen und Strahlungs-feld," *Nuovo Cimento 15* (1938), 108-14.

42   H. A. Bethe, "The Electromagnetic Shift of Energy Levels," *Phys. Rev. 72* (1947), 339-41, reprinted in Schwinger (ed.), *Quantum Electrodynamics* (Note 4).

43   J. B. French and V. F. Weisskopf, "The Electromagnetic Shift of Energy Levels," *Phys. Rev. 75* (1949), 1240-8; Norman M. Kroll and Willis E. Lamb, Jr., "On the Self-Energy of a Bound Electron," *Phys. Rev. 75* (1949), 388-98, reprinted in Schwinger (ed.), *Quantum Electrodynamics* (Note 4).

44   See Schwinger (ed.), *Quantum Electrodynamics* (note 4).

45   Julian Schwinger, "On Quantum Electrodynamics and the Magnetic Moment of the Electron," *Phys. Rev. 73* (1948), 416, reprinted in Schwinger (ed.), *Quantum Electrodynamics* (Note 4). J. E. Nafe, E. B. Nelson, and I. I. Rabi, "The Hyperfine Structure of Atomic Hydrogen and Deuterium," *Phys. Rev. 71* (1947), 914-15; H. M. Foley and P. Kusch, "On the Intrinsic Moment of the Electron," *Phys. Rev. 73* (1948), 412, reprinted in Schwinger (ed.), *Quantum Electrodynamics* (Note 4).

46   There are simpler field theories in which the renormalization can be carried out exactly. See, for example, James Glimm and Arthur Jaffe, "Positivity of the ϕ_3^4 Hamiltonian," *Fortschritte der Physik 21* (1973), 327-76.

47   See M. Veltman, "Gauge Field Theories," in *International Symposium on Electron and Photon Interactions at High Energies, 1973*, ed. by H. Rollnick and W. Pfeil (Amsterdam: North-Holland, 1974), pp. 438-47; and S. Coleman, "The 1979 Nobel Prize in Physics," *Science 206* (1979), 1290-2.

4 The development of meson physics in Japan

SATIO HAYAKAWA

Born 1923, Niihama, Japan; Ph.D., University of Tokyo, 1951; specialist in high-energy astrophysics, theoretical and experimental; Nagoya University, Nagoya, Japan.

Why was it that elementary particle physics developed in Japan, in spite of its isolation? This question is often asked out of curiosity by scientists in developed countries, as well as by scientists in countries that are still developing who are concerned about their own prospects. It was one of the questions studied recently by a Japanese-American joint research program on the history of elementary particle physics, a report of which provided the basis for this chapter.[1]

The story of meson physics in Japan began with an unsuccessful attempt in 1933 by Hideki Yukawa to account for the nuclear force in terms of an electron-exchange model. At the same time, Yoshio Nishina and Sin-itiro Tomonaga were making a correct calculation of electron-positron pair production.[2] In the 1930s, an active group in Osaka and Kyoto, under the young leadership of Yukawa, developed the celebrated meson theory he invented, and a group under Tomonaga's leadership in Tokyo carried out notable work on electromagnetic interactions of particles. Both of these theoretical groups were in close contact with experimental groups, headed by Seishi Kikuchi in Osaka and by Nishina in Tokyo, studying nuclear physics and cosmic rays. The experimental studies stimulated theoretical work by contributing to such topics as proton-proton scattering and the discovery of the cosmic-ray meson.

By 1940, serious concern was felt about the discrepancy between the observed properties of cosmic-ray mesons and those predicted by Yukawa's theory. Yukawa himself was involved in the relativistic formulation of quantum field theory. These problems were discussed at a series of informal meetings during 1941-4 of the so-called Meson Club, resulting in two-meson theories of Shoichi Sakata, Yasutaka Tanikawa, Takeshi Inoue, and Seitaro Nakamura and the super-many-time theory of Tomonaga.

When I was a student at the University of Tokyo during 1942-5, I learned about these works from publications, particularly from a collection of papers on meson theory prepared for seminars given by Yukawa, from the proceedings of meetings of the Science Research Council, and from the proceedings of the 1943 meeting of the Meson Club. I also had much time for reading the original literature when I stayed out of ruined Tokyo in the latter half of 1945. I had opportunities to hear directly from those who had participated in these activities, because I attended seminars given by Tomonaga to students in their final year and worked in 1945 under Osamu Minakawa at the Cosmic Ray Laboratory of the Central Meteorological Bureau. I obtained much information personally from these two teachers and their friends at the Institute of Physico-Chemical Research (Riken, in abbreviated Japanese). I am particularly indebted to Mituo Taketani and Hidehiko Tamaki, who edited the mimeographed proceedings of the Meson Club. I further benefited by personal contacts with other physicists: Nakamura, at the University of Tokyo; Kikuchi, with whom I shared an office at Cornell University during 1950-1; Yuzuru Watase, who pioneered in cosmic-ray research in 1936 and offered me a position at Osaka City University during 1949-53; Yukawa, Minoru Kobayasi, and Inoue, at Kyoto University during 1954-9; and Sakata and Yataro Sekido, of Nagoya University. I wrote a paper on the history of nuclear and cosmic-ray physics in Japan, based on literature surveys and personal information. I presented it at the International Colloquium on Theoretical Physics held in Peking in 1966, and an extended version appeared in my book *Cosmic Rays*, a part of which was condensed in Chapter I of *Cosmic Ray Physics*.[3]

That is the background of this chapter; however, this is not a new version of older work. My earlier version was somewhat revised as a result of the joint Japanese-American program on the history of elementary particle physics in Japan. A more radical revision was made by referring to Yukawa's manuscripts and laboratory diary, which were recently discovered in a corner of the physics library of Kyoto University and were put in order by Rokuo Kawabe, Michiji Konuma, and

Ziro Maki.[4] It is my pleasant duty to acknowledge these people as well others who participated in the joint Japanese-American program. Without their contributions this chapter could not have been written.

Quantum field theory and nuclear physics

It is known that the meson theory of Yukawa was preceded by his unpublished attempt to explain the nuclear force by the exchange of electrons between proton and neutron. This is noteworthy because it was one of the earliest attempts to deal with the creation and annihilation of fermions and to introduce a field to mediate a force between nucleons. Motivated by Werner Heisenberg's nuclear model,[5] Yukawa tried to describe β decay as the creation of an electron associated with the transformation of a neutron into a proton. This required the quantization of the electron field using anticommutation relations. At that time, he guessed that the electron field would mediate the nuclear force. However, he was not satisfied by this theory because of its long force range and its difficulties with spin and statistics. He did not publish this work, although he started to write a paper entitled "On the Problem of Nuclear Electrons I," with a revised version entitled "Preliminary Studies on the Theory of Electron, Proton and Neutron." He also gave a talk on this subject at the annual meeting of the Physico-Mathematical Society of Japan held in Sendai on April 3, 1933.

In the same year, Nishina and Tomonaga published a short note (independently of J. Robert Oppenheimer and Milton Plesset and of Walter Heitler and Fritz Sauter) to point out that the positron discovered by Anderson should be created together with an electron by a γ ray in the nuclear Coulomb field.[2,6] In this note they did not explicitly use the quantized electron field, but they did assume the transition of an electron from a negative to a positive energy state. This work was elaborated in collaboration with Sakata,[7] taking into account the Coulomb wave functions of the electron and positron. Nishina, Tomonaga, and Hidehiko Tamaki then calculated the annihilation of a positron with an orbital electron (one quantum annihilation), correcting a wrong factor given by Enrico Fermi and George Uhlenbeck.[8]

These works can be regarded as outstanding, especially if one considers that in those days, as pointed out by Takabayasi, Japanese physicists were more concerned with absorbing the progress in the Western world than with original work. The achievements of quantum mechanics were known to Japanese physicists through the literature, and we find translations of important papers as early as 1927. The flavor of Western physics was imported by those few Japanese, such as Kikuchi and Yoshikatsu Sugiura, who had spent some years in Europe and by visitors such as Otto Laporte, Paul A. M. Dirac, and Heisenberg; the latter two were invited in 1929 by Riken. At Kyoto University, where Yukawa and Tomonaga were students during 1926-9, only occasional lectures on quantum mechanics were given: in 1926 by Toshima Araki, a professor of astrophysics, and later by Sugiura and Laporte. The first regular course on quantum mechanics was given in 1931 by Matsuhei Tamura, although a course in optics by Masamichi Kimura contained a brief account of the Schrödinger equation. Kimura arranged for Nishina to visit Kyoto and deliver a series of lectures in 1931. (The situation in the more traditional atmosphere of Tokyo University was more backward.) It was under these circumstances that Yukawa and Tomonaga studied quantum field theory and nuclear physics on their own and tried to do some original work.

In 1934, Yukawa was already interested in the relativistic formulation of quantum mechanics, as indicated by the program of the annual meeting of the Physico-Mathematical Society on April 4, 1934. This interest, apparently stimulated by a work of Dirac, would reappear in Yukawa's later work.[9]*

The first paper on meson theory

At a talk by Yukawa on the electron exchange between nucleons, Nishina commented that the exchange of a Bose particle could overcome the difficulty of violating the conservation of spin and statistics. Soon afterward, the paper on β decay by Fermi was published, and it first came to the notice of Japanese physicists without Wolfgang Pauli's idea of the neutrino.[10] Then came the papers by Igor Tamm and by D. Iwanenko demonstrating that the exchange of an electron-neutrino pair would result in too weak a nuclear force.[11]

In early October of 1934, Yukawa conceived the idea that a Bose particle of mass of about 200 m_e, strongly interacting with nucleons, could be the quantum of the nuclear force. He spoke of this idea in Osaka at a colloquium of Kikuchi's group and then at a meeting of the Osaka branch of the Physico- Mathematical Society. Then, on November 1, he started writing the first draft of a paper that on November 17 was read as the last among seven 10-minute papers at a regular monthly meeting of the Physico-Mathematical Society. After rewriting

---

* *Ed. note:* See also Chapter 18 in this volume.

the paper four times, he submitted it to the *Proceeding of the Physico-Mathematical Society of Japan*. Received on November 30, 1934, it was published in the first issue of 1935.[12]

In his paper he considered the proton and the neutron as two states of the nucleon, which would be degenerate in the absence of interaction.* The interactions are mediated by a field analogous to the electromagnetic field, which results in the analogue of the Coulomb force. Quantization of the field yields the "heavy quantum" of mass about 200 m_e, just as quantization of the electromagnetic field yields the light quantum of mass zero. The field is vector (like the electromagnetic potential), but only the fourth component is used to derive the nuclear force. The exchange of the charged heavy quanta between proton and neutron gives a potential

$$J(r) = - g^2 e^{-\lambda r}/r,$$

where λ is related to the mass of the heavy quantum by

$$m_U = \hbar\lambda/c \simeq 200\ m_e,$$

and g is the coupling constant, whose value is

$$g^2/\hbar c \simeq 0.1.$$

Yukawa further attempted to explain the β decay as being due to the interaction of the heavy quantum with the electron-antineutrino pair. The coupling constant g' for this interaction is related to the Fermi coupling constant as

$$g_F = 4\pi g g'/\lambda^2.$$

He concluded this paper with a comment on the difficulty of creating the heavy quantum, because this would require a high energy, then available only in the cosmic rays.

Experimental activities in the early 1930s

Experimental physics also developed in the middle 1930s, although slowly, at Riken and Osaka University, where theoretical and experimental physicists were in close contact. At Riken, Nishina asked Ryokichi Sagane and Masa Takeuchi to construct a counter-controlled cloud chamber. They learned the relevant techniques mainly from the literature and partly from Takeo Shimizu of Tokyo University, who

---

* *Ed. note*: This degeneracy is not actually discussed in Yukawa's paper, where the actual masses of neutron and proton are used.

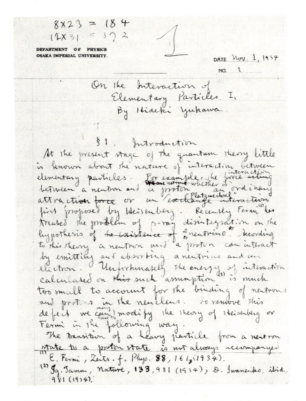

First page of the manuscript of the first article proposing the existence of the meson, in Yukawa's own hand (credit: Ziro Maki and Michiji Konuma from Yukawa Hall Archival Library, Kyoto).

had constructed a cloud chamber with a high expansion rate at the Cavendish Laboratory. They reported success on November 17, 1933, at the semiannual meeting of Riken, shortly after Patrick Blackett and Giuseppe Occhialini. Torao Ichimiya joined the cloud chamber group to measure the momenta of cosmic-ray particles using a magnet powered by a dc power supply for charging submarine batteries at a navy base, and Sagane left to take part in the construction of a cyclotron. Nishina also organized another cosmic-ray group to measure the intensity with ionization chambers and counters; Chihiro Ishii was responsible for this group, which worked in collaboration with the Meteorological Bureau.

Seiji Nishikawa, also at Riken, was stimulated by Fermi's work on neutron physics and, collaborating with Nishina, imported radium to make neutron sources. For further study, a Cockroft-Walton machine of 200 keV was constructed in 1934 and another of 600 keV later. Minakawa and Shizuo Miyake carried out experiments on neutron scattering and neutron-induced reactions.

At Osaka University, Kikuchi built a 600-keV Cockroft-Walton accelerator in 1934 and, in collaboration with Hiroo Aoki (later named Kumagai) and Kōzi Husimi, measured γ rays resulting from neutron capture. In 1935, Watase began cosmic-ray experiments using the counter-coincidence technique and, in collaboration with Junkichi Itoh, obtained the shower transition curve. This group was then asked to work on the construction of a 40-inch cyclotron.

Tomonaga often consulted with these experimental physicists while he continued theoretical work, such as that on the pair production by charged particles in the nuclear Coulomb field with Nishina and Kobayasi.[13]

Yukawa was also interested in experiments, and he published papers on the slowing down of neutrons and on counter efficiencies. In 1935 he was much concerned with nuclear physics, and he worked with Sakata on the internal pair-production process and on K capture.[14] However, he carried the development of meson theory no further in 1935.

Unpublished work before the second paper

In 1936, Yukawa began to develop his theory of heavy quanta. On November 28, 1936, he presented a five-minute paper entitled "On the Interaction of Elementary Particles. II" at the Osaka branch meeting of the Physico-Mathematical Society and then started to write a manuscript on various processes associated with the heavy quanta, the first section in English, the second in Japanese.

In Section 1 he gave a program to improve on the first meson paper, as follows: (1) relativistically invariant formalism, (2) introduction of the Majorana force, and (3) modification of the interaction of the meson with electron and neutrino so as to obtain the β-decay interaction of George Gamow and Edward Teller. For the first program he introduced a four-vector and a relativistic six-vector.

In Section 2 he tried to formulate scalar meson theory, following Pauli and Victor Weisskopf.[15] Without completing the formulation, however, he turned to discussing observable features of this hypothetical particle. First, he pointed out that the trajectory of the heavy quantum U could be distinguished from the trajectories of a proton or an electron by its curvature in a magnetic field, combined with its rate of ionization or its range. He made numerical estimates to determine if cloud chamber tracks obtained by Carl Anderson and Seth Neddermeyer might be those of heavy quanta.[16]* They had operated a cloud chamber with a magnetic field of 7,900 G at Pike's Peak. In one picture they found an event in which five particles of positive charge and one with unidentified sign were produced in a 0.35-cm-thick lead plate by a nonionizing ray. One of the particles ejected vertically upward had a range of about 5 cm in the gas and a radius of curvature of about 20 cm. In a long figure caption they expressed a hesitation to identify this track with a new particle of intermediate mass because of uncertainty regarding the contribution of multiple scattering to the curvature. Four positive particles ejected downward were called positrons, because they were lightly ionizing, although the total energy of 1,000 MeV given to these "positrons" would be too high. Having been stimulated by this paper, Yukawa showed that the upward particle track was consistent with a U particle of mass 200 m_e, and he suspected that the four downward particles could also be U particles.

Here I insert a story on the discovery of mesons in cosmic rays. Laurie Brown has remarked that Anderson and Neddermeyer were discouraged by Oppenheimer from publishing their results on the existence of intermediate-mass particles on the basis that the radiation theory was unreliable at energies above 137 $m_e c^2$. At a colloquium by Oppenheimer at M.I.T., their results were told to Jabez Street and E. C. Stevenson, who also had obtained a track that could be interpreted as due to an intermediate-mass particle. Then both groups decided to publish their respective results of the discovery of the meson.[17] An independent discovery of the meson was made by Nishina, Takeuchi, and Ichimiya with the cloud chamber mentioned earlier.[18] They submitted a letter to the *Physical Review* on August 28, 1937, earlier than the letter of Street and Stevenson, but the Japanese letter was sent back because it was too long. That resulted in its publication being delayed for several months.†

It is interesting that Blackett and John G. Wilson obtained results almost identical with those of Neddermeyer and Anderson by measur-

---

* *Ed. note*: This article is not the one of 1937 announcing the discovery of the cosmic-ray meson.

† *Ed. note:* The Japanese paper[18] "received August 28, 1937" was published in the December 1 issue of 1937. The Street and Stevenson letter (not the abstract[17]) was sent in on October 6 and appeared November 1, 1937.

ing the momentum loss of cosmic-ray particles by absorbers in their cloud chamber.[19] However, Blackett and Wilson interpreted their results as being due to the breakdown of quantum electrodynamics at energies above 137 $m_e c^2$. Once the existence of the meson was established, a number of tracks already obtained were found to be due to mesons; the earliest track was obtained by Paul Kunze.[20]

Immediately after the foregoing discovery, Yukawa submitted a paper, received July 5, 1937, stating that the intermediate-mass particle discovered in cosmic rays could be identified with his heavy quantum.[21] Oppenheimer and Robert Serber also took the same interpretation for granted.[22]* It is worth noting that no physicists in the Western world had earlier paid attention to Yukawa's work, and Yukawa, for his part, had not seriously considered the hard component of cosmic rays. The existence of the hard component was established at the 1934 London conference.[23]

Williams noted that the hard component might consist of protons, because of their small radiation, but he questioned this interpretation because negative particles were as abundant as positive particles.[24] Had this paper been noticed by Yukawa, he would have mentioned this fact as supporting his theory.

Yukawa's idea was at first ignored in the West. Heitler (1937) and Lothar Nordheim (1938) carefully analyzed the Pfotzer curve of cosmic-ray intensity versus atmospheric depth. The cascade theory predicted too low an intensity at lower altitudes, and the difference was attributable to the hard component. They then suggested, without referring to Yukawa's papers, that heavy electrons of intermediate mass could be responsible for the hard component.[25] It is well known that neither Neddermeyer and Anderson nor Street and Stevenson knew of Yukawa's heavy quanta. Homi J. Bhabha was the first to discuss cosmic-ray phenomena explicitly taking into account Yukawa's theory.[26]

Turning back to Yukawa's unpublished manuscript, we find that in late 1936 he was deeply concerned with the interactions of heavy quanta. He estimated their cross sections for bremsstrahlung to be much smaller than those of electrons. He then mentioned the charge-exchange scattering of a heavy quantum by a nucleus, whereby a neutron would change into a proton, or vice versa. A nucleus thus excited could emit γ rays, heavy quanta, or β rays, which might initiate showers. He calculated the cross section for capture of U particles in

* *Ed. note*: However, see the discussion with Serber in Chapter 17.

Walter Heitler (left) and Leon Rosenfeld at Bohr Institute conference, 1934 (credit: Paul Ehrenfest, Jr.).

flight and found that it was small at relativistic energies but large in the
nonrelativistic region. He mentioned that the cross section would depend on the charge of U particles. Only qualitative remarks were made
on production processes, such as the pair creation by γ rays and the
creation by nucleon-nucleus collisions.

Yukawa rewrote the manuscript and dated it January 11, 1937, but
he kept it. At about this time Sakata began to collaborate with Yukawa, and they elaborated scalar theory. Even at this stage Yukawa
wanted to publish the second paper by himself; he presented the
paper at Riken on August 19, 1937, with particular attention to the
experimentally discovered particle, which in his note he named the
"cosmon." In the third manuscript, probably written after a joint
paper given on September 25, 1937, at the Osaka branch meeting of
the Physico-Mathematical Society, he referred to Sakata's work on β
decay.

Finally, on November 9, 1937, he abandoned his unpublished manuscript and instead submitted a paper jointly with Sakata.[27] This is the
second paper on the meson.

Shoichi Sakata watches Hideki Yukawa write a poem in the auditorium at Nagoya University at the celebration of the 25th anniversary of the two-meson theory (credit: Satio Hayakawa).

Second and third papers

The second and the third papers, which dealt with scalar and vector fields, were written almost in parallel. The program resulting in these papers had been prepared in mid-1936, as revealed in the unpublished manuscript of Yukawa. Some of the questions raised were the sign of the proton-neutron force, the force between like nucleons, the anomalous magnetic moments (i.e., the deviation, due to meson effects, from the value expected for spin-1/2 particles without interaction, the Dirac magnetic moment; for the neutron, the latter is zero), and β decay.

Yukawa and Sakata found that in scalar theory (mesons of zero spin and positive parity) the nuclear force in the spin triplet state is repulsive; thus the deuteron would not be bound.[27] Between like nucleons the force is much weaker than that between unlike particles, because two heavy quanta have to be exchanged in order to conserve charge.

Vector theory, formulated according to Yukawa's program, was incorporated with Taketani's idea that a vector particle, virtually dissociated from a nucleon, carries a magnetic moment that contributes to the anomalous moment an amount of the right magnitude to explain the observed value.[28]

After some calculations on vector field theory, Yukawa, Sataka, and Taketani in November 1937 wrote the first draft of the third paper as far as the third section.[27] It was completed and read on January 22, 1938, at the Osaka branch meeting of the Physico-Mathematical Society and was received for publication on March 15, 1938.

The vector theory raised new problems, such as a term varying as $1/r^3$ in the nuclear force. The spontaneous decay of the U particle was first calculated in this paper, following a suggestion of Bhabha.[29] They obtained the lifetime 5×10^{-7} sec.

Yukawa's group and the fourth paper

In April 1938, Kobayasi left Riken and joined Yukawa's group at Osaka University. The activities of the group have been described in Yukawa's laboratory diary.[4] At a colloquium held April 21, 1938, whose participants included Daiske Okayama and Zaimoku Hai (first-year postgraduate students) and Tanikawa (a third-year student), Yukawa presented a program for further development of the meson theory. Seven persons* agreed to share the work in the following way: Yukawa and Hai were to work on nuclear forces, Kobayasi and Okayama on cosmic rays, Sakata and Tanikawa on β disintegration, and Taketani on the formalism and other things. Taketani then gave a talk on the magnetic moment and the self-energy. Colloquia were held twice a week and occasionally were visited by Husimi and others. Actually, even before this series of colloquia started, Yukawa had already written a preliminary draft of the fourth paper.

At the Osaka branch meeting of the Physico-Mathematical Society on May 28, 1938, they presented four successive papers: on nuclear forces by Yukawa, on β disintegration by Sakata, on the production and energy loss of U particles by Kobayasi, and on nuclear magnetic moments by Taketani. Kemmer's papers on the charge independence of nuclear force and on the formalism and that of Fröhlich, Heitler, and N. Kemmer were discussed on June 4, June 11, and June 18, respectively. Papers on cosmic rays by Heitler and by Lothar Nordheim and G. Nordheim were discussed on July 25 and on September 15, after completion of the fourth paper.[30] In writing the fourth paper, Sakata, Kobayasi, and Taketani contributed notes on their respective parts, and Yukawa completed the manuscript in July. The paper was received for publication on August 2, 1938.[27] Taketani was arrested by

---

* *Ed. note*: Familiarly known as the "'seven samurai.'"

the security police on September 13.* The group moved to Kyoto in May 1939, because Kyoto University offered a professorship and a lectureship to Yukawa and Sakata, respectively, but Kobayasi stayed at Osaka University to form a subgroup with new students, including Ryoyu Utiyama.

From the record of colloquia one can see how research was incorporated with education and how foreign work interacted with the fourth paper. Senior members reported on papers directly connected with their own work: Yukawa on leptons; Sakata on spinor equations; Kobayasi on vector field theory; Taketani on the electromagnetic properties of nuclei.[31] They also spent whole afternoons presenting their original work: Yukawa on deuteron problems; Sakata on β decay; Kobayasi on the creation of U particles; Taketani on the magnetic moments. New postgraduates, Okayama and Hai were asked to read papers of fundamental importance, such as those by Heisenberg and Pauli on the quantum theory of wave fields.[32] Discussions of papers by Kemmer and associates came after the four senior members gave talks on their own parts. Some talks given afterward by Kobayasi on the absorption of U particles and by Taketani on the magnetic moments indicate some extension and revision based on the British papers.

After submitting the paper, Yukawa seems to have become interested in the question whether or not the quantum field theory was really applicable to high-energy phenomena associated with mesons. This can be seen from the discussions held on the classical description of particles by Dirac and by Heisenberg, presented by Yukawa and by Kobayasi, respectively.[33] They were deeply concerned with the question whether or not cosmic-ray and nuclear phenomena could be understood in terms of Yukawa's meson alone.

Decay of mesons
By 1938, Yukawa's theory of the meson became widely appreciated. The meson was considered to play an important role in cosmic rays, as elucidated by H. Euler and Heisenberg, although Heisenberg told Tomonaga in 1937 that the foundation of Yukawa's theory did not

---

* *Ed. note:* Taketani was arrested for his antimilitarist and Marxist views. He was held until April 1939, when he was released "into the custody of Yukawa." See Taketani Mituo, "Methodological Approaches in the Development of the Meson Theory of Yukawa in Japan," in *Science and Society in Modern Japan, Selected Historical Sources*, ed. by Nakayama Shigeru, David L. Swain, and Yagi Eri (Cambridge, Mass.: M.I.T. Press, 1974).

seem clear.[1,34] In the article by Euler and Heisenberg, cosmic-ray phenomena observed in the atmosphere were interpreted on the basis of the meson and the cascade theory: Mesons produced at high altitudes, possibly by energetic photons in cascade showers, were identified as the hard component, which penetrates the atmosphere down to the ground level; electrons produced by these mesons through knock-on processes and by spontaneous decay formed the main part of the soft component at low altitudes.

The lifetime of the meson in cosmic rays was obtained in several ways. Blackett interpreted the temperature effect, derived by Arthur Compton and Louis Turner from their measurement of the cosmic-ray intensity over a wide latitude range spanning the equator, in terms of the spontaneous decay of mesons occurring over their passage through the atmosphere.[35] A number of measurements of the intensity of cosmic rays having passed through the same quantity of matter but at different densities gave an appreciable difference of intensities that could be explained by the decay of mesons, as summarized by Rossi.[36] All these results indicated the lifetime of the cosmic-ray meson to be about 1μsec.*

On the other hand, the lifetime of Yukawa's meson was given in the fourth paper as 1.3×10^{-7} sec, correcting an error of a factor of 2 in the third paper and adopting $m_U = 200\ m_e$.[27] Because this was appreciably shorter than the lifetime of cosmic-ray mesons, the lifetime was calculated also using the Konopinski-Uhlenbeck (rather than the Fermi) interaction, although the calculation could be carried out only to first order in the couplings. More quantitative study, with proper account taken of the nuclear matrix elements, showed that the lifetime should be 10^{-8} sec rather than 10^{-6} sec.[37] Various attempts were made to get rid of this discrepancy.

Fermi and Otto Halpern and Harvey Hall asked if the difference in the intensities of cosmic rays passing through materials of different densities might be attributed to a difference in the ionization loss rate, but their theories of the density effect gave too small an effect.[38]

Yukawa and Okayama questioned whether or not absorption of mesons in matter could modify the interpretation of meson decay. They calculated the time for a meson to stop in a given medium and found that a meson would be captured before it could decay in a dense medium such as lead.[39] Although this argument gave an effect opposite

---

* *Ed. note:* See Chapter 11 for more details.

to the density effect on ionization, this work provided an important clue for later development. In this paper they also remarked that few mesons had shown decay electrons.

Stimulated by the work of Yukawa and Okayama, Tomonaga and Araki called attention to the effect of the nuclear Coulomb field that should attract negative mesons, causing them to be absorbed, but repel positive ones, causing them to decay. To support this prediction, they pointed out that the decay events thus far observed had yielded only positrons, and no electrons.[40] It is interesting that on the basis of the absence of decay particles in an experiment, Montgomery and associates suspected the existence of a meson different from cosmic-ray mesons.[41] Because the paper by Tomonaga and Araki was published in the *Physical Review*, it called the attention of Western physicists to the problem and stimulated experiments on meson decays; as a result, M. Conversi and associates obtained indisputable evidence for the weak nuclear interaction of cosmic-ray mesons.[42]

In an attempt to get rid of the lifetime discrepancy, Sakata proposed a different scheme in which the meson decays through a virtual nucleon pair:

$$U^{\pm} \rightarrow P(N) + \bar{N}(\bar{P}) \rightarrow e^{+}(e^{-}) + \nu(\bar{\nu})$$

(also proposed by Tomonaga in 1939). This is the scheme that is adopted at present, and it assumes a direct interaction between nucleon and lepton.* In Sakata's scheme, the lifetime for nuclear β decay was determined solely by the direct nucleon-lepton coupling, whereas that of spontaneous meson decay depended also on the meson-nucleon interaction. This gave a long lifetime for the spontaneous decay, which matches the presently known partial lifetime for π-e decay.

Because this scheme required a cutoff procedure, Sakata made a quantitative calculation of the lifetimes for four types of mesons using Yukawa's scheme.[43] The result for the pseudoscalar meson with pseudovector coupling gave the mass dependence, which was later applied to obtain the branching ratio for π-μ and π-e decays.

The process of decay through a nucleon pair was applied by Sakata and Tanikawa to the decay of a postulated neutral meson into γ rays.[44] The theory drew on the Furry theorem (i.e., the invariance under charge conjugation), predicting two and three photon decays of spin-zero and spin-one mesons, respectively.

---

* *Ed. note:* In the "'standard" Salam-Weinberg-Glashow model, the nucleons are replaced by quarks.

Interactions of mesons with matter

If the meson interacted strongly with nucleons, as predicted by Yukawa's theory, then the cross section for meson-nucleon scattering should have been about 10^{-26} cm^2. The production of the hard component of cosmic rays at high altitudes and the production of penetrating showers by this hard component were considered supportive of the strong interaction of mesons. However, the scattering of cosmic-ray particles in a lead plate was explainable, without the strong interaction, in terms of multiple Coulomb scattering, except for a single event.[45] If this one event were due to the nuclear interaction of a meson, the cross section would be only 10^{-28} cm^2. This cross section was taken for granted, although the event is now thought to be the scattering of a proton. The efforts then made to reduce the theoretical meson-nucleon cross section will be described in the next section.

Clearer evidence for the weakness of the nuclear interaction of cosmic-ray mesons would have been the penetration of cosmic rays down to great depths underground. It is strange that nobody seems to have taken this fact seriously in those days. The intensity of cosmic rays was measured down to a depth of 140 kg-cm^{-2} and the intensity-depth curve was found to change its slope at about 50 kg-cm^{-2}.[46] In Japan, Nishina and associates observed cosmic-ray intensity at a depth of 140 kg-cm^{-2} in the Shimizu Tunnel during 1939-40 and later moved the station to a deeper place, at 300 kg-cm^{-2}.[47] However, they were mainly interested in interactions produced by rays reported, but as yet unidentified.[48]

It was also considered that a change in the slope of the intensity-depth curve might be due to an anomalous interaction, beyond the validity of the existing theory. Oppenheimer and his collaborators suspected that the cross sections for electromagnetic interactions of vector mesons would increase with energy and thus be responsible for the break in the intensity-depth curve.[49] The first half of the preceding statement is correct for bremsstrahlung, but Tomonaga showed that there was no appreciable difference between the knock-on cross sections of scalar and vector.[50] The interpretation of the intensity-depth curve in terms of electromagnetic interactions alone was later demonstrated by Hayakawa and Tomonaga, taking the π-μ decay effect into account, whereas Robert Marshak and Hans Bethe overlooked the latter point in their two-meson theory, having kept to the inference of Oppenheimer and associates.[51]

If the meson were of vector nature, the frequency of bursts observed in a heavily shielded ionization chamber would have been observed. Such bursts were interpreted by Robert F. Christy and Shuichi Kusaka in terms of large cascade showers originating from electromagnetic interactions of cosmic-ray mesons.[52] The frequency-size relation observed by Marcel Schein and Piara Gill appeared to reject the vector nature of mesons and favor the meson spin to be zero or 1/2.[53] This work was reported by Kusaka on a visit to Japan in 1941, and it influenced the development of meson theory during the war.

Efforts during the war to develop a new meson theory

Japanese physicists tried various methods to overcome the difficulties in understanding the properties of cosmic-ray mesons, and their efforts can be traced in the records of informal meetings, usually held after the semiannual meetings of Riken. For example, we find a vivid record in Yukawa's laboratory diary.

The first two meetings, those on June 12 and December 13, 1941,* were called "theory meetings." The meetings held on April 24 and June 13, 1942, have the name "illusion meetings," which is a pun on the Japanese word for meson. Those on December 12, 1942, on June 19, and on September 26 and 27, 1943 were named "meson meeting," "informal meeting on mesons," and "discussion meeting on mesons," respectively. The last one was a two-day meeting, and its proceedings were published in mimeographed form. The final one, on November 18 and 19, 1944, was sponsored by the Science Research Council and was called "elementary particle theory meeting." Each of the first six meetings was attended by about 20 physicists, whereas about 50 attended each of the last two, which were of a more formal character. Because there were regular attendees who formed a core throughout these meetings, they are called meetings of the Meson Club.

From one of Yukawa's articles, one can see how the Japanese group went about analyzing difficulties in those days.[54] In this paper, Yukawa questioned whether or not relativistic quantum field theory was firmly founded, given the divergence difficulties, but he nevertheless tried to understand the properties of the meson within the framework of the existing theory. Because the scalar and pseudovector theories were ruled out, as they gave repulsive forces in the ^3S state of the deuteron, he discussed the predictions of vector and pseudoscalar theories in comparison with experiment. Both gave qualitatively correct results for

---

* *Ed. note:* The attack on Pearl Harbor notwithstanding.

the anomalous magnetic moment and the nuclear force, although a cutoff procedure was required to avoid divergence in the former, and $1/r^3$ terms appeared in the latter. Pseudoscalar theory was preferred over vector theory, because the former explained the frequency of cosmic-ray bursts, gave the right sign for the quadrupole moment of the deuteron, and allowed a reduced lifetime of the meson, by using Sakata's scheme of meson decay.[43,52] However, neither vector nor pseudoscalar could explain the small scattering cross section observed in cosmic rays.

He then considered whether or not the mixture of vector and pseudoscalar fields proposed by C. Møller and Léon Rosenfeld could avoid these difficulties.[55] By an appropriate choice of coupling constants, the singularities in the nuclear forces would cancel, and the other divergence difficulties might also be avoided. Vector mesons, of lifetime about 10^{-8} sec, should decay at the high altitudes where they are produced, whereas pseudoscalar mesons, of lifetime about 10^{-6} sec, should survive to the lower atmosphere. However, the cross sections for meson-nucleon scattering were found to be large for both types.[56] An alternative possibility suggested by Kobayasi was that vector theory could give a small cross section if only the longitudinal component were used.[57] He mentioned also that Tomonaga's method could reduce the cross section.[58]

In the last section of his paper Yukawa discussed the problems related to the neutral meson. Nishina and Birus failed to observe neutral particles in cosmic rays, but this was not regarded as evidence against the neutral meson, because neutral mesons should decay very quickly in the upper atmosphere.[44,59] Marshak proposed a pair theory to account for the nuclear force between like nucleons by the exchange of two fermions of opposite charges.[60] This did not require a neutral meson, but the scattering cross section was found to be even larger.

To summarize: There were four directions taken by Japanese meson physicists in those days: (1) identifying the type of meson needed to give a consistent understanding of nuclear and cosmic-ray phenomena, (2) developing a suitable approximation to deal with the strong meson-nucleon coupling, (3) resolving the discrepancies between Yukawa's theory of nuclear forces and cosmic-ray phenomena, and (4) formulating a relativistic and divergence-free quantum field theory. Of these, problem 3 was not discussed explicitly in Yukawa's 1942 paper in *Zeitschrift für Physik*.[54] Let us discuss these problems now in more detail.

Identifying the type of meson was the principal subject of the article, which was received January 9, 1942. It seems to have been based in part on talks at the first Meson Club meeting, by Tanikawa on scattering, and by Sakata on decay. An extended treatment of the pseudoscalar meson was given by Gentaro Araki at the seventh meeting.

The existence of the neutral meson was suggested at the second and seventh meetings by Taketani, to explain the soft component of the cosmic rays at high altitudes.[61] In work that stemmed from the analysis of cosmic-ray phenomena by Tamaki, the intensity of the soft component was derived by careful application of cascade theory and was found to be lower by a factor of about 2 than the observed intensity at the maximum. The directional distribution also required the production of a soft component of low energy. The discrepancies were explained by γ rays arising from the decay of neutral mesons.

The second problem, developing approximations suitable for strong nuclear interactions of mesons, was attacked by Tomonaga, who believed that perturbation theory was not permissible for meson-nucleon interactions. The strong coupling theory had been worked out by Gregor Wentzel; Tomonaga invented intermediate coupling theory and presented it at the first meeting.[62] He employed the configuration space method that he had developed earlier and combined it with the Hartree approximation.[58] This method is now called the Tamm-Dancoff approximation, which leaves out the name of the first person to introduce the method. Tomonaga demonstrated that the meson-nucleon scattering cross section could be reduced to 10^{-28} cm^2, the value accepted in those days. Tatuoki Miyazima and Tomonaga then applied this method to multiple meson production and to nuclear forces.[63]

The fourth meeting was devoted to the third problem, that of the two-meson theory; at the seventh meeting, full accounts were given by Sakata and Tanikawa of their respective versions. In Yukawa's laboratory diary, this idea first appeared on May 13, 1942, at a colloquium given by Sakata at Kyoto University. Papers on this subject were read by Inoue and by Nakamura at the Kyoto branch meeting of the Physico-Mathematical Society on July 11, 1942.

The two-meson theory assumed that a new meson (M), different from the meson (Y) introduced by Yukawa, exists and is the meson observed in cosmic rays. In addition to the primary interactions in Yukawa's theory,

$$P(N) \rightleftarrows N(P) + Y^{\pm}, \ P(N) \rightleftarrows P(N) + Y^0, \qquad (g)$$

$$Y^{\pm} \rightarrow e^{\pm} + \nu(\bar{\nu}), \qquad (g')$$

there is another interaction: If M is a boson,

$$Y^\pm \rightleftharpoons M^\pm,$$

or if M is a fermion,

$$Y^\pm \rightarrow M^\pm + n, \qquad\qquad (\gamma)$$

where n is a neutral fermion (e.g., a neutrino). The possibility that M is heavier than Y was considered, but we shall discuss only the case where M is lighter than Y. For the decay of $M^\pm$, we must again distinguish the boson and fermion cases. For $M^\pm$ boson, a possible decay mode is

$$M^\pm \rightarrow e^\pm + \nu(\bar\nu),$$

whereas in the fermion case we might expect

$$M^\pm \rightarrow e^\pm + \nu(\bar\nu) + n. \qquad\qquad (g'\gamma)$$

If these decay processes take place with $Y^\pm$ as an intermediate boson, then the decay rate is proportional to $(g'^2/\hbar c)(\gamma^2/\hbar c)$.* The nuclear scattering of $M^\pm$ takes place also through $Y^\pm$, so that the cross section is proportional to $(g^2/\hbar c)(\gamma^2/\hbar c)$. Because Yukawa's theory and cosmic-ray experiments give discrepancies of 10^{-2} in these two processes, the coupling constant γ has the value

$$\gamma^2/\hbar c \simeq 10^{-2}.$$

This leads to a very short lifetime for $Y^\pm$ of about 10^{-21} sec, very different from what we know about π-μ decay at present. Another difference is that the decay of Y^0 into two n's is faster than its two-γ decay.

Experimental results available in those days were insufficient to establish the two-meson theory, but the members of the Meson Club seem to have regarded it as one of the likely means of eliminating the difficulties in Yukawa's original theory. The theory was first published in 1942 in Japanese, and after the war in English, by Sakata and Inoue, using for $M^\pm$ the Fermi meson version, and by Tanikawa, using for $M^\pm$ the Bose meson.[64]

The two-meson theory developed in Japan was unknown in the West until Sakata and Inoue's paper in English was published. Independently, Marshak and Bethe proposed a two-meson theory on the basis

---

* *Ed. note:* The strong coupling constant of $Y^\pm$ to nucleons is g; the coupling constant of $Y^\pm$ to leptons is g'; γ is the coupling constant of $Y^\pm$ to $M^\pm$ in the boson case and of $Y^\pm$ to $M^\pm + n$ in the fermion case.

of the weak nuclear interactions of cosmic-ray mesons, established by Marcello Conversi, Ettore Pancini, and Oreste Piccioni.[42,51] The strength of meson-nucleon interactions was therefore assumed to be as weak as that for nuclear β decay, thus resulting in a very small cross section for the nuclear scattering of M and the lifetime of Y-M decay of about 10^{-8} sec. However, they adopted the pair boson theory for Y, and the boson choice for M, thus predicting a two-body decay of M and a low excitation energy for the nuclear capture of Y^-.[*]

The two-meson theory later found a firm basis with the discovery of the π-μ decay.[65] There have been several amendments in detail, based on experiment. The fermion nature of $M^\pm$ has been established, with $Y^\pm$ and $M^\pm$ identified as $\pi^\pm$ and $\mu^\pm$, respectively. Identification of ν with ν_e and n with ν_μ was confirmed in 1962.

Turning finally to the fourth problem, that of a consistent quantum field theory, Yukawa had suspected that the difficulties might be due to an incompleteness of the formalism. The quantum field theory in those days satisfied the requirement of relativistic invariance, but the probability amplitude was not defined in a covariant way. From time to time, since his first meson paper, Yukawa had asked himself (as described earlier) if one could reconstruct quantum field theory in a totally covariant way. He gave talks and published papers in Japanese on this problem on several occasions.[66] In doing so he always drew a circle in four-space, and this became famous as Yukawa's circle.[†] The third meeting was solely dedicated to discussing Yukawa's idea.

The answer to this question was given by Tomonaga, in his super-many-time theory, and is identical with the covariant field theory developed later by Schwinger.[67] In this theory, Tomonaga showed that the probability amplitude is defined on a spacelike surface, not on a circle in four-space, and that it develops in time according to the Schrödinger equation in the interaction representation. The formal aspects of this theory were discussed on the second day of the eighth meeting, and Tomonaga and his students then applied this formalism to quantum electrodynamics, providing a powerful framework for renormalization theory.

One might ask how it was possible to work on such fundamental problems in the last phase of the war. Most physicists were, of course, mobilized to work for military research, organized by the Science Re-

---

[*] *Ed. note:* This is because in a pair theory the capture of a Y can only occur with emission of a Y^0.

[†] *Ed. note:* See Chapter 18 in this volume.

Hideki Yukawa and Richard Feynman during Feynman's visit to Kyoto in the summer of 1955. Left to right: Minoru Kobayasi, Koichi Mano, Yukawa, Feynman, Satio Hayakawa, Mrs. Yukawa (credit Satio Hayakawa).

search Council. But they wanted to work as much as possible on physics problems, as may be seen from articles published in university journals by Kobayasi and Husimi.

In the February 21, 1945, issue of the *Tokyo University Journal* Kobayasi wrote "Self-Energy of Elementary Particles," in which he discussed the importance of the self-energy problem and methods for subtracting infinities in quantum electrodynamics and meson theory; there was no mention of the war. In the March 15, 1945, issue of the *Osaka University Journal*, Husimi wrote "Prospects of the Fundamental Theory of Physics and Super-Many Time Theory," describing the discussions at the eighth meeting of the Meson Club. He introduced the questions raised by Yukawa about covariant field theory and the answer given by Tomonaga, providing an introduction to relativity and quantum field theory. He then sketched a paper by Satoshi Watanabe, using poetic and philosophical expressions, and he concluded by praising this as a restful interlude during the difficult war period, but mentioned the involvement of Yukawa, Tomonaga, and Watanabe in military research.

Notes

1 L. M. Brown, M. Konuma, and Z. Maki (eds.), *Particle Physics in Japan, 1930-1950*, (Kyoto: Research Institute for Fundamental Physics, 1980), a collaboration under auspices of the Japan Society for the Promotion of Science and the U.S. National Science Foundation.

2 Y. Nishina and S. Tomonaga, "On the Creation of Positive and Negative Electrons," *Proc. Phys.-Math. Soc. Japan 15* (1933), 248-9.

3 S. Hayakawa, *Cosmic Rays* (Tokyo: Chikuma Shobō, 1972), in Japanese; S. Hayakawa, *Cosmic Ray Physics* (New York: Wiley, 1969).

4 R. Kawabe and M. Konuma, paper presented at the annual meeting of the Physical Society of Japan, March 27-30, 1980.

5 W. Heisenberg, "Ueber den Bau der Atomkerne," *Z. Physik 77* (1932), 1-11, and *78* (1932), 156-64.

6 J. R. Oppenheimer and M. S. Plesset, "On the Production of the Positive Electron," *Phys. Rev. 44* (1933), 53-5; W. Heitler and F. Sauter, "Stopping of Fast Particles with Emission of Radiation and the Birth of Positive Electrons," *Nature 132* (1933), 892.

7 Yoshio Nishina, Sin-itiro Tomonaga, and Shoichi Sakata, "On the Photoelectric Creation of Positive and Negative Electrons," *Sci. Pap. Inst. Phys. Chem. Res. 24, Suppl. 17* (1934), 1-5.

8 Y. Nishina, S. Tomonaga, and H. Tamaki, "On the Annihilation of Electrons and Positrons," *Sci. Pap. Inst. Phys. Chem. Res. 24, Suppl. 18* (1934), 7; Enrico Fermi and George E. Uhlenbeck, "On the Recombination of Electrons and Positrons," *Phys. Rev. 44* (1933), 510-11.

9 P. A. M. Dirac, "The Lagrangian in Quantum Mechanics," *Phys. Zeit. U.S.S.R. 3* (1933), 64-72.

10 E. Fermi, "Versuch einer Theorie der β-Strahlen. I," *Z. Physik 88* (1934), 161-71; English translation in Charles Strachan (ed.), *The Theory of Beta Decay* (Oxford: Pergamon, 1969), pp. 107-28.

11 I. Tamm, "Exchange Forces between Neutrons and Protons and Fermi's Theory," *Nature 133* (1934), 981; D. Iwanenko, "Interaction of Neutrons and Protons," *Nature 133* (1934), 981-2.

12 Hideki Yukawa, "On the Interaction of Elementary Particles. I," *Proc. Phys.-Math. Soc. Japan 17* (1935), 48-57.

13 Y. Nishina, S. Tomonaga, and M. Kobayasi, "On the Creation of Positive and Negative Electrons by Heavy Charged Particles," *Sci. Pap. Inst. Phys. Chem. Res. 27* (1935), 137-78.

14 Hideki Yukawa and Shoichi Sakata, "On the Theory of Internal Pair Production," *Proc. Phys.-Math. Soc. Japan 17* (1935), 397-407; Hideki Yukawa and Shoichi Sakata, "On the Theory of the β-Disintegration and the Allied Phenomenon," *Proc. Phys.-Math. Soc. Japan 17* (1935), 467-79, and (Suppl.) *18* (1936), 128-30.

15 W. Pauli and V. Weisskopf, "Ueber die Quantisierung der skalaren relativistischen Wellengleichung," *Helv. Phys. Acta 7* (1934), 709-31.

16 Carl D. Anderson and Seth H. Neddermeyer, "Cloud Chamber Observations of Cosmic Rays at 4300 Meters Elevation and Near Sea Level," *Phys. Rev. 50* (1936), 263-71.

17 Seth H. Neddermeyer and Carl D. Anderson, "Note on the Nature of Cosmic Ray Particles," *Phys. Rev. 51* (1937), 884-6; J. C. Street and E. C. Stevenson, "Penetrating Corpuscular Component of the Cosmic Radiation," *Phys. Rev. 51* (1937), 1005 (abstract).

18 Y. Nishina, M. Takeuchi, and T. Ichimiya, "On the Nature of Cosmic-Ray Particles," *Phys. Rev. 52* (1937), 1198-9.

19 P. M. S. Blackett and J. G. Wilson, "The Energy Loss of Cosmic Ray Particles in
Metal Plates," *Proc. Roy. Soc. (London) A160* (1937), 304-23.

20 Paul Kunze, "Untersuchung der Ultrastrahlung in der Wilsonkammer," *Z. Physik
83* (1933), 1-18.

21 Hideki Yukawa, "On a Possible Interpretation of the Penetrating Component of the
Cosmic Rays," *Proc. Phys.-Math. Soc. Japan 19* (1937), 712-13.

22 J. R. Oppenheimer and R. Serber, "Note on the Nature of Cosmic-Ray Particles,"
Phys. Rev. 51 (1937), 1113.

23 *International Conference on Physics, London 1934. Vol. I. Nuclear Physics,* ed. by J.
H. Awberry (Cambridge University Press and the Physical Society, 1935). See especially P. Auger and L. Leprince-Ringuet, "Some Measurements of Cosmic Radiation at High Altitudes," pp. 195-8, and B. Rossi, "Some Results Arising from the
Study of Cosmic Rays," pp. 233-47.

24 E. J. Williams, "Nature of the High Energy Particles of Penetrating Radiation and
Status of Ionization and Radiation Formulae," *Phys. Rev. 45* (1934), 729-80.

25 W. Heitler, "On the Analysis of Cosmic Rays," *Proc. Roy Soc. (London) A161*
(1937), 261-83; L. W. Nordheim, "A New Analysis of Cosmic Radiation Including
the Hard Component," *Phys. Rev. 53* (1938), 694-706.

26 H. J. Bhabha, "Penetrating Component of Cosmic Radiation," *Proc. Roy. Soc.
(London) A164* (1938), 257-94.

27 H. Yukawa and S. Sakata, "On the Interaction of Elementary Particles. II.," *Proc.
Phys.-Math. Soc. Japan 19* (1937), 1084-93; H. Yukawa, S. Sakata, and M. Take-
tani, "On the Interaction of Elementary Particles. III.," *Proc. Phys.-Math. Soc.
Japan 20* (1938), 319-40; H. Yukawa, S. Sakata, M. Kobayasi, and M. Taketani,
"On the Interaction of Elementary Particles. IV.," *Proc. Phys.-Math. Soc. Japan 20*
(1938), 720-45.

28 M. Taketani, "Quantization of Proton and Neutron Fields and Their Magnetic Moments," *Kagaku 7* (1937), 532-3 (in Japanese).

29 H. J. Bhabha, "Nuclear Forces, Heavy Electrons and the β-Decay," *Nature 141*
(1938), 117-18.

30 N. Kemmer, "Quantum Theory of Einstein-Bose Particles and Nuclear Interaction,"
Proc. Roy. Soc. (London) A166 (1938), 127-53; H. Fröhlich, W. Heitler, and N.
Kemmer, "On the Nuclear Forces and the Magnetic Moments of the Neutron and
the Proton," *Proc. Roy. Soc. (London) A166* (1938), 154-77; W. Heitler, "Showers
Produced by Cosmic Radiation," *Proc. Roy. Soc. (London) A166* (1938), 529-42; L.
W. Nordheim and G. Nordheim, "On the Production of Heavy Electrons," *Phys.
Rev. 54* (1938), 254-65.

31 Yukawa: E. Majorana, "Symmetrical Theory of Electrons and Positrons," *Nuovo
Cimento 14* (1937), 171-84; G. Racah, "Symmetry between Particles and Anti-Parti-
cles," *Nuovo Cimento 14* (1937), 322-8. Sakata: B. L. van der Waerden, "Spinor-
analyse," *Göttinger Nachrichten* (1929), 100-9; O. Laporte and G. E. Uhlenbeck,
"Application of Spinor Analysis to the Maxwell and Dirac Equations," *Phys. Rev.
37* (1931), 1381-97.

32 W. Heisenberg and W. Pauli, "Zur Quantendynamik der Wellenfelder," *Z. Physik
56* (1929), 1-61; W. Heisenberg and W. Pauli, "Zur Quantentheorie der Wellenfelder," *Z. Physik 59* (1930), 160-90.

33 P. A. M. Dirac, "Classical Theory of Radiating Electrons," *Proc. Roy. Soc. (London) A167* (1938), 148-69; W. Heisenberg, "Die Grenzen der Anwendbarkeit der
bisherigen Quantentheorie," *Z. Physik 110* (1938), 251-66.

34 H. Euler and W. Heisenberg, "Theoretische Gesichtspunkte zur Deutung der kosmischen Strahlung," *Ergeb. Exakt. Naturw. 17* (1938), 1-69.

35 P. M. S. Blackett, "On the Instability of the Baryon and the Temperature Effect of Cosmic Rays," *Phys. Rev. 54* (1938), 973-4.

36 Bruno Rossi, "The Disintegration of Mesotrons," *Rev. Mod. Phys. 11* (1939), 296-303.

37 Hideki Yukawa and Shoichi Sakata, "The Mass and Mean Life-Time of the Mesotron," *Proc. Phys.-Math. Soc. Japan 21* (1939), 138-40; Hideki Yukawa and Shoichi Sakata, "Mass and Mean Life-Time of the Meson," *Nature 143* (1939), 761-2; L. W. Nordheim, "Lifetime of a Yukawa Particle," *Phys. Rev. 55* (1939), 506; H. A. Bethe and L. W. Nordheim, "On the Theory of Meson Decay," *Phys. Rev. 57* (1940), 998-1006.

38 E. Fermi, "The Ionization Loss of Energy in Gases and in Condensed Materials," *Phys. Rev. 57* (1940), 485-93; O. Halpern and H. Hall, "Energy Losses of Fast Mesotrons and Electrons in Condensed Materials," *Phys. Rev. 57* (1940), 459-60.

39 H. Yukawa and D. Okayama, "Note on the Absorption of Slow Mesotrons in Matter," *Sci. Pap. Inst. Phys. Chem. Res. 36* (1939), 385-9.

40 S. Tomonaga and G. Araki, "Effect of the Nuclear Coulomb Field on the Capture of Slow Mesons," *Phys. Rev. 58* (1940), 90-1.

41 C. G. Montgomery, W. E. Ramsey, D. H. Courie, and D. D. Montgomery, "Slow Mesons in the Cosmic Radiation," *Phys. Rev. 56* (1939), 635-9.

42 M. Conversi, E. Pancini, and O. Piccioni, "On the Disintegration of Negative Mesons," *Phys. Rev. 71* (1947), 209-10.

43 Shoichi Sakata, "Connection Between the Meson Decay and the Beta Decay," *Phys. Rev. 58* (1940), 576; Shoichi Sakata, "On the Theory of the Meson Decay," *Proc. Phys.-Math. Soc. Japan 23* (1941), 283-91; Shoichi Sakata, "On Yukawa's Theory of the Beta-Disintegration and the Lifetime of the Meson," *Proc. Phys.-Math. Soc. Japan 23* (1941), 291-309; Shoichi Sakata, "On the Lifetime of the Pseudoscalar Meson," *Proc. Phys.-Math. Soc. Japan 24* (1942), 843-4.

44 Shoichi Sakata and Yasutaka Tanikawa, "The Spontaneous Disintegration of the Neutral Mesotron (Neutretto)," *Phys. Rev. 57* (1940), 133.

45 J. G. Wilson, "Absorption of Penetrating Cosmic Ray Particles in Gold," *Proc. Roy. Soc. (London) A172* (1939), 517-29.

46 Volney C. Wilson, "Nature of Cosmic Rays Below Ground," *Rev. Mod. Phys. 11* (1939), 230-1.

47 Y. Nishina, Y. Sekido, Y. Miyazaki, and T. Masuda, "Cosmic Rays at a Depth Equivalent to 1400 Meters of Water," *Phys. Rev. 59* (1941), 401.

48 J. Barnóthy and M. Forró, "Variation of Cosmic Ray Intensity with Sidereal Time," *Nature 138* (1936), 325.

49 J. R. Oppenheimer, H. S. Snyder, and R. Serber, "The Production of Soft Secondaries by Mesotron," *Phys. Rev. 57* (1940), 75-81.

50 Sin-itiro Tomonaga, "Ueber den Zusammenstoss des Mesotrons mit Elektronen," *Sci. Pap. Inst. Chem. Res. 37* (1940), 399-413.

51 S. Hayakawa and S. Tomonaga, "Cosmic Rays Underground. I," *Progr. Theoret. Phys. (Kyoto) 4* (1949), 287-96; "Cosmic Rays Underground. II," *Progr. Theoret. Phys. (Kyoto) 4* (1949), 496-501; R. E. Marshak and H. A. Bethe, "On the Two-Meson Hypothesis," *Phys. Rev. 72* (1947), 506-9.

52 R. F. Christy and S. Kusaka, "The Interaction of γ-Rays with Mesotrons," *Phys. Rev. 59* (1941), 405-14; R. F. Christy and S. Kusaka, "Burst Production by Mesotrons," *Phys. Rev. 59* (1941), 414-21.

53 Marcel Schein and Piara S. Gill, "Burst Frequency as a Function of Energy," *Rev. Mod. Phys. 11* (1939), 267-76.

54 H. Yukawa, "Bemerkungen über die Natur des Mesotrons," *Z. Physik 119* (1942), 201-5.

55 C. Møller and L. Rosenfeld, "On the Field Theory of Nuclear Forces," *Kgl. Danske Videnskab. Selskab, Mat.-Fys. Medd. 17* (1940), 1-72.

56 Y. Tanikawa and H. Yukawa, "On the Scattering of Mesons by Nuclear Particles," *Proc. Phys.-Math. Soc. Japan 23* (1941), 445-54.

57 M. Kobayasi, "On the Meson Theory of the Penetrating Component of Cosmic Radiation," *Proc. Phys.-Math. Soc. Japan 23* (1941), 891-914.

58 Sin-itiro Tomonaga, "Bemerkung über die Streuung der Mesotronen am Kernteilchen," *Sci. Pap. Inst. Phys. Chem. Res. 40* (1942), 73-86.

59 Karl Birus, Yataro Sekido, and Yukio Miyazaki, "Ein Umwandlungseffekt neutraler Mesotronen," *Sci. Pap. Inst. Phys. Chem. Res. 38* (1941), 353-9; Yoshio Nishina and Karl Birus, "Neutrale Mesotronen in der Höhenstrahlung?" *Sci. Pap. Inst. Phys. Chem. Res. 38* (1941), 360-70.

60 R. E. Marshak, "Heavy Electron Pair Theory of Nuclear Forces," *Phys. Rev. 57* (1940), 1101-6.

61 The English-language version, published later, is as follows: Mituo Taketani, "On the Neutral Meson," *Progr. Theoret. Phys. (Kyoto) 3* (1948), 349-55.

62 Gregor Wentzel, "Zur Problem des statischen Meson-feldes," *Helv. Phys. Acta 13* (1940), 269-308; Sin-itiro Tomonaga, "Zur theorie des Mesotrons. I," *Sci. Pap. Inst. Phys. Chem. Res. 39* (1941), 247-66.

63 Tatuoki Miyazima and Sin-itiro Tomonaga, "Zur Theorie des Mesotrons. II," *Sci. Pap. Inst. Phys. Chem. Res. 40* (1942), 21-67; Tatuoki Miyazima and Sin-itiro Tomonaga, "On the Mesotron Theory of the Nuclear Forces," *Sci. Pap. Inst. Phys. Chem. Res. 40* (1943), 274-310.

64 Shoichi Sakata and Takesi Inoue, "On the Correlations between Mesons and Yukawa Particles," *Progr. Theoret. Phys. (Kyoto) 1* (1946), 143-50; Shoichi Sakata and Takesi Inoue, "On the Relation between the Meson and the Yukawa Particle," *Bulletin Phys.-Math. Soc. Japan 16* (1942), 232-4 (Japanese version of the preceding); Yasutaka Tanikawa, "On the Cosmic-Ray Meson and the Nuclear Meson," *Progr. Theoret. Phys. (Kyoto) 2* (1947), 220-1.

65 C. M. G. Lattes, H. Muirhead, G. P. S. Occhialini, and C. F. Powell, "Processes Involving Charged Mesons," *Nature 159* (1947), 694-7.

66 Hideki Yukawa, "On the Foundation of the Theory of Fields – New Theory of Particles, Part 3," *Kagaku (Science) 12* (1942), 249-55, 282-6, 322-6 (in Japanese), English translation in *Scientific Works*, ed. by Yasutaka Tanikawa (Tokyo: Iwanami Shoten, 1979), pp. 386-414.

67 S. Tomonaga, "On a Relativistically Invariant Formulation of the Quantum Theory of Wave Fields," *Rikon-Iho 22* (1943), 545 (in Japanese), English version in *Progr. Theoret. Phys. (Kyoto) 1* (1946), 27-42.

Part III

Discoveries of particles

5 The early stage of cosmic-ray particle research[1]

DMITRY SKOBELTZYN

Born 1892; graduated St. Petersburg University, 1915; use of cloud chamber for nuclear physics and cosmic rays; State Prize, 1951; P. N. Lebedev Physical Institute, U.S.S.R. Academy of Sciences, Moscow.

This chapter is a brief account of personal recollections pertaining to the prehistory of cosmic-ray research, written by probably the oldest living participant in the scientific events of that period.

In 1927, I published photographs of secondary electron tracks produced by a beam of γ rays in a Wilson cloud chamber placed in a magnetic field.[2] On two of these photographs the tracks of one or two unusually high-energy particles, not connected with the γ-ray beam, were discovered. Subsequent observations revealed the relatively frequent appearance of similar tracks and a very striking peculiarity: the appearance of groups of simultaneous correlated tracks.*

I began research on γ rays (and subsequently cosmic rays) in late 1923 in the laboratory of my father, a professor of physics at the Leningrad Polytechnical Institute. It started spontaneously under the impact of a very important discovery of that time, the Compton effect, and a fortunate idea of my own, namely, to investigate in a Wilson chamber the recoil-electron tracks of γ rays. My first photographs were taken without a magnetic field present.

---

* *Ed. note*: For a discussion of Skobeltzyn's first observations and reactions to them, see Norwood Russell Hanson, *The Concept of the Positron* (Cambridge University Press, 1963), especially pp. 136-9 and 216-17. See also Skobeltzyn's "note added in proof" (Note 20).

The use of a magnetic field combined with a cloud chamber, which turned out to be so fruitful, was not an invention. I used such a field (1,500-2,000 G) as an auxiliary means to deflect the tracks of secondary β rays produced by γ rays in the walls of the Wilson chamber. The spurious background of such tracks hindered observation of the recoil-electron tracks produced in the gas of the chamber.

At that time I was not interested in cosmic rays. I was, however, aware of the work being done in this field. One of my older university colleagues, Leo Mysovski, performed many important experiments in studying cosmic rays, called at that time *Höhenstrahlung*. To Mysovski and his associate Tuvim belongs, for instance, credit for discovery of the "barometric effect" exhibited by this radiation. The highlights in this field of research at that time were the very important results of the experiments by Robert Andrews Millikan and G. Harvey Cameron concerning the absorption of cosmic rays in mountain lakes and Millikan's (unfounded) hypothesis on the nature and origin of cosmic rays as very hard "ultra γ rays" produced as a product of synthesis of various light nuclei, such as He, O, Si, etc.

It appears that before my Wilson photographs were published nobody had tried to observe the secondary β particles of such hypothetical ultra γ rays. In this connection, I shall quote a paper by a renowned and very competent experimenter, Walther Bothe, that appeared in early 1923.[3] One finds the following astonishing statement: "Man muss offenbar schliessen, dass ein β-Strahl, dessen Geschwindigkeit der Lichtgeschwindigkeit nahekommt, nach der Wilsonschen Methode nicht oder nur sehr schwer zu erkennen ist."*

Luckily for me, I could not be impressed by such a pessimistic pronouncement, because before I began my observations with the Wilson chamber I had had the opportunity to study thoroughly two fundamental papers by Niels Bohr on the theory of ionization produced by fast β particles.[4] Bothe probably overlooked this important contribution of Bohr, which seems to have been overshadowed by his papers containing the famous quantum postulates that appeared practically simultaneously.

After my first casual observation on the appearance of cosmic-ray tracks in a Wilson chamber[2] [the term *cosmic ray* (which had not yet been adopted) was not mentioned by me on that occasion, nor even *Höhenstrahlung*], a full report on the corresponding results of my work

---

* *Ed. note*: "One must apparently conclude that a β ray whose speed approaches the speed of light is not recognizable by the Wilson method, or only with great difficulty."

was published in 1929.[5] But most of the pertinent facts and photographs were presented by me earlier in the course of a discussion at an informal conference on γ- and β-ray problems held at Cambridge, July 23 to 27, 1928, under the patronage of Ernest Rutherford. One session of the conference was scheduled for a discussion on β-ray problems (it seems that no paper was read at that session). During that discussion, I demonstrated a collection of photographs of cosmic-ray tracks, and, I dare say, it produced some impression on the audience. Incidentally, I began with a comment on the Bohr theory of the ionization produced by very fast β particles.

Immediately after my impromptu remarks, Hans Geiger took the floor to announce that Bothe and Werner Kolhörster were working on a method to register cosmic rays by the coincidence of pulses in two wire counters and that they hoped to be able to study the penetrating power of the rays by this method.

In connection with this communication of Geiger on July 25 (evoked, I believe, by my own presentation), I call attention to the following dates. The announcement by Geiger and Müller of their invention of wire counters was published the 7th of the same month (July 7, 1928).[6] A very brief communication of Bothe and Kolhörster, informing that as a result of their coincidence observations with recently invented wire counters there were observed ionizing particles penetrating 1 cm of lead, was dated November 2, 1928.[7] A detailed communication of my results, published in *Zeitschrift für Physik*, is dated February 5, 1929, and that of Bothe and Kolhörster is dated May 1929.[5,8*]

It is well known what extremely important results followed from further development of the technique of using a Wilson chamber plus a magnetic field. The next step that appeared natural was to use a magnetic field of much higher strength. There were many reasons why I myself never tried to do this. In 1929-31, I was working at the Curie laboratory in Paris. Pierre Auger, who had been working at the neighboring institute of Jean Perrin, asked me (probably at the beginning of 1931) if I intended to undertake such investigations. I answered in the negative, whereupon he told me that he would try to perform that kind of experiment. Soon thereafter, he showed me his installation that was ready for operation. However, his attempt turned out to be unsuccessful and probably was dropped by him in the fall of 1931 when (as we

---

* *Ed. note*: These are dates of submission, as given in the articles themselves.

shall see later) it was disclosed that Carl David Anderson had already obtained some thousand beautiful pictures of cosmic-ray tracks in a strong magnetic field (13,000 G).

It seems that something was wrong with the cloud chamber that Auger had constructed. I was told in the spring of 1931 that his Wilson chamber, when put into operation, showed no cosmic-ray tracks whatsoever. I obtained this information from a fellow of the Curie laboratory staff (Georges Fournier), who concluded from this that my observations of 1927-9 were erroneous. By that time, however, my results had already been corroborated.[9] Until the end of my sojourn in Paris (1931), I had no further occasion to meet Auger himself, and afterward I never heard from him what had gone wrong with his cloud chamber in a strong magnetic field.

In November of 1931, when I was already in Leningrad, Millikan visited Europe and gave sensational lectures at Paris and Cambridge, showing a collection of Anderson's photographs. The main content of these lectures was published by Millikan, with Anderson as coauthor, in May 1932.[10] The tracks observed by Anderson were ascribed to high-energy protons produced by cosmic-ray photons. In November and December of 1931, I received letters from Marie Curie and Frédéric Joliot-Curie from Paris and L. Harold Gray from Cambridge, who mentioned that they attended Millikan's lectures and in a few lines informed me of the results communicated by him. Somewhat more detailed information was given in the letter of Gray, with whom at that time I was in correspondence, discussing some problems of γ-ray research. He wrote (November 27): "I *think* [emphasis added] that in every case the proton tracks were more dense than those of electrons." That probably was his imagination. Millikan showed to his audience some 11 pictures of good tracks in the range of 20 to 80 MeV energy, say $E \simeq 50$ MeV on the average.

Now, the ionization density (specific ionization) depends mainly on the velocity of the particle, or on the quantity E/mc^2, where m is the proper mass of the particle. With this quantity given, the absolute value of the mass m of a particle of given charge has but little importance. It follows that the ionization produced by a proton of 50 MeV is practically the same as that of an electron of about 25 keV. It is impossible to confuse the specific ionization of such a slow electron with the ionization of a fast ("relativistic") one having several MeV or more. However, the positive tracks on pictures demonstrated by Millikan did not differ essentially from electron tracks on the same pictures, with energies of about 50 MeV. Millikan and his audience overlooked this inconsistency. After I received Joliot's letter, I wrote him straightforwardly my views on the subject and suggested that something was wrong with Millikan's photos or their interpretation.

In their note of May 1932, Millikan and Anderson repeated the interpretation Millikan had made in Europe. They asserted that the fast protons were a product of the interaction of ultra γ rays with nuclei, and they even saw in the phenomenon a new confirmation of Millikan's old hypothesis on the origin of cosmic rays (ultra γ rays as a product of synthesis of certain nuclei). Progress by Anderson in the deciphering of his experimental evidence was slow. Only in September 1932 (a year after a stock of more than 1,000 Wilson photos in a strong magnetic field had been obtained) did he make reference, in a brief and very cautiously drafted note, to the specific ionization of the positively curved tracks, and he concluded the existence of positively charged particles, the mass of which "must be small compared with the mass of a proton."[11] The next paper of Anderson appeared in February 1933; it was entitled "The Positive Electron," and it contained more definite statements.[12] In writing it, he was already aware of the results of the outstanding work of Patrick Blackett and Giuseppe Occhialini (he refers to a press report about this work).[13]

The paper of Blackett and Occhialini was received for publication in the *Proceedings of the Royal Society of London* on February 7, somewhat earlier than that of Anderson in *Physical Review* (February 28). It now appears strange, perhaps, that in discussing their experiments the authors of both papers did not make any attempt to connect their results with the Dirac theory of the positron. Paul A. M. Dirac's work of 1930 was certainly known to them (to the Cambridge physicists, anyway). However, it is true that Dirac himself was inclined to identify the positive particles of his theory with protons.[14] Anderson, in his paper, suggested some far-reaching hypotheses that appeared strange, as, for instance, the transformation of protons into electrons, induced by cosmic rays. (Such speculation considered now on the background of modern subnuclear physics may appear, perhaps, less strange?)

During some months after the discoveries of Anderson and Blackett-Occhialini, the experimental evidence obtained simultaneously by many experimenters in the field of γ-ray research showed that the concepts of the Dirac theory were adequate. The conclusions drawn from this evidence were summarized by Blackett at a session of the Solvay Congress (October 22-9, 1933). The discussion that followed is certainly of historical interest. In the record of this discussion, one finds, incidentally, the following characteristic remark of Rutherford: "It seems to a certain degree regrettable that we had a theory of the positive electron before the beginning of experiments. . . . I would be more pleased if the theory had appeared after the establishment of the experimental facts" (translated from the French text by the author).[15]

I have dwelt on the facts related in the preceding paragraphs because they can give one an idea of the psychological barriers lying in the way of discovering the first of a sequence of many generations of new particles that soon followed.

There is another important line of development that followed (with some delay) my first observation of cosmic-ray tracks. I refer to the appearance of groups of simultaneous tracks (up to four in my subsequent observations) and "showers" (up to 20 particles) on photographs of Blackett and Occhialini (1933) taken with their counter-controlled Wilson chamber.[13] This phenomenon immediately attracted the attention of physicists working in the field. But its importance in leading to perhaps the most interesting chapter in the history of this branch of science had probably not been foreseen. In fact, it was a precursor of modern high-energy physics. For a relatively long time, however, its nature remained a puzzle.

Two aborted attempts to solve this puzzle followed soon. A note by Auger and myself on this problem appeared in July 1929.[16] Later (early 1932), Werner Karl Heisenberg published his version.[17] How meager was the experimental evidence needed to answer the question as to the nature of the phenomenon can be seen from the fact that on five pages of Heisenberg's paper my observations of 1929 were quoted 11 times – quite unusual. It was unfortunate for the author to have published his paper as early as he did. During the same year (1932), the discovery of the positron was announced, and a year later pictures of showers taken by Blackett and Occhialini appeared. Both these events changed the whole situation radically.

The suggestion of Auger and myself was that the groups of tracks could appear as a result of simultaneous production of several Compton electrons by ultra γ radiation. We even saw in this effect a reason to reject the interpretation of the nature of primary cosmic rays as high-energy β rays, an idea put forward by Bothe and Kolhörster on the basis of their measurements of the penetrating power of cosmic-ray particles.[8] Heisenberg, on the other hand, based his deductions on the aforementioned hypothesis of Bothe and Kolhörster. According to his scheme, simultaneous tracks of an observed group were tracks of many δ rays generated by one and the same high-energy β particle.*

A long road of hard work was trodden by many theoreticians of the highest caliber (and to a lesser degree by experimenters) to arrive at an understanding of the nature of the phenomenon.

·The names of Walter Heitler, Hans Albrecht Bethe, J. Robert Oppenheimer, C. F. von Weizsäcker, and many others should be mentioned in this connection. The solution to the puzzle (the first version of the cascade theory of showers) came in 1937 in a paper by Homi J. Bhabha and Heitler.[18]

For subsequent developments, the discovery by Auger, just before the war, of extensive atmospheric cosmic-ray showers (at first called Auger showers) was of great importance.[19] He observed coincidences in two counters placed up to about 300 m apart (in a horizontal plane). This occurred at the time that I began my work at the P. N. Lebedev Institute in Moscow.

During the last years (1944-5) of World War II, cosmic-ray research was resumed at the Lebedev Institute. I suggested to George Zatsepin that he begin observation of atmospheric (later known as extensive) showers and try to obtain by this method information on the highest energies of cosmic-ray particles. The first step toward this goal was to use an array of two pairs of counters separated by a distance as long as possible. By using fourfold coincidences, instead of twofold as in Auger's experiments, one could diminish the influence of spurious background due to chance coincidences. The experiments performed by Zatsepin and associates were successful. They became the starting point for the development of more complicated and more refined devices in various branches of research, and they constituted a very exciting chapter in the history of modern high-energy physics.

During the following decades, work in this field was pursued by a highly qualified team of research fellows from the Lebedev Institute and Moscow University, and it continues to this day.

---

* *Ed. note*: δ rays are secondary electrons, usually of short range, ejected from atoms by fast charged particles.

Notes

1 This chapter was originally prepared for the book *Early History of Cosmic Ray Studies*, edited by Y. Sekido and H. Elliot (Dordrecht, Holland: Reidel, 1981). We are grateful to the editors of that book and to Professor Skobeltzyn for permission to adapt the chapter for inclusion in this volume.

2 D. Skobeltzyn, "Die Intensitätsverteilung in dem Spektrum der γ-Strahlen von RaC," *Z. Physik 43* (1927), 354-78.

3 W. Bothe, "Untersuchungen an β-Strahlenbahnen," *Z. Physik 12* (1923), 117-27.

4 N. Bohr, "On the Theory of the Decrease of Velocity of Moving Electrified Particles on Passing through Matter," *Phil. Mag. 25* (1913), 10-31; N. Bohr, "On the Decrease of Velocity of Swiftly Moving Electrified Particles in Passing through Matter," *Phil. Mag. 30* (1915), 581-612.

5 D. Skobeltzyn, "Ueber eine neue Art sehr schneller β-Strahlen," *Z. Physik 54* (1929), 686-702; D. Skobeltzyn, "Sur le mécanisme de rayonnement ultrapénétrant (rayons cosmiques)," *Compt. Rend. 195* (1932), 315-17.

6 H. Geiger and W. Müller, "Elektronenzählrohr zur Messung schwächster Aktivitäten," *Naturwiss. 16* (1928), 617-18.

7 W. Bothe and W. Kolhörster, "Eine neue Methode für Absorptionsmessungen an sekundären β-Strahlen," *Naturwiss. 16* (1928), 1045.

8 W. Bothe and W. Kolhörster, "Das Wesen der Höhenstrahlung," *Z. Physik 56* (1929), 751-77.

9 L. M. Mott-Smith and G. L. Locher, "A New Experiment Bearing on Cosmic-Ray Phenomena," *Phys. Rev. 38* (1931), 1399-408.

10 Robert A. Millikan and Carl D. Anderson, "Cosmic-Ray Energies and Their Bearing on the Photon and Neutron Hypotheses," *Phys. Rev. 40* (1932), 325-8.

11 Carl D. Anderson, "The Apparent Existence of Easily Deflectable Positives," *Science 76* (1932), 328-9.

12 Carl D. Anderson, "The Positive Electron," *Phys. Rev. 43* (1933), 491-4.

13 P. M. S. Blackett and G. P. S. Occhialini, "Some Photographs of the Tracks of Penetrating Radiation," *Proc. Roy. Soc. (London) A139* (1933), 699-727.

14 P. A. M. Dirac, "A Theory of Electrons and Protons," *Proc. Roy. Soc. (London) A126* (1930), 360-5.

15 *Structure et propriétés des noyeaux atomiques, rapports et discussions du 7ᵐᶜ Conseil de Physique Solvay, 1933* (Paris: Gauthier-Villars, 1934), quotation on pp. 177-8.

16 Pierre Auger and D. Skobeltzyn, "Sur la nature des rayons ultrapénétrants (rayons cosmique)," *Comp. Rend. 189* (1929), 55-7.

17 W. Heisenberg, "Theoretische Ueberlegungen zur Höhenstrahlung," *Ann. Physik 13* (1932), 430-52.

18 H. J. Bhabha and W. Heitler, "The Passage of Fast Electrons and the Theory of Cosmic Showers," *Proc. Roy. Soc. (London) A159* (1937), 432-58.

19 Pierre Auger and Roland Maze, "Extension et pouvoir pénétrant des grandes gerbes de rayons cosmiques," *Compt. Rend. 208* (1939), 1641-3.

20 Note added in proof by D. V. Skobeltzyn concerning *The Concept of the Positron* (Cambridge University Press, 1963) by N. R. Hanson: The interpretation of my results of 1926-7 by the author of this book, based on his examination of plates printed in *Zeitschrift für Physik* (Note 1), is nothing else but sheer nonsense. It is absurd to pretend that such conclusions as his can be drawn on the basis of such material. To persuade him that he is wrong, I sent him the original print of a stereoscopic pair of photos that could show him clearly his error. Everybody who has the least habit of exploring such photos would be certainly satisfied. But he persisted in his interpretation. The main point of my remarks is the following: Professor Hanson was inspired by a story told him by Professor Dirac. The story in itself is correct. Dirac remembered a private meeting held in the evening in a room of one of the colleges at Cambridge University (I remember him attending it) where I had been invited by someone (if I am not mistaken, by the late Dr. Cockroft) to speak about some of my recent observations on positron phenomena. The results of these

observations of mine later turned out to be controversial. I presented a short communication on the subject at a plenary session of the Congress on Physics in London some days before visiting Cambridge with other participants of the congress. A curious point that makes the whole story anecdotal is that Dirac, remembering after a time of about 20 years correctly about the event itself, has been mistaken in indicating its date as something that happened in 1926-7 (see the book by Hanson, p. 136). In fact, it occurred in October of 1934, at the time of the Congress on Physics organized by late Professor R. Millikan (incidentally, my short communication in the course of discussion is published in the proceedings of the congress). This shift by six or seven years of the event quoted by Professor Hanson makes his deductions pertaining to it senseless.

6 Some reminiscences of the early days of cosmic rays

H. VICTOR NEHER

Born 1904, Page, Kansas; Ph.D., California Institute of Technology, 1931; cosmic rays; California Institute of Technology (emeritus).

In February of 1932, Robert A. Millikan and Ira S. Bowen came into my laboratory to discuss the possibility of me joining them in a new series of measurements of cosmic rays.[1] I had done my thesis work on the scattering of high-energy electrons. Nevill F. Mott had just published his relativistic, quantum mechanical scattering formula for electrons in which he predicted a polarization of the scattered electrons. I thought it would be interesting to test his prediction, but to do so I needed a much more sensitive detector for the double scattering experiment involved.* Of the several possible small current detectors known at that time, a Lindemann type of electrometer with a statically balanced crossarm on a fused-quartz torsion fiber seemed the best possibility. I had undertaken the job of developing such a detector. Jack Workman, then a National Research Fellow at Caltech, knew some of the tricks of making quartz fibers. I was still very much of a novice in February 1932, but it turned out that these techniques were just what were needed to make the instrument that Millikan and Bowen had in mind. This, then, is how I became involved in cosmic-ray research. I told Millikan that I would like to see my research on electron scattering continued. "Oh," he said, "we'll find someone to carry it on." And thus Bill Pickering, as a senior, became involved.

---

* *Ed. note*: Mott's formula was not confirmed until 1940. See Allan Franklin, "The Discovery and Nondiscovery of Parity Nonconservation," *Studies in the History and Philosophy of Science 10* (1979), 201-57.

But the scattering experiment was dropped within a year because of the press of cosmic-ray work.

What Millikan and Bowen were proposing was to make a series of experiments on cosmic rays using airplanes, ships, and balloons at various latitudes. What was needed was a self-recording ionization chamber and detector that was insensitive to vibration and tilt. I told them I would think about it and let them know.

With such a device there are two possible arrangements: (1) The detecting device can be placed exterior to the ionization chamber. (2) The detecting device can be placed inside the ionization chamber. There are obvious advantages to the second arrangement, which was the method used by Millikan for the instruments in his underwater experiments in mountain lakes. Having the detector inside obviates the possible difficulties one might have with humidity causing leakages over insulators and effects due to the polarization of dielectrics. One can easily realize how extraneous effects can be serious, because the typical electric currents that one needs to measure at sea level are of the order of 10^{-14} A. If an accuracy of 1% is desired, then these extraneous effects must correspond to currents of less than 10^{-16} A. Another distinct advantage of this simple system is that the sensitivity is not dependent on auxiliary potentials and further is not sensitive to small changes in the single potential used. But the Wulf type of electroscope used in Millikan's ion chamber was very subject to vibration and tilt, and besides was not very sensitive, even in a 2-liter-volume ion chamber with 30 atm of air pressure.

After a few days I took my ideas to Millikan, and he said to go ahead. I also had some ideas on the recording end of things, and he went along with these, too. He then laid out a time schedule. He wanted first to make some airplane flights at March Field, near Riverside in southern California, during the summer. Then he wanted to go to northern Canada to make similar flights. An intermediate latitude would be Spokane, Washington. Next we would go to Seattle and take readings on board ship going south to Los Angeles. He proposed that I then take a ship from Los Angeles, making measurements going south to Peru, make airplane flights there, and also go into the Peruvian mountains, ending up by making measurements on Lake Titicaca. I would then take a ship from Peru to New York through the Panama Canal, making measurements at sea level.

He then went on to relate his experiences on a boat trip with G. Harvey Cameron going south from Los Angeles to Peru in 1926. Millikan expected that there might be a latitude effect, even though he then thought that the majority of the primaries were photonic in nature. A charged component would be anticipated because of the secondary radiation produced due to Compton collisions as the original radiation traveled through space.

On the trip to Peru, Millikan and Cameron were some 600 miles south of Los Angeles before they succeeded in getting two of their three instruments set up and working properly. As we learned later, the decrease in cosmic-ray intensity at sea level sets in rather rapidly at about the latitude of Los Angeles as one goes south, and by the time they had covered 600 miles most of the latitude effect was over with. In analyzing the results of their sea-level measurements on that trip, they arrived at the conclusion that there was no latitude effect, with an estimated uncertainty of 6%.

Arriving at Mollendo, Peru, they went inland to a Bolivian mountain lake. Millikan said they actually measured a lower cosmic-ray intensity, by 10% to 20%, at these altitudes than they had measured with the same instruments at northern latitudes. But because of the shielding of surrounding mountains, as well as the possible effects of air leaks of the ionization chambers, they decided that within the experimental uncertainties there was no latitude effect at these mountain altitudes either. He then went on to say that Arthur Compton had received a grant to make a world survey of cosmic-ray intensities at sea level as well as mountain altitudes. "We welcome this kind of independent work," he said. "I'm sure we will all agree in time."

The time schedule that he laid out frightened me a bit, but I was promised a lot of help. Two to three machinists were kept busy in the shops, and I concentrated on the quartz electroscope system. It took a couple of months to arrive at a satisfactory design. The gold-coated fused-quartz system consisted of a quartz torsion fiber about 1 cm long and 4 μ in diameter. The crossarm was also about 1 cm long and 10 μ in diameter. One end of the crossarm was bent over, and this was viewed with an objective lens that cast an image on the photographic film. The other end of the crossarm was snipped off until the arm was balanced statically. A stationary crossarm repelled the movable arm as electric charge was applied. A means was provided to produce an initial twist in the torsion fiber so that the deflection did not commence until a predetermined potential was reached. The design was such that once the motion started, the deflection was close to linear with change of electric charge.

The original systems were assembled using a cement that polymerized (i.e., hardened) with the application of heat. Later, techniques were developed to fuse all of the quartz parts together. These systems had a capacitance of about 0.5×10^{-12} F and could be made to give sensitivities such that α particles could easily be counted. (Someone has said that a system made of quartz like this either gives the right answer or is broken.)

Dr. Millikan (or "The Chief," as he was affectionately known by the faculty as well as the graduate students) was very much interested in these developments and often had useful suggestions to make. He was, of course, busy running the institute and had a busy social life. Appointments with him were often at night. He would say, "I'll see you tonight in your room at 11 o'clock." Sure enough, he would be there, and usually on time, and might stay until 2 o'clock in the morning. Sometimes he would come after a formal affair, still dressed in tuxedo.

One night he came to help test one of the early versions of the quartz system. It was mounted in a brass cylinder and could be viewed with an eyepiece and objective lens. A flashlight lamp provided illumination. When we wanted to test the system for vibration, we used his 1928 Chevrolet; he did the driving while I observed, changing the coupling between the vibrating car and the instrument by placing more or less human flesh in between the two. It seemed to behave better than I had expected. He wanted to see for himself. While he looked through the eyepiece, I started the car. Not being familiar with its workings, I let out the clutch; it grabbed, and the car lurched forward. The eyepiece struck the bridge of his nose; the injury was not too serious, but it left a scar that remained the rest of his life.

By the last week of July we had assembled a working instrument with a 2,000-cm^3 ion chamber having an internally mounted quartz-system detector and a recorder with a clock-driven film. A built-in 45-deg-angle partially reflecting mirror permitted visual readings to be taken simultaneously if desired. A bimetallic coiled strip also gave a temperature record on the film. The last instrument that Millikan had used in his underwater measurements in mountain lakes and the one that had 30 atm of air pressure was used to calibrate the new instrument. Millikan had determined the constants of this instrument with some precision.

Two quantities need to be known for such an instrument to determine the ionization due to any source. These are (1) the constant one needs to multiply the readings by to arrive at the ionization one would get in a standard volume of air under standard conditions of temperature and pressure and (2) the zero of the instrument. If all external sources of radiation are removed, there still remains a residual ionization mainly due to α particles emitted by the inner wall of the ionization chamber. By taking two values of ionization and comparing the two instruments, the constants of the new instrument can be determined in terms of the old, taken as a standard.

Millikan's instrument was used as a standard for over 35 years. It is still usable and is in the archives at Caltech. An independent determination of the absolute values of ionization by Michael George in 1967 showed that the calibrations of the instruments used through the years, as derived from Millikan's instrument, agreed to within 1%.

After comparing the two instruments in Pasadena, we decided to compare them also out on the surface of Lake Arrowhead in southern California. This lake, at an elevation of about 5,000 ft, is about 125 ft deep at its center and had been used by Millikan in his underwater work, where the readings were compared at various depths with those taken at Muir and Gem lakes in the high Sierra Nevada Mountains.

We anchored the rowboat in the middle of the lake. In the boat, the new instrument was left running for the night. Also in the rear of the boat were the six 100-lb pieces that went together to form the lead shield. When we looked out on the lake the next morning, there was no boat. A windstorm had come up during the night, and it had sunk. Our hopes also sank. Using a grappling hook, we dragged the lake for three days. We pulled up old automobile tires, fishing gear, etc., but no instrument.

Arrangements had been made to start a series of airplane flights at March Field at the end of July, and obviously these had to be canceled. Millikan sent a telegram to Colonel Arnold (later General H. H. "Hap" Arnold) telling him of the bad news and asking to have the flights postponed for a few weeks until we could build another instrument. At that time the air force was the Army Air Corps, a part of the U.S. Army. Arnold sent back a telegram: "Serves you right for playing with the Navy."

Now we really had to get busy. In just 33 days we had duplicated the instrument and were out at March Field. Using Curtiss Condor bombers, we made five flights, some with the lead shield and some without. The pilots' instructions were to fly at four given altitudes for 1 hr each, the highest being 22,000 ft. Accompanied by Ike Bowen, we then went by train to northern Manitoba, Canada, where the Royal Canadian Air

Force supplied a plane. Three flights were made from Cormorant Lake, north of The Pas. I then took the instrument with its lead shield to the top of Pike's Peak in Colorado, while Bowen and Millikan made balloon flights from South Dakota, using instruments that Bowen had had made that summer. Measurements were made on Pike's Peak in a hut with the instrument inside its shield. After a week, Millikan arrived, and we took the train for Spokane, where similar airplane flights were made. Next on the agenda was to go to Seattle, where we boarded a passenger ship for Los Angeles. On this trip the new instrument was run inside the lead shield, and the older one was operated with no shield.

Some comments might be made about using airplanes to make cosmic-ray ionization measurements. The chief uncertainty was the effect of the radioactive materials in the airplane's instrument dials. In all cases we located our instrument as far aft as possible. To find the effect of the dials, readings were made with the airplane on the ground. These involved measurements in the aircraft with the pilot in his seat and also measurements at the same location with the aircraft absent. By this means the effects of the dials and the ground radiation could be separated and determined. There was always the uncertainty of the airplane's absorption of the local radiation, but the skin of these airplanes was either fabric or thin aluminum, and this was probably a small factor. To determine the air pressure during a flight, we used a recording barometer calibrated by the Bureau of Standards, and we checked it at intervals.

After we returned to Pasadena, the data were worked up. Even with some uncertainties about corrections for the airplane dials, it was quite obvious that there was a latitude effect of cosmic rays at airplane altitudes between southern California and Canada. We did not have much time to think about these results though, because I was scheduled to go south by ship from Los Angeles to Peru. Most of the four weeks of intervening time was spent in making calibrations and making a new quartz system. It seemed foolish to leave with just one instrument, with no preparation for failure. A new system was made, but because of lack of time the crossarm was not balanced. A pressure gauge and a cylinder of old compressed air were also taken along; the air had to be old to allow for decay of radon.

After we left Los Angeles harbor, all went well for a day or so. Then symptoms began to develop that all was not well. Finally, on the second day, the quartz system broke. This was a horrible thing to happen. Millikan, six years earlier, had missed this strategic region, and now we had done the same. I had the other quartz system, but the seas were too rough to permit the delicate operation of balancing the crossarm. I waited until we anchored at Mazatlan. This was in a rather quiet part of the bay, but the swells made the ship roll. Nevertheless, I managed to snip off, bit by bit, the back part of the crossarm, using a pair of cuticle shears and a magnifying glass.

It was a day after leaving Mazatlan before data were again being collected. The constant of the new system was estimated. Its zero was probably close to the value of the old one. The data showed no latitude effect on the remaining part of the trip to Panama.

On arriving at Panama I had to send Millikan a cable. But what should I say? Millikan was not one to depart from the reputation of his Scottish ancestry, and he had instructed me to be as brief as possible. I decided to send a cable saying there was no latitude effect, and I followed this with a letter describing what had happened.

We changed ships at Panama and continued on to Mollendo, Peru. In changing ships there is always the uncertainty of different absorbing materials in the walls and overhead. There is sufficient difference so that readings cannot be readily compared. On reaching Mollendo, I sent another cable indicating again that there was no latitude effect from Panama down. We had stopped at Lima before reaching Mollendo, and I had talked with the South American representative, Captain Harris of Panama Grace Airways, about putting our instrument in one of their planes. He was a Caltech graduate and was very willing to help out. Their Ford trimotor planes were supercharged to get over the Andes.

From Mollendo we took the train to Arequipa. It was there that I mounted the instrument and recording barometer in the tail of a Panama Grace airplane. On the way to Lima our maximum altitude was 19,000 ft, which for a Ford trimotor was not too bad. The cabin was not pressurized, and no oxygen was available. On the return flight, on another plane, we flew at other altitudes to give more points on the ionization-pressure curve.

After making measurements in the mountains at 14,700 ft, we went to Lake Titicaca, at 12,500 ft. There measurements were made on a small cargo ship, with the instrument unshielded on the deck. We then boarded a ship at Mollendo and took data all the way to New York via the Panama Canal. Going through the Caribbean and on up the east coast of the United States it became clear that there was a latitude effect. The following message was sent to Millikan on December 27,

1932, when we were at 43° geomagnetic north: "Seven percent change returning. Concealed before by broken system and different ships."

I saw Millikan at the Atlantic City meeting of the American Physical Society. The original plan was to return by train to Pasadena. Because of the trouble in going south from Los Angeles to Panama, it was decided to take a ship back through the canal. Arrangements were also made with the Army Air Corps to make an airplane flight at Panama.

It need only be added that the data from New York through the canal and up to Los Angeles were consistent with the data taken from Mollendo to New York when geomagnetic latitudes are used. Also, the results of the airplane flight at Panama were consistent with the flights made in Peru. It was necessary to wait until returning to Pasadena to calibrate the instrument used on the trip so that an accurate comparison could be made between airplane flights made in Peru and Panama and those made at northern latitudes.

Concerning the latitude effect at sea level, it was Professor Jacob Clay of Holland who in 1928 first reported a latitude effect. In going from Amsterdam to Batavia he found a decrease of 45%. We were aware of Clay's work, but this large latitude effect seemed unreasonable. In later years the value found between these points was about 12% in summer and several percent larger in winter. As Ike Bowen remarked, "If one is concerned with errors, Millikan's estimated uncertainty of 6 percent for his South American trip in 1926 was considerably less than Clay's."

The larger latitude effect found at airplane altitudes gave added impetus to the project of making balloon flights, where one could expect to make measurements near the top of the atmosphere. Consequently, a number of lightweight instruments were made that were used at latitudes as far north as Saskatoon, Canada, and as far south as southern India, which is close to the geomagnetic equator. These measurements were made in 1936 and 1937.

In taking the difference between the ionization-depth curves for different latitudes and longitudes, it was possible to determine how much total energy was brought in to the earth by charged particles. By using the calculations of G. Lemaitre and M. S. Vallarta it was possible to find the average magnetic rigidity of these particles. Furthermore, by assuming the nature of the charged particles involved, it was possible to make an estimate of the numbers of primaries within a certain energy range, that is, arrive at a rough differential energy distribution. The big question at that time, of course, was, What kind of particles?

The 1930s and 1940s were surely the days and years of innocence, as Bruno Rossi has emphasized in Chapter 11. I suppose there has never been a period in the history of physics when so many new phenomena have come to light in so short a time in a single field of investigation. The original workers could only proceed from the then-known behavior of photons or charged particles. Many wrong guesses were made about the nature of the absorption mechanisms in the atmosphere, as well as the nature of the primaries. Also, some right guesses were made for the wrong reasons. A typical example of the latter is the experiment of Walther Bothe and Werner Kolhörster in 1929.[2] Using a piece of gold 4.1 cm thick between two Geiger counters, used to count coincidences, they concluded that the primaries were charged particles. We now know that most of these penetrating particles that they detected were μ mesons, which, of course, were secondaries.

In this same sense, I suppose, the early interpretation that the primaries were photons, by the way in which the radiation was absorbed in the atmosphere, could be said to be wrong for the right reasons as then understood.

There were several keys that eventually unlocked the doors to the mysteries of the ways cosmic rays behaved. One of the first breakthroughs came with the work of Hans Bethe and Walter Heitler in 1934 on the absorption of electrons.[3] This led to an understanding of the way in which the so-called nonpenetrating radiation was absorbed in the atmosphere. Soon after, in 1936, came the discovery of the μ meson by Carl D. Anderson and Seth Neddermeyer. This cleared up the mystery of the penetrating part of the radiation found in the atmosphere. Finally came an understanding of the role that nuclear collisions of high-energy particles played in the production of the various kinds of secondaries in the atmosphere. Although there was indirect evidence before 1940 that the primaries consisted largely of protons, direct evidence that they consisted mainly of positively charged high-energy particles waited until the late 1940s, when photographic emulsions showed that not only protons but also α particles and heavier nuclei were the primary constituents of the incoming cosmic rays.

A very brief history of Millikan's interpretation of cosmic-ray experiments as they indicated the nature of the primary radiation may be stated as follows: The absence of a measured latitude effect and the nature of the ionization-depth curve found for the earth's atmosphere argued for a photonic nature of the primaries. In seeking a source of the photons, Millikan hit upon the idea that the photons resulted from a fusion process, that is, the building up of the more complex nuclei from hydrogen. Using the currently available theories, the absorption curve seemed to support this idea. There then followed a series of experimental results such as the latitude effect, the longitude effect, balloon flights at different latitudes and longitudes, direct measurement of particle momenta in the cloud chamber, etc., that, combined with developments on the theoretical side, showed conclusively that the atom-building process could not give the required particle energies. To account for these higher energies, Millikan suggested that atom annihilation in the universe might be the mechanism. It had to be a mechanism that would give rise predominantly to charged particles, but it must also give more positive particles than negative. Millikan rejected the idea of incoming protons because he saw no way for them to give the atmospheric absorption curves that were found. To him, a proton of a given initial energy had a definite range determined by the ionization losses along its path. His last attempt to analyze the situation is described in the last chapters of his revised book *Electrons (+ and −), Protons, Photons, Neutrons, Mesotrons and Cosmic Rays,*[4] published in January 1947. He was then almost 79 years old. In this book he reiterated his stand that the primaries consist mostly of electrons, with positive electrons predominating. The remaining energy that is not affected by the earth's magnetic field he believed to consist of photons.

To those of us who knew Millikan, he was a remarkable individual. During the 1920s and early 1930s he was intimately involved with directing the research that went on in the Norman Bridge Laboratory of Physics, and he would assign many of the graduate students their research problems. He taught a first-year graduate course so that he might become acquainted with the strong and weak points of each student.

Millikan had a personality that attracted many people, including newspaper reporters, who for the benefit of their readers picked up any tidbits that served their purpose, *The New York Times* not excluded. Contention that developed in the early 1930s between Millikan and Compton was certainly fueled by the reporters. I think that most of us who knew Millikan regarded these events as aberrations that did not represent the Millikan that we knew.[5]

Like Newton, Millikan often talked of the giants on whose shoulders we stand. To many, Millikan may not qualify as a giant, but there is no doubt that he has had a lasting influence on the scientific development of this country as well as many other countries.

Notes

1 See also the account in Daniel J. Kevles, *The Physicists. The History of a Scientific Community in Modern America* (New York: Alfred A. Knopf, 1978), pp. 240-3.

2 W. Bothe and W. Kolhörster, "Das Wesen der Höhenstrahlung," *Z. Physik 56* (1929), 751-77.

3 H. Bethe and W. Heitler, "Stopping of Fast Particles and Creation of Electron Pairs," *Proc. Roy. Soc. (London) A146* (1934), 83-112.

4 Robert A. Millikan, *Electrons (+ and −), Protons, Photons, Neutrons, Mesotrons and Cosmic Rays* (Chicago: University of Chicago Press, 1947).

5 See, for example, my essay "Millikan – Teacher and Friend," *Am. J. Phys. 32* (1964), 866-77.

7 Unraveling the particle content of cosmic rays[1]

CARL D. ANDERSON,
with HERBERT L. ANDERSON

Carl D. Anderson: Born 1905, New York City; Ph.D., California Institute of Technology, 1930; Nobel Prize, 1936, for discovery of the positron; one of the discoverers of the muon (with S. Neddermeyer); California Institute of Technology (emeritus).

Herbert L. Anderson: Born 1914, New York City; Ph.D., Columbia University, 1940; assisted Enrico Fermi in producing the first nuclear chain reaction; built the 170-in. synchrocyclotron at University of Chicago, 1947-51, and found the first nucleon isobar; currently Professor Emeritus, University of Chicago and Senior Fellow, Los Alamos National Laboratory.

Introduction by Herbert L. Anderson

Carl Anderson's story is one of those American success stories that Horatio Alger might have written about the boy who makes good. He also exemplifies a maxim that's a favorite of mine, ascribed to Pascal: "Fortune favors the prepared mind." Carl Anderson is a man who was prepared, in the right way at the right time, for some remarkable opportunities that he didn't hesitate to take.

Carl Anderson received his Ph.D. at Caltech in 1930, at which time Robert A. Millikan, who was in a sense his patron, strongly suggested that Anderson build a cloud chamber to look at cosmic rays. Previous work had turned up some curious puzzles. Millikan had the idea that by looking at the cosmic rays with a cloud chamber one might see many interesting things. Carl Anderson built a cloud chamber with the

Ed. note: Carl Anderson was unable to attend the symposium. His presentation was read by Herbert Anderson, who added some further information included here in his introduction.

highest magnetic field of any in the world. The size of the chamber was
17 × 17 × 6 cm. Its strong magnetic field went up to 25,000 G. Before
the end of its first year, 1931, Anderson had it working and had already collected enough pictures to make it evident that a whole new
world of physics was being opened up.

The following year, at age 27, Anderson discovered the positron. In
1936, at age 31, he was awarded the Nobel Prize. That did not slow
him up. He had already begun to investigate the possibility of other
strange phenomena. In the following year he discovered, with Neddermeyer, the particle we now call the muon. Three years later, in 1939,
Caltech decided that it was time to appoint him a full professor.

An article by Watson Davis in *Science News Letter* in 1931 showed
some of Anderson's early results with the cloud chamber. Almost the
first time he turned the cloud chamber on he could see an electron and
a positively charged particle, which at that time, being conservative, he
said was probably a proton. But it became clear later that given the
low degree of ionization and the curvature, it had to be a positron. At
the time, Millikan was off in Europe giving lectures on cosmic rays.
Anderson sent him a batch of photographs of the cloud chamber results, and they were displayed all over Europe, giving this work wide
exposure.

The *Scientific American* had an article on this work, but it came a
little late; it was printed in 1935 and was written by Edward U.
Condon, who tried to answer the question, What are positrons? His
attempt to answer this question illustrates the difference between experimentalists and theorists. For the experimental side, the author
displayed photographs showing how the track of a positron going
through a lead plate increases its curvature because of a loss of energy, as only a particle with mass like that of an electron would. Yet
the direction of the curvature in the magnetic field shows it to be
positively charged. What did he do about the theoretical contribution? There are no pretty pictures of the experimental apparatus or
experimental results. The only thing you can do for a theorist is to
publish his picture. Dirac's theory of the positron took on great importance because of this experimental demonstration.

I have one final remark. I thought it might be interesting to see what
Robert Millikan said in his autobiography about the work in Pasadena
during his tenure at Caltech. In this book of some 300 pages I found
one reference to Carl Anderson. It simply said that the work on cosmic
rays was carried on by some of his younger colleagues, Carl Anderson,

Carl Anderson and positron photograph, 1931 (credit: *Science News Letter V.20* (Dec. 19, 1931) p. 387).

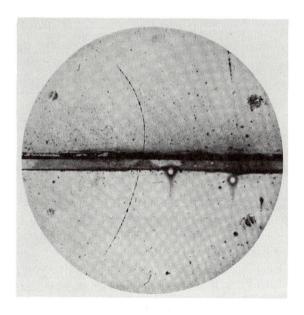

Positron passing through lead plate. From C. Anderson, in *Phys. Rev. 43* (1933), 419. Reprinted in E. U. Condon, "What Are Positrons?," *Scientific American 153* (August 1935) 71.

P. A. M. Dirac, 1935. From E. U. Condon, "What Are Positrons?," *Scientific American 153* (August 1935) 72.

Seth Neddermeyer, H. Victor Neher, William Pickering, and half a score of graduate students. They discovered a few new particles that had important roles to play in the development of nuclear physics and nuclear power. That effort was supported by a grant from the Carnegie Corporation that amounted to an average of $15,000 per year. That is the work we shall now consider.

Contribution by Carl D. Anderson

I understand that my task is to report on the part played by cosmic-ray research in providing new information on the particles of physics, until such work was gradually driven into oblivion by the advent of the new accelerators. I have not had access to a library except under very trying and difficult circumstances, and hence this chapter was prepared at my home, based almost wholly on a limited amount of material and my own recollections. This has inevitably tended to make for omissions in the chapter, to make it less objective and more personal than it otherwise might or should have been. I regret the omission of the contributions provided by the extensive programs of research using Geiger counters and ionization chambers and the extremely important results from the photographic emulsion techniques originally devised by Cecil Powell and later continued by many other investigators. In particular, I regret the omission of the innovative and important series of experiments carried out by Bruno Rossi beginning about 1930, based mainly on the adaptation of Geiger counter techniques to a study of the complex character of the cosmic radiation. His results, as well as those of other investigators, had a strong influence on my thinking and were very helpful in interpreting some of our own results.

The name of Dr. Robert A. Millikan appears early in this chapter. Later I shall describe some differences of opinion with respect to the interpretation of certain cosmic-ray data that occurred between Dr. Millikan and me, and later on between Dr. Millikan and Seth Neddermeyer, who was my first graduate student and afterward my close collaborator for several years. I wish to emphasize that these differences with Dr. Millikan were very minor, and in no way have they affected the respect and admiration that I have always had for this great man, to whom I owe so much.

At about the end of 1929, when it became clear to me that I was likely to receive my Ph.D. degree at Caltech in June 1930, I made an appointment to see Dr. Millikan. The purpose of my visit was to see if it were at all possible for me to spend one more year at Caltech as a postdoctoral research fellow. My reason for doing so was twofold: to carry out an experiment I had in mind and to learn something about quantum mechanics.

After a brief discussion with Dr. Millikan, in which I described the experiment and my desire to study quantum mechanics, he informed me that this would not be possible. The gist of his remarks was that, having had both my undergraduate and graduate training at Caltech, I was very provincial and should plan to continue my work at some other institution under a National Research Council fellowship, about the only fellowship available at that time for postdoctoral studies. Thus, I had no choice but to apply for the fellowship, and I wrote to Arthur H. Compton at the University of Chicago. I received a cordial reply and began planning for my sojourn at Chicago, an idea that appealed to me more and more as time went on.

My thesis work as a graduate student at Caltech consisted of studying, by means of a Wilson cloud chamber, the space distributions in various gases of photoelectrons produced by x rays. At the time (1927-30) that I was doing this work, Dr. Chung-yao Chao, working in a room close to mine, was using an electroscope to measure the absorption and scattering of γ rays from ThC''. His findings interested me greatly. At that time it was generally believed that the absorption of "high-energy" γ rays (2.6 MeV from ThC'') was almost wholly by Compton collisions, as governed by the Klein-Nishina formula. Dr. Chao's results showed clearly that both the absorption and scattering were substantially greater than calculated by the Klein-Nishina formula. A detailed explanation of these anomalous effects was not possible from his experiments because one could not obtain detailed information from an electroscope. My proposed experiment was to study the interaction of ThC'' γ rays with matter using a cloud chamber operated in a magnetic field: to view the secondary electrons produced in a thin lead plate inserted in the cloud chamber, to measure their energy distribution, and to see what further light could be thrown on Chao's results.

While still a graduate student at Caltech, I obtained the ThC'' source used by Chao and attempted to photograph the secondary electron tracks produced by its γ rays in the same cloud chamber I had previously used for the x-ray studies. I found the tracks produced by the ThC'' difficult to photograph; the tracks were very thin because of the much higher speed of the ThC'' electrons, as compared with those from the x rays (80 keV), which ionized heavily and therefore produced a heavy and easily photographed track.

After trying various things, I happened one day to pick up a bottle containing ethyl alcohol and pour some of it into the water normally used as a condensing liquid in the cloud chamber. This immediately produced tracks of much higher visibility that were very easy to photograph. To my knowledge, it was the first time a mixture of alcohol and water, rather than pure water, was used in a cloud chamber. With this problem solved, I was ready to design and build the equipment for the ThC'' studies, consisting of a cloud chamber operated in a magnetic field.

It is my firm conviction that had this experiment been carried out, the positive electron would have been discovered at an earlier date than its actual discovery, for about 10% of the electrons emerging from the lead plate would have had a positive charge. It is, of course, now well known that Chao's excess absorption was caused by pair production, and the excess "scattering" by the γ rays produced from the annihilation of positive and negative electrons.

Cosmic rays

One day I received a call from Dr. Millikan asking me to see him in his office. The gist of his comments on this occasion was that he wanted me to spend one more year at Caltech and build an instrument to measure the energies of the electrons present in the cosmic radiation (more about this instrument later). By this time, Chicago was clearly my first choice, and I used all the arguments that he had previously presented for not staying at Caltech. He replied that all these arguments were valid and cogent, but that my chances of receiving an NRC fellowhsip would be much better after one more year at Caltech. He was a member of the NRC fellowship selection committee at the time.

Again, I seemed to have no choice in the matter. Without further ado I began work on the design of the instrument he had proposed for the cosmic-ray studies. It was to consist of a cloud chamber operated in a magnetic field. This equipment, however, would require a very powerful magnetic field, for the cosmic-ray electrons were expected to have energies in the range of at least several hundred million electron volts, rather than the approximately 1 MeV for the ThC'' experiment.

The magnet cloud chamber

The existence at that time of the Guggenheim Aeronautical Laboratory on the Caltech campus helped to dictate the design of the new magnet cloud chamber instrument, for the laboratory was equipped with a 450 kW dc generator used for supplying power to a wind tunnel. Under overload conditions, the generator was capable of safely delivering 600 kW for extended periods of time.

The magnet, as actually designed and built, was essentially a pair of air-core solenoids capable of operating at 600 kW. Cold-rolled steel was used to form a frame to support the solenoid coils rigidly enough to withstand the rather large forces expected. Only the pole pieces were made of high-quality permeable iron, and one of these contained a large square hole in order to permit the cloud chamber at the center to be phtographed. When operated at full power, the magnetic field through the whole of the cloud chamber was slightly over 25,000 G. The solenoid coils were wound with copper tubing, and tap water was used for cooling, making possible continuous operation at full power.

Funding was in short supply and was the underlying factor that determined how one built scientific equipment in those days. For example, to build a magnet of conventional design would have been completely out of the question.

First results

The first results from the magnet cloud chamber were dramatic and completely unexpected. There were approximately equal numbers

Cloud chamber and magnet in Aeronautics Laboratory, 1931 (credit: Carl Anderson).

of particles of positive and negative charges, in sharp contrast to the Compton electrons expected from simply the absorption of high-energy photons. Dr. Millikan was on a visit to England at the time the first results were obtained, and I sent him a group of 11 photographs.[*] The accompanying letter describing the photographs revealed my own excitement; the concluding sentences of the letter were as follows:

> A hundred questions concerning the details of the effects immediately come to mind. Such questions as the loss of energy by high energy electrons, loss of energy by high energy protons, presence or absence of heavier nuclei of high energy, energy distribution among the particles in the case of double or triple ejections, momentum relations, etc. It promises to

---

[*] *Ed. note*: For reactions to Millikan's discussion of Anderson's photographs in England and France, see Chapter 5 in this volume.

be a fruitful field, and no doubt much information of a very fundamental character will come out of it.

It was, of course, important to provide unambiguous identification of these unexpected particles of positive charge, and this could best be done by gathering whatever information was possible on the mass of the particles, inasmuch as the photographs clearly showed that in all cases these particles carried a single unit of electric charge. Experimental conditions were such that no information as to a particle's mass could be ascertained except in those cases in which the particle's velocity was appreciably lower than the velocity of light, which was true for only a small fraction of the events. Only a few of the low-velocity particles were clearly identified as protons.

One of the first tasks undertaken with the first photographs, in fact the original purpose of the experiment, was to determine an energy distribution of the particles by means of the curvature they showed in traversing the powerful magnetic field. My original measurements showed an energy distribution extending from very low energies (~ 100 MeV) up to above 1 BeV, with the great majority of particles having energies in the range of several hundred million electron volts.

At about this time, Neddermeyer joined me as my first graduate student, and I assigned him the task of continuing the curvature measurements, paying particular attention to obtaining as precise measurements as possible for those of highest energy (i.e., in the range above 1 BeV). As will be seen later, the results of these energy measurements were to lead to some very interesting discussions with Dr. Millikan.

Positive electrons

As more data were acccumulated, however, a situation began to develop that had its awkward aspects, in that practically all of the low-velocity cases involved particles whose masses seemed to be too small to permit their interpretation as protons. The alternative interpretations in these cases were that these particles were either electrons (of negative charge) moving upward or some unknown lightweight particles of positive charge moving downward. In the spirit of scientific conservatism, we tended at first toward the former interpretation (i.e., that these particles were upward-moving negative electrons). This led to frequent, and at times somewhat heated, discussions between Professor Millikan and myself, in which he repeatedly pointed out that everyone knows that cosmic-ray particles travel downward, not upward, except in extremely rare instances, and that therefore these particles must be downward-moving protons. This point of view was very difficult to accept, however, because in nearly all cases the specific ionization of these particles was too low for particles of proton mass.

To resolve this apparent paradox, a lead plate was inserted across the center of the chamber in order to ascertain the direction in which these low-velocity particles were traveling and to distinguish between upward-moving negatives and downward-moving positives. It was not long after the insertion of the plate that a fine example was obtained in which a low-energy lightweight particle of positive charge was observed to traverse the plate, entering the chamber from below and moving upward through the lead plate. Ionization and curvature measurements clearly showed this particle to have a mass much smaller than that of a proton and, indeed, a mass entirely consistent with an electron mass. Curiously enough, despite the strong admonitions of Dr. Millikan that upward-moving cosmic-ray particles were rare, this indeed was an example of one of those very rare upward-moving cosmic-ray particles.

Soon additional instances of lightweight positive particles traversing the plate were observed; in addition, events in which several particles were simultaneously emitted from a common source were observed. Clearly, in both types of cases the direction of motion was known, and it was therefore possible to identify the presence of several more lightweight positive particles whose mass was consistent with that of an electron but not with that of a proton.

After the existence of positrons was clearly indicated, the question naturally arose as to how they came into being. Just what was the mechanism responsible for their production?

It has often been stated in the literature that the discovery of the positron was a consequence of its theoretical prediction by Paul A. M. Dirac, but this is not true. The discovery of the positron was wholly accidental. Despite the fact that Dirac's relativistic theory of the electron was an excellent theory of the positron, and despite the fact that the existence of this theory was well known to nearly all physicists, including myself, it played no part whatsoever in the discovery of the positron.

It was not immediately obvious to me just what the detailed mechanism was in the production of positrons. Did they somehow acquire their positive charge from the nucleus? Could they be ejected from the nucleus when there were presumably no positrons present in the nucleus? The idea that they were created out of the radiation itself did not occur to me at that time, and it was not until several months later when Patrick M. S. Blackett and Giuseppe P. S. Occhialini suggested the pair-creation hypothesis that this seemed the obvious answer to the production of positrons in the cosmic radiation. Blackett and Occhialini suggested the pair-production hypothesis in their paper published in the spring of 1933, in which they reported their beautiful experiments on cosmic rays using the first cloud chamber that was controlled by Geiger counters.

Soon after that, experiments in which γ rays were used showed that a pair of electrons, one positive and one negative, could be created in the Coulomb field of a nucleus in such a way that the energy required to create the mass of the pair, $2\,mc^2$, and their kinetic energies as well, was supplied by the incident radiation, thus giving quantitative support to the pair-creation hypothesis. The positron thus represents the first example of a particle consisting of antimatter. It is now generally believed that all particles have their corresponding antiparticles; in fact, several have been identified.

If one goes back a few years, say to just after the Dirac theory was announced, it is interesting to speculate on what a sagacious person working in this field might have done. Had he been working in any well-equipped laboratory, and had he taken the Dirac theory at face value, he could have discovered the positron in a single afternoon. The reason for this is that the Dirac theory could have provided an excellent guide as to just how to proceed to form positron-electron pairs out of a beam of γ-ray photons. History did not proceed in such a direct and efficient manner, probably because the Dirac theory, in spite of its successes, carried with it so many novel and seemingly unphysical ideas, such as negative mass, negative energy, infinite charge density, etc. Its highly esoteric character was apparently not in tune with most of the scientific thinking of that day. Furthermore, positive electrons apparently were not needed to explain any other observations. Clearly, the proton was the fundamental unit of positive charge, and the electron the corresponding unit of negative charge. This kind of thinking prevented most experimenters from accepting the Dirac theory wholeheartedly and relating it to the real physical world until after the existence of the positron was established on an experimental basis, although the Dirac theory has since proved to be a great milestone in early twentieth-century physics.

The discovery of the positron is also an example of a situation that is so often present in physics, in which the same discovery is made, or could easily have been made, in experiments simultaneously under way but carried out for quite different purposes. One such example was the famous experiment of Walther Bothe and H. Becker in which a light nucleus such as Be was bombarded by α particles from a radioactive source. This experiment was first performed in 1930 by Bothe and Becker and later repeated by a number of investigators. As was shown later, this single simple experiment produced neutrons, positrons, and induced radioactivity.

Energy measurements

Let us return to the matter of the energy measurements of the cosmic-ray particles. I think it should be said at this point that Millikan, in his previous studies of cosmic rays in which he used electroscopes, had found what he interpreted as a "banded structure" in the absorption coefficients of the cosmic rays as they passed through the atmosphere. This led him to his atom-building theory of the origin of the cosmic radiation and to the conclusion that the primary cosmic-ray beam consisted of photons in the energy range of several hundred million electron volts.[*] To explain the origin of cosmic rays, he postulated a process by which electrons and protons in outer space would somehow combine and coalesce to form nuclei of atoms, with the "packing-fraction" energy released as photons forming the primary cosmic-ray beam that impinges on the earth's atmosphere. He thus expected the magnet cloud chamber experiments to reveal secondary electrons produced in Compton collisions by the primary photons constituting the incoming cosmic-ray beam. According to his hypothesis, the energies of the electrons observed should be in the general energy range of about a hundred million electron volts, but not to exceed some 400 to 500 MeV.

On many occasions Neddermeyer and I would meet with Millikan to discuss energy measurements and their interpretation. Millikan was a very busy man, and although he was not officially president of Caltech, he performed that complex function and many more. Not the least of the demands on him was raising money. Thus, because of Millikan's many duties, our meetings often occurred late at night in the labora-

---

[*] *Ed. note*: See Daniel J. Kevles, *The Physicists* (New York: Alfred A. Knopf, 1978), pp. 179-80.

tory after the conclusion of one of his many evening social engagements. He would remove his necktie, open his collar, and relax during these discussions.

Although the atom-building hypothesis did not appeal to Neddermeyer and me, it seemed to be very firmly fixed in Millikan's mind. Millikan seemed steadfastly to think in terms of the atom-building hypothesis, which did not permit energies above 400 MeV. I remember that on one occasion Neddermeyer was relating energy measurements he had made on a series of tracks and he came to one over 1 BeV. Millikan virtually hit the ceiling and gave Neddermeyer a rather tough third-degree-type questioning. Both Neddermeyer and I tried to argue with Millikan, but it seemed impossible to change the direction of his thinking – his mind's momentum seemed close to infinite. It was only after many of these meetings that Millikan readily accepted energies in the range of several billion electron volts.

Paradoxes

During the months that followed, Neddermeyer and I accumulated much more data and at least for a while believed the bulk of the high-energy particles to be electrons about equally divided between positive and negative charges. But doubts soon began to develop, and it was only through the discovery of the meson that these doubts were finally resolved.

The discovery of the meson, unlike that of the positron, was not sudden and unexpected. Its discovery resulted from a two-year series of careful, systematic investigations all arranged to follow certain clues and to resolve some prominent paradoxes that were present in the cosmic rays.

The gist of the matter was as follows. Neddermeyer and I were continuing the study of cosmic-ray particles using the same magnet cloud chamber in which the positron was discovered. In these experiments it was found that most of the cosmic-ray particles at sea level were highly penetrating in the sense that they could traverse large thicknesses of heavy materials like lead and lose energy only by the directly produced ionization, which amounted to something like 20 MeV per centimeter of lead. A principal aim of the experiments was to identify these penetrating cosmic-ray particles. They had unit electric charge and were therefore presumably either positive or negative electrons or protons, the only singly charged particles known at that time.

There were difficulties, however, with any interpretation in terms of

Robert Millikan (center) visits Seth Neddermeyer (right) and Carl Anderson on the summit of Pike's Peak, 1935 (credit: Carl Anderson).

known particles, as was pointed out as early as 1934 in a paper presented to the International Conference on Physics held in London that year.[2]

The most important objection to their interpretation as protons lay in the fact that the energy of the electron secondaries produced by the direct impact of these particles as observed in a cloud chamber contained too many "knock-on" secondaries of high energy to correspond with the known energy spectrum if particles as massive as protons were producing these secondaries.* On the other hand, the spectrum was just that to be expected if the particles producing the secondaries were much lighter than protons. Furthermore, to interpret these particles as protons would mean assuming the existence of protons of negative charge, because these sea-level particles occurred equally divided between negative and positive charges, and at that time there was no evidence for the existence of protons of negative charge.

There were difficulties also in interpreting these sea-level penetrat-

---

* *Ed. note*: Particles of similar masses tend to share their energy equally in "knock-on" collisions.

Carl Anderson and control panel for cloud chamber in trailer on Pike's Peak, 1935 (credit: Carl Anderson).

ing cosmic-ray particles as positive and negative electrons. The most important objections to their being electrons arose from three considerations. First, theoretical calculations by Hans Bethe, Walter Heitler, and Fritz Sauter on the energy loss of electrons led to the conclusion that high-energy electrons should lose large amounts of energy through the production of radiation, which the penetrating particles in question were observed not to do. Second, we had found individual cases of electrons that did, in fact, show large energy losses through radiation, in some cases 100 MeV or more per centimeter of lead. Clearly, in these cases the electrons showed a behavior quite different from that of the penetrating particles. Third, the so-called highly absorbable component of the cosmic rays and the existence of electron showers could find an appealing explanation in terms of electrons if electrons did, in fact, suffer large radiative losses at high energies, as demanded by the aforementioned theory.

This, then, was the situation in 1934 in which the sea-level penetrating particles had this paradoxical behavior. They seemed to be neither electrons nor protons. We tended, however, to lean toward their interpretation as electrons, and we "resolved" the paradox in our informal discussions by speaking of "green" electrons and "red" electrons – the green electrons being the penetrating type, and the red the absorbable type that lost large amounts of energy through the production of radiation.

Evidence of an entirely new type was soon obtained. In experiments carried out on the summit of Pike's Peak in 1935, a number of cases of cosmic-ray-produced nuclear disintegrations were observed from which many protons were ejected, but showing also in a few cases particles that, from ionization and curvature measurements, were lighter than protons and heavier than electrons. These observations were not conclusive evidence in themselves for the existence of a new type of particle, but they did tend to lend support to this assumption in view of the other difficulties involved in interpreting the data in terms of known particles.

The next year or so brought further evidence on all the preceding points and only tended to strengthen the paradox further. The hypothesis that the penetrating particles were protons was further weakened by the observation of many cases of particles that did not suffer appreciable radiative collisions and still could not be as massive as protons, as evidenced by the ionization-curvature relations of their cloud chamber tracks. These cases could, however, be interpreted as electrons, but only if electrons ceased to radiate appreciably above a certain energy, say 100 MeV.

The crux of the matter was whether electrons above a certain energy did or did not experience a large energy loss through radiative impacts. In other words, the paradoxical character of our data could be removed if one assumed that the Bethe-Heitler theory, although correct for electrons of energies below a few hundred million electron volts, in some way became invalid for electrons of high energy, thus permitting high-energy electrons to have a much greater penetrating power and thus perhaps permitting the interpretation of the highly penetrating sea-level cosmic-ray particles as positive and negative electrons. This was the view held by Blackett, but it did not seem to fit most of the facts as known at that time.

New particles

In the summer of 1936, Neddermeyer and I were quite firmly convinced that all the data on cosmic rays as known at that time nearly forced on us the conclusion that the penetrating sea-level particles could be neither electrons nor protons and must therefore consist of particles of a new type. Let me give two quotations from our previous papers that strongly support these views. That the interpretation of the penetrating particles as protons seemed untenable:

> It was shown above that the number and distribution in energy of the negatrons ejected as secondaries by particles traversing plates of lead and carbon agree with those to be expected theoretically for extranuclear encounters if the incoming particles possess electronic mass. On the other hand, the assumption that the incoming particles possess protonic mass and have the curvature distribution in the magnetic field given above, would lead to an electron-secondary energy distribution noticeably different from that actually found.[3]

That the penetrating particles could not readily be interpreted as electrons:

> This large absorbability of electrons and photons is difficult to reconcile with the highly penetrating character of a large fraction of the sea-level particles on the view that the latter are electrons.[4]

Evidence for the existence of new particles of intermediate mass was first presented in a colloquium at Caltech on November 12, 1936; a brief summary of this colloquium was sent out by Watson Davis of *Science Service* on November 13, 1936. A brief report also appeared in *Science*, November 20, 1936 (page 9 of the supplement). Perhaps the first reference in the "literature" to the new particles was the last sentence in my Nobel lecture on the positron delivered in Stockholm on December 12, 1936. In the more than 40 years since the delivery of that address I have received no reaction at all from it; so I will quote that sentence here: "These highly penetrating particles, although not free positive and negative electrons, will provide interesting material for future study."

The first formal publication relating to the new particles appeared in the spring of 1937.[5] However, before publishing this paper, we wished to obtain as convincing proof as possible that the theory was indeed valid at energies sufficiently high to require that the penetrating particles could not be electrons but must be particles of a new type. To do this, we took an additional 6,000 photographs in which we measured separately the losses in energy of "single" and "associated" particles in traversing a platinum plate of 1-cm thickness inserted in the cloud chamber. In this paper our conclusion was "that there exist particles of unit charge, but with a mass (which may not have a unique value) larger than that of a free electron and much smaller than that of a proton."

At about this time, Jabez Street and E. C. Stevenson reported experimental results from which they concluded that "from the data in the table it is evident that the penetrating particles cannot be described as electrons obeying the Heitler theory nor can an appreciable fraction be protons."[6] That the validity of the theory was not generally accepted is clear from a quotation from a paper by Blackett and J. G. Wilson at about the same time:

> Since it has long been clear that the radiation formula . . .
> must break down at high energies, though the exact value of
> the energy at which the breakdown started was not known,
> attempts have been made by Evan James Williams (1934), by
> J. Robert Oppenheimer (1935), and by Lothar Wolfgang
> Nordheim (1936), to modify the formula so as to reduce the
> calculated energy loss.[7]

Nomenclature

At first the new particles were known by various names, such as baryon, Yukawa particle, x-particle, heavy electron, etc. One day Neddermeyer and I sent off a note to *Nature* suggesting the name *mesoton* (meso for intermediate). At the time, Millikan was away, and after his return we showed him a copy of our note to *Nature*. He immediately reacted unfavorably and said the name should be *mesotron*. He said to consider the terms *electron* and *neutron*. I said to consider the term *proton*. Neddermeyer and I sent off the *r* in a cable to *Nature*. Fortunately or not, the *r* arrived in time, and the article appeared containing the word *mesotron*. Neither Neddermeyer nor I liked the word, nor did anyone else that I know of. Further discoveries (e.g., the pion) and the passage of time have greatly improved the matter of nomenclature as related to particles.

There is an interesting story that will not be related here concerning the origin of the name *positron*. I do not like the name particularly well, and although I have never discussed it with Professor Dirac, my feeling is that he may find it wholly inelegant.

In discussing the discovery of the meson I have not so far mentioned anything about the theoretical aspects of the situation. We saw previously how the Dirac theory predicted the existence of positrons, although it played no role in their discovery. The discovery of mesons, similarly, was based on experimental measurements and procedures, with no guide from any theoretical predictions.

As with the positron, this need not have been the case, for before the discovery of the meson had been finally achieved, a novel idea was published in a Japanese journal by Hideki Yukawa. Reasoning by analogy with quantum electrodynamics, he made the suggestion that perhaps nuclear forces, which are not electromagnetic in character, could be described in terms of a particle carrier of these nuclear forces, analogous to the photon being the carrier of electromagnetic forces. Nuclear forces, however, differ from electromagnetic forces in that they possess only a short range of action. This means that if nuclear forces are described in terms of a particle carrier, this particle carrier must have a finite rest mass, unlike the photon of zero rest mass, which is appropriate to the long-range electromagnetic forces. Yukawa estimated from the known range of nuclear forces that this carrier should have a rest mass about 200 times that of an electron.

This novel suggestion by Yukawa was unknown to the workers engaged in the experiments on the meson until after the meson's existence was established. Although Yukawa's suggestion preceded the experimental discovery of the meson, he published it in a Japanese journal that did not have general circulation in the United States. It is interesting to speculate on just how much Yukawa's suggestion, had it been known, would have influenced the progress of the experimental work on the meson. My own opinion is that this influence would have been considerable, even though Dirac's theory, which was much more specific than Yukawa's, did not have any effect on the discovery of the positron. My reason for believing this is that for a period of almost two years there was strong and accumulating evidence for the meson's existence, and it was only the caution of the experimental workers that prevented an earlier announcement of its existence. I believe that a theoretical idea like Yukawa's would have appealed to the people carrying out the experiments and would have provided them with a belief that maybe, after all, there was some need for a particle as strange as a meson, especially if it could help explain something as interesting as the enigmatic nuclear forces.

It was clear almost from the beginning, however, that the Yukawa particle and the cosmic-ray meson could not possibly be the same particle. The Yukawa particle was invented to explain the strong nuclear forces, whereas even in the very early experiments the cosmic-ray mesons seemed to ignore nuclear forces completely and to interact with matter only through electromagnetic forces.

For example, as described at the London conference in 1934, a total of 2,437 traversals of carbon and lead plates in the cloud chamber showed no evidence of a nuclear reaction. In this instance, 1,215 cm of carbon and 663 cm of lead had been traversed. Thus, these data provided strong support for the conclusion that the experimental mesons could not be quanta of the nuclear force field. Other experiments carried out in later years provided proof for this conclusion: for example, the important experiments of Marcello Conversi, Ettore Pancini, and Oreste Piccioni, in which they observed the capture and decay of negative and positive mesons stopped in an iron absorber.[8]*

It was not until the discovery of the pion in 1947 by Powell and his group that a particle was known that could be identified with the Yukawa particle, although the subsequent discoveries of a host of other particles have shown the extremely complex character of nuclear forces, a problem still unsolved.

The B-29 airplane

During the war years, I worked on the artillery rocket research and development project based at Caltech. This program, the brainchild of Charles C. Lauritsen, under his leadership and with the able assistance of William A. Fowler and Thomas Lauritsen, grew into a large and successful operation and made a substantial contribution to the war effort. My own part in the program was the responsibility for adapting the firing of various types of Caltech rockets from suitable army and navy aircraft. It was through this work that I made close contacts with the proper military authorities to obtain the use of a B-29 airplane in 1946 for cosmic-ray experiments.

The plan was to continue the cosmic-ray studies in the same magnet cloud chamber formerly used in Pasadena and on Pike's Peak, but in the airplane at an altitude as high as possible. The B-29 was stripped of all armaments at the factory and modified for operation at extreme altitudes. It was equipped with optional electric generators on the

---

* *Ed. note*: See Chapter 13 in this volume.

engines capable of providing 750 A at 28 V, giving a magnetic field strength in the cloud chamber of 7,500 G.

My collaborators in the B-29 experiments were Raymond V. Adams, R. Ronald Rau, Ram C. Saxena (graduate students), and Paul E. Lloyd (postdoctoral research fellow). I do not remember the date of our maiden flight, but I do remember that as I was climbing aboard the B-29, the pilot asked me what our route and destination were to be. This took me aback, because in no previous airplane flight had this question come up. I quickly answered, however, by suggesting we fly first to Mt. Shasta, circle it once, then to San Diego, circle it once, then to Pike's Peak, circle it once, and then back home to the naval ordnance test station, China Lake, California, where the B-29 was stationed. The cloud chamber operated perfectly throughout the flight.

In all, 35 flights lasting about 5 hr each were made at altitudes between 30,000 and 40,000 ft,[9] and a few flights were made at higher altitudes when the B-29 was able to fly that high. The highest altitude we ever reached was 41,000 ft.

At this point I cannot resist saying that although the B-29 flights were successful and gave considerable new information on the cosmic radiation, they could have accomplished much more. Now, of course, I am speaking of hindsight. Many of the engineering problems associated with the B-29 flights were formidable, and, unfortunately, they received much of our attention. Had we forgotten the B-29 and spent a week in the nearby High Sierra arguing only about cosmic rays and physics, we might have done better. All the clues were present and were published, one of the most important being the experiments by Lajos Janossy, in which he used counter arrays separated by various thicknesses of lead chosen to select nuclear collisions of high energy.

Such a week, solely devoted to physics, might have given us a proper goal, that is, to study the nucleonic component of cosmic rays. The modification required in the B-29 equipment would have been very minor and would have taken no more than an hour or so to do. We needed only to have added a small block of lead about 20 cm thick and an additional counter and required triple rather than double coincidences. This would have selected nuclear events, and undoubtedly would have given us hundreds of examples of the new unstable particles, heavy mesons and hyperons, subsequently discovered by George D. Rochester and Clifford C. Butler. The relative intensity of high-energy nuclear events in the cosmic rays at 30,000 ft as compared with sea level must be in the vicinity of several hundred to one. In any case, this is an example of a superb piece of experimental equipment not used to maximum advantage.

Strange particles

While the B-29 experiments were in progress, Robert B. Leighton and Eugene W. Cowan joined me in a program of building a variety of cloud chambers for further studies of cosmic rays in Pasadena and at White Mountain, California. This program was active for several years, until the advent of the accelerator era made the use of cosmic rays for particle studies more and more difficult. I will not attempt to cover the work carried out during this period of time except to refer to one early experiment.

In 1947, Rochester and Butler[10] reported observing in a cloud chamber two cases of "forked tracks," one charged and one neutral, which they interpreted as examples of the spontaneous decay of particles of a new type. No further cloud chamber photographs of events such as these were observed for about two years; then, in a series of 11,000 photographs made at Pasadena and White Mountain, we observed 34 additional cases of forked tracks and were led to the same conclusion as that drawn by Rochester and Butler, namely, the existence of unstable particles of a new type.[11] From these 34 cases we obtained some information as to the nature of the decay products, and we found the lifetime of the neutral particles to be about 3×10^{-10} sec. Various further investigations were carried out in an attempt to elucidate the properties of the hyperons and heavy mesons, their modes of decay, lifetimes, associated production, relative abundance, etc. Robert W. Thompson of the University of Indiana was also active during this period and was especially successful in his precision work on the neutral θ meson.

However, the ever-encroaching larger and larger accelerators clearly indicated the end of the period when cosmic rays could be useful in studies of particle physics. To exemplify this threat, I should like to give two quotations from Robert Marshak's article on the Rochester conferences.[12] Commenting on the 1952 conference, Marshak said,

> The machine results were beginning to overtake the cosmic-ray results, certainly in quantity, albeit to a lesser extent in importance (Anderson, Leighton and Thompson were coming through strong and clear on the two unstable particles), and the theorists were contributing their wisdom with regard to

selection rules in particle reactions, with particular reference to isospin invariance.

And, on the 1956 conference, he wrote,

> It was the year when machine results were coming in so fast that Leighton was led to remark that "next year those people still studying strange particles using cosmic rays had better hold a rump session of the Rochester Conference somewhere else."

However, undaunted by the irresistible encroachment of the accelerators, Cowan built a complex arrangement of 8 flat ionization chambers and 12 flat cloud chambers of a total height about 20 ft, designed for investigations at energies above those obtainable in any accelerator, and he continued his studies of cosmic-ray particle events until 1971.[13]

Notes

1 This chapter is based in part on an earlier paper: C. D. Anderson, "Early Work on the Positron and Muon," *Am. J. Phys. 29* (1961), 825-30. The editors wish to thank John S. Rigden, editor of the *American Journal of Physics*, for permission to adapt a portion of Dr. Anderson's 1961 article.

2 Carl D. Anderson and Seth H. Neddermeyer, "Fundamental Processes in the Absorption of Cosmic-Ray Electrons and Photons," in *International Conference on Physics, London, 1934, Vol. 1*, ed. by J. H. Awberry (London: Physical Society, 1935), pp. 171-87.

3 *Ibid.*, p. 182.

4 Carl D. Anderson and Seth H. Neddermeyer, "Cloud Chamber Observations of Cosmic Rays at 4300 Meters Elevation and Near Sea-Level," *Phys. Rev. 50* (1936), 263-71.

5 Seth H. Neddermeyer and Carl D. Anderson, "Note on the Nature of Cosmic-Ray Particles," *Phys. Rev. 51* (1937), 884-6.

6 J. C. Street and E. C. Stevenson, "Penetrating Corpuscular Component of the Cosmic Radiation," *Phys. Rev. 51* (1937), 1005.

7 P. M. S. Blackett and J. G. Wilson, "The Energy Loss of Cosmic Ray Particles in Metal Plates," *Proc. Roy. Soc. (London) 160* (1937), 304-23.

8 M. Conversi, E. Pancini, and O. Piccioni, "On the Disintegration of Negative Mesons," *Phys. Rev. 71* (1947), 209-10.

9 Carl D. Anderson, Raymond V. Adams, Paul E. Lloyd, and R. Ronald Rau, "On the Mass and the Disintegration Products of the Mesotron," *Phys. Rev. 72* (1947), 724-7; Raymond V. Adams, Carl D. Anderson, Paul E. Lloyd, R. Ronald Rau, and Ram C. Saxena, "Cosmic Rays at 30,000 Feet," *Rev. Mod. Phys. 20* (1948), 334-49.

10 G. D. Rochester and C. C. Butler, "Evidence for the Existence of New Unstable Elementary Particles," *Nature 160* (1947), 855-7.

11 Robert B. Leighton, Carl D. Anderson, and Aaron J. Seriff, "The Energy Spectrum of the Decay Particles and the Mass and Spin of the Mesotron," *Phys. Rev. 75*

(1949), 1432-7; A. J. Seriff, R. B. Leighton, C. Hsiao, E. W. Cowan, and C. D. Anderson, "Cloud-Chamber Observations of the New Unstable Cosmic-Ray Particles," *Phys. Rev. 78* (1950), 290-1.

12 Robert E. Marshak, "The Rochester Conference: The Rise of International Cooperation in High Energy Physics," *Bull. At. Scientists 26* (1970), 92-8.

13 E. W. Cowan and M. K. Moe, "Cosmic Rays as a High-Energy Particle Source," *Rev. Sci. Instr. 38* (1967), 874-8; E. W. Cowan and K. Matthews, "Cosmic-Ray Interactions at 10^{11}-10^{13} eV," *Phys. Rev. D4* (1971), 37-45.

8   The intriguing history of the μ meson

GILBERTO BERNARDINI

Born 1906, Florence, Italy; *laurea*, Scuola Normale; a central figure in Italian cosmic-ray school; director of research for CERN, 1962-5; first president of the European Physical Society; Scuola Normale Superiore, Pisa.

Discovery of the μ meson in cosmic rays

The existence of a particular ionizing fraction of the cosmic rays that we now recognize as muons was first revealed through an experiment by Walther Bothe and Werner Kolhörster in 1929.[1] They showed that the cosmic rays at sea level were not, as had been thought, composed mainly of high-energy photons and Compton recoil electrons produced by them higher in the atmosphere. Their experiment showed that at sea level, 75% of the cosmic rays were ionizing particles able to traverse a block of gold of 4 cm thickness.

This entirely unexpected result was obtained by comparing the numbers of simultaneous discharges (coincidences) in two Geiger-Müller counters, placed vertically one above the other, when the block was alternately inserted between the counters and then removed for an equal time interval. With the gold block between the counters, the number of coincidences was reduced by only 25%.

Already at that time, although it was still possible to make various assumptions about the probable behavior of photons and their Compton recoil electrons, it was difficult to explain such a small reduction in the number of coincidences. This milestone cosmic-ray experiment in elementary particle physics did not, however, prove the corpuscular composition of the primary cosmic radiation. What it did show was the high penetrating power through dense matter of cosmic-ray particles

155

arriving at, or generated near, sea level. It also posed the problem of their origin.

A decisive step toward the solution of this problem was taken by Bruno Rossi and his associates over a period of several years.[2] One of his many outstanding results confirmed the high penetrating power and gave information on the energy spectrum of the cosmic-ray particles. Using a coincidence device in a vertical plane, Rossi showed that 60% of those cosmic rays able to cross a 25-cm lead shield were also capable of passing through 1 m of vertically aligned lead bricks.

These rays, thought to consist of individual ionizing particles arriving at sea level from high altitudes with great residual energy (as was later established by a number of other experiments), have since that time been known as the "penetrating component" of cosmic rays. However, these pioneering experiments were not able to identify these particles among those known at the time, nor could they lead to any conclusion as to their primary or secondary origin.

A part of the answer came from the discovery of "electron showers" by Patrick M. S. Blackett and Giuseppe Occhialini in 1933, using a cloud chamber in a strong magnetic field that was partitioned by a lead plate and expanded under the control of a counter coincidence device.[3] Each of these showers consisted of a cluster of positive and negative electrons, coming usually from the lead partition, and often originated by a cosmic-ray particle hitting the chamber from above, as established by the counter control. Theoretical interpretation was provided simultaneously by Homi J. Bhabha and Walter Heitler and by J. F. Carlson and J. Robert Oppenheimer, on the basis of the quantum electrodynamics cross sections calculated by Hans A. Bethe and Heitler.[4]

The gratifying consistency between theory and the experimental results on showers ruled out the idea that the penetrating component could consist of highly energetic electrons; the probability that an electron of any energy whatsoever would have been capable, either directly or through secondaries, of triggering Rossi's counters separated by 1 m of lead was totally negligible. Furthermore, in accordance with the behavior predicted by the Bethe-Heitler theory, the penetrating component would have had to consist of particles of mass at least of the order of 100 times that of the electron.[5]

Meanwhile, experiments were done for the purpose of surveying the cosmic-ray intensity, as well as that of the penetrating component itself, at different altitudes and at different latitudes.[6] This research was planned with a view to detecting and measuring the effect of the earth's magnetic field, according to a theory advanced by Carl Störmer (originally in connection with the aurora borealis) and applied to cosmic rays especially by G. Lemaitre and M. S. Vallarta.[7] It became clear (from the east-west asymmetry) that the cosmic rays reaching the upper atmosphere from interplanetary space were positively charged, or at least mostly so.

The discovery of the positron by Carl Anderson in 1932 in the cosmic rays did not imply that the particles approaching the earth from outer space were also positive electrons.[8] In fact, it was already clear that even extremely energetic electrons of either sign would be absorbed by forming a chain of showers coming down from the upper layers of the atmosphere.* Therefore, because they were not positrons, it was assumed that the cosmic-ray primaries were mainly protons, but it was not possible to extend this conclusion (which is still consistent with present astrophysical knowledge) to the penetrating particles arriving at sea level. How, then, could one explain the presence among them of both signs of the electronic charge, already indicated by many cloud chamber pictures?

The enigma was solved by research carried out by Anderson and Seth Neddermeyer with a counter-controlled cloud chamber.[9] Using as absorber a platinum plate 1 cm thick (in which there is a negligible probability for an electron to pass through without radiating) with an appropriate magnetic field applied across the chamber, their main result was the finding that at sea level there exist two clearly distinguishable types of cosmic-ray particles. One showed large variable energy losses, in good agreement with the theoretical predictions for electrons, and often associated with visible showers; the other type consisted of particles that did not radiate appreciably, while exhibiting ionization energy losses compatible with a unit charge. Thus, these latter particles were identified with Rossi's penetrating component, but some of them, those of lower momentum, should have ionized at least three times more strongly had they been protons. Instead, their ionization was more like that of electrons of the same momentum.

Hence, Anderson and Neddermeyer suggested that "there exist particles of unit charge but with a mass (which may not have a unique value) larger than that of a normal free electron and much smaller than that of a proton." They were unable to establish a more definite value

* *Ed. note*: The "radiation unit" (i.e., the mean free path for an electron to radiate a photon) is 340 m in standard air (37.7 g-cm^{-2}).

for the mass of the new particles because they were unable to trigger the expansion of their cloud chamber by low-energy particles.

However, Jabez C. Street and E. C. Stevenson had an apparatus that could select low-energy tracks, and by using great care in the counting of droplets of the "cloud" formed along a certain particle's track, and comparing it with that of a proton, they were able to estimate the particle's velocity. Combined with the magnetically analyzed momentum, this gave a mass value of about 130 electron masses.[10]

The new particle was called by various names, including *yukon*, *mesotron*, and *meson*, the last two names being suggested by its intermediate mass value. The name *yukon*, on the other hand, came from identifying it as the particle predicted by Hideki Yukawa in 1935, to serve as the mediator of the strong nuclear force.[11] Aside from the mass value, which the Yukawa theory demanded to be about 200 electron masses, in order to agree with the range of nuclear forces, the cosmic-ray evidence seemed to favor an instability of the particle. This instability was also implied by the Yukawa theory, in which the meson was supposed to interact with electron and neutrino in order to account for, as Yukawa thought, the nuclear β decay.[12]

It was pointed out by Kuhlenkamff that the spontaneous decay of free mesons produced in the upper layers of the atmosphere by the primary cosmic rays could explain the so-called anomalous absorption of cosmic rays in air.[13] The effect consisted in greater "absorption" of mesons in a given (large) layer of the atmosphere than in an equivalent thickness (measured in grams per square centimeter) of dense material, such as carbon or lead.[14] According to Kuhlenkamff, this was to be regarded as proof of decay of cosmic-ray particles moving through the atmosphere, whereas this effect would be far smaller in the geometrically thin layer of dense material.

Enrico Fermi was skeptical about this conclusion, showing theoretically that in a dense material polarized by the extended, relativistically contracted Coulomb field of a fast charged particle, the ionization loss rate would be less than in a diffuse medium.[15] However, he found that not more than half of the anomaly could be attributed to this cause, and he concluded that the decay of the mesotrons had probably been observed, although indirectly.

Mean lifetime of the muon

As mentioned earlier, the cosmic-ray meson (i.e., the muon) became identified with the nuclear force meson of Yukawa, which was at the same time supposed to explain the nuclear β decay. According to this assumption, such a meson had to disintegrate into an electron and a neutrino. The corresponding mean lifetime of a meson at rest was variously estimated[16] in a range from 5×10^{-7} sec to 10^{-8} sec, which was at least an order of magnitude smaller than any of the values estimated by measurements of the anomalous absorption in the atmosphere that came out around 1940.[13,14] However, the differences were not considered to be serious, in view of the experimental and theoretical uncertainties, and in fact there turned out to be a welcome agreement between the average of these values and the mean life at rest of the cosmic-ray mesons that in 1941 were directly measured by Franco Rasetti.[17]

Rasetti's experiment had such a great influence on the orientation of research on the behavior of muons between 1940 and 1950 that it seems appropriate to describe it here in some detail. The experiment was stimulated by cloud chamber pictures,[18] obtained by E. J. Williams and G. E. Roberts, showing the decay of a cosmic-ray meson into an electron.

Rasetti used a rather refined coincidence device to select for observation mesons that were near the end of their range. They were then stopped in an absorber that was thin enough to allow most of the decay electrons to emerge and be detected, either directly or by means of secondary electrons they might produce. Figure 8.1 shows schematically the arrangement of the counters and the related electronic coincidence circuits. The vertically aligned counters ABCD selected the mesons capable of traversing the 10 cm of lead interposed between A and B and arriving at the absorber Σ placed between D and the triple counter S. Mesons that stopped in Σ discharged the sequence of counters A to D, but not any of the S counters. In this way, mesons stopping in Σ were selected electronically from among the others crossing the setup.

When an electron resulting from the decay of a stopped meson emerged from Σ, it could trigger one of the six Ge counters, but an electronic circuit arranged that a signal from those Ge counters was registered only when the discharge occurred within a microsecond after the arrival of the meson that had stopped. Two similar electronic circuits operated similarly, with time delays of 2 and 15 μsec. Rasetti observed that a great many decay electrons were emitted after a delay of between 2 and 15 μsec, and after long observation time, in part with iron as absorber and in part with aluminum, a mean lifetime value τ_μ was found: $\tau_\mu = (1.5 \pm 0.3) \times 10^{-6}$ sec.

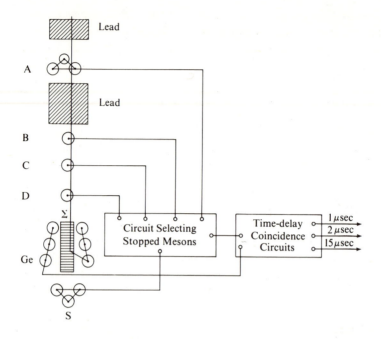

Figure 8.1. Schematic of the apparatus used by Franco Rasetti in 1941 to measure the mean life of the cosmic-ray meson. From W. Heisenberg (ed.), *Cosmic Radiation* (N.Y.: Dover, 1946), 88.

By the summer of 1943, Rossi and his young associates had measured τ_μ with methods similar to Rasetti's, but with a new electronic device that was able to measure continuously the decay time of each individual stopped meson,[13] and they obtained a result very close to the best current value.

These experiments on muon decay, as well as the cloud chamber pictures of Williams and Roberts, had clearly shown that in its decay the electric charge of the muon was transferred to an electron; but what other particle, or particles, shared energy and momentum with the electron was not easy to determine. The most straightforward assumption would have been a radiative decay of the muon into an electron and a photon, which did not necessarily violate any conservation law known 40 years ago. However, this reaction was excluded (although not until the late 1940s) by experiments showing that no shower-producing photon was among the decay products of the muon.[19]

Then J. Steinberger studied the energy distribution of the electrons emitted in the decay of stopping muons and found it to be continuously variable, consistent with a decay into three light particles.[20] Assuming them to be an electron and two neutrinos, one wrote (and has written since then) $\mu \rightarrow e + \nu + \bar{\nu}$, where ν and $\bar{\nu}$ are, respectively, neutrino and antineutrino. Because each of the three decay products has spin 1/2, the muon must also have half-integral spin. As we shall see later, the half-integral spin was considered to be necessary after it was discovered that it originates (for the most part) in the decay of the π meson.

Muon interactions

In 1932, Werner Heisenberg, prompted by the neutron discovery and by the conditions imposed by his indeterminacy principle, advanced the hypothesis that electrons had to be excluded as nuclear constituents and that nuclei should be thought of as made up of neutrons and protons.[21] They were able to coexist and interact in nuclear matter as two different states of the same particle: the nucleon.* A new variable, not related to space and time, was assumed to constitute the difference between the neutron and proton states. Because of its formal behavior, identical with Pauli's matrices for spin 1/2, this new variable came to be called "isotopic spin," and it may be considered as the first example of an internal or intrinsic variable characterizing the elementary particles and their interactions.

The exclusion of electrons from nuclear matter was cheating, up to 1933 when the β-decay theory was advanced by Fermi as "an exercise" in the second quantization and its creation and annihilation operators.[22] He assumed that inside a decaying nucleus a neutron is directly transformed into a proton by the simultaneous emission of an electron and a neutrino. This was the ordinary β process. A similar one, with the emission of positrons, observed only in nuclei produced by artificial disintegration, was explained shortly afterward by Gian Carlo Wick as the conversion of a proton into a neutron, with the creation of an antineutrino and a positive electron.[23]

The Heisenberg interpretation of nuclear forces and the Fermi theory referred to quite different interactions, and it was not necessary that they be unified by the Yukawa meson and its supposed β decay. Identification of this meson with the cosmic-ray muon, and the mass value provided by Street and Stevenson, was specifically supported by

---

* *Ed. note*: This is, from the historical viewpoint, an oversimplified version of Heisenberg's theory. See, for example, Joan Bromberg, "The Impact of the Neutron: Bohr and Heisenberg," *Historical Studies in the Physical Sciences 3* (1971), 307-41.

the results of Rasetti. Heisenberg remarked that only about half of all stopped mesons in Rasetti's experiment decayed by emitting electrons, whereas the other half were absorbed.[24] This was theoretically justified by Sin-itiro Tomonaga and Toshima Araki, who computed the probability for the capture of a Yukawa meson at rest by an atomic nucleus.[25] They found a drastic difference between negative and positive mesons. Whereas slow positive mesons were practically immune to capture because of Coulomb repulsion, the average time for capture for the negative ones was considerably less than the mean life for decay. It was thus concluded that the 50% capture observed by Rasetti was to be regarded as another indication supporting identification of the cosmic-ray meson with the Yukawa meson.

The problem was then to check the validity of all these indications, but for this purpose a simpler and faster method than the cloud chamber was required, one able to separate the negative from the positive cosmic-ray mesons in such numbers as to provide good statistics, whether they decayed or were captured. Luigi Puccianti, an old professor in Pisa, one of my teachers more than 50 years ago, taking into account the energies and penetration of these mesons, suggested a method that was actually very simple. He firmly believed that inside magnetized iron the magnetic field is B and not H, as our group then demonstrated by some preliminary tests.

The first setup of that experiment, carried out with Marcello Conversi, is shown schematically in Figure 8.2, where the circles indicate counters in coincidence. The rectangles A,A' (and also B,B') represent magnetized iron blocks whose horizontal fields are oppositely directed. Particles crossing these blocks, even if their incidence was more or less tilted vertically, were, according to the sign of charge, made either to converge or to diverge. Thus, the blocks served as "magnetic lenses." Different possible trajectories are shown in Figure 8.3, which refers to work done with Conversi, Wick, and others in the years between 1940 and 1945.[26]

The same magnetic lenses were then used by Conversi, Ettore Pancini, and Oreste Piccioni[27] to make the positive and the negative cosmic-ray mesons converge alternately on an absorber. In this way they discovered, in 1947, that these mesons did not behave at all as Yukawa's nuclear particle was supposed to; they found that the capture of muons was a rather unlikely process. In fact, in light absorbers such as carbon, a negative muon, presumably captured in an internal atomic orbit, almost always decayed before being captured by the nucleus.

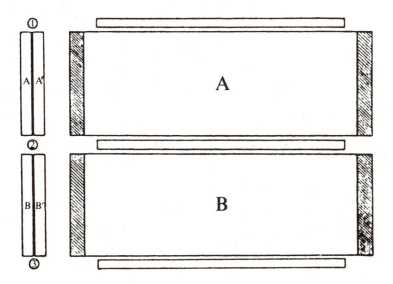

Figure 8.2. Magnetized iron blocks used as a "magnetic lens" for penetrating charged particles of the cosmic rays. From *Ricerca Scientifica, Consiglio Nazionale Richerche* (1944).

Only with absorbers of high atomic number, such as iron, did the time required for nuclear capture fall below 2 μsec; then capture became more probable than decay. These results, as then discussed by Fermi, Edward Teller, and Victor Weisskopf, definitively excluded any identity of the muon with the Yukawa meson or "pion" (as it may be better to call it from now on).[28]

Because Piccioni will discuss the experiment in Chapter 13, I have mentioned it only in order to emphasize its great importance. I should add that the existence of two kinds of mesons was soon directly shown by C. M. G. Lattes, Occhialini, and Cecil Powell using specially prepared Ilford photographic plates exposed at a mountain-top laboratory in the Pyrenees.[29] On these plates they found several traces of a charged particle that by grain density and range measurements showed a mass around 200 times the electron mass, but was observed to decay, one of the products having almost the mass of the primary particle. Thus, both the primary and the charged secondary particle were mesons, but they were related by an entirely unexpected interaction.

Yukawa had supposed that his meson decayed into an electron and a neutrino, not into a muon and a neutrino. He did not know about this

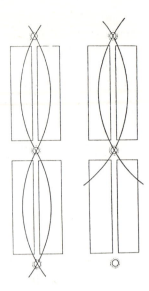

Figure 8.3. Qualitative shape of trajectories for two arrangements of the magnetized iron blocks shown in Figure 8.2. From G. Bernardini et al. in *Phys. Rev. 68* (1945), 110.

interaction, and it was neither required nor even compatible with his theory of the nuclear β decay. However, within a short time the two mesons linked by the decay observed by Lattes, Occhialini, and Powell were assumed to be the pion and the muon, the latter being the meson studied for years in the cosmic rays.

Since then, the problem posed by the muon has been to explain its raison d'être. The frustration associated with this problem, which lasted for decades, may be summarized in these words of Murray Gell-Mann and E. P. Rosenbaum: "The muon was the unwelcome baby on the doorstep, signifying the end of days of innocence."[30]

To recount briefly the progress achieved after the muon was distinguished from all other known elementary particles, we must consider its own mode of decay and its behavior as a Dirac particle. Concerning the former, some direct measurements made by 1949 of the energy and angular distributions of the electrons emitted in the decay showed that this decay was not a two-body process.[31] These results were interpreted by assuming that two neutral particles were emitted with the electron in the decay, thought to be neutrinos, because experiments showed they were not photons. So the reaction was supposed to be (as indeed it is)

Left to right: Hideki Yukawa, Cecil Powell, Mrs. Powell, and Mrs. Yukawa at dinner at a restaurant in Kyoto in 1956 (credit Iwanami Shoten).

$$\mu \rightarrow e + \nu_1 + \nu_2,$$

but without any other specific assumption concerning the neutrinos ν_1 and ν_2.

So, in itself, the weak interaction of the muon was not at all puzzling. Far from it: It was of fundamental help in extending our understanding of the overall weak interaction. The first step in this direction was to exclude the hypothesis that ν_1 and ν_2 might be a neutrino and its antineutrino, because in that case, among other possibilities, the two could annihilate and convert into a photon. But photons in the muon decay have been searched for, and not found, with an upper limit of 2×10^{-8} for the branching ratio. This led to the hypothesis that there might exist two kinds of neutrinos: one coupled to the electron and the other to the muon.[*]

This is now a well-established fact, as shown by an experiment in

[*] *Ed. note*: Although Professor Bernardini has provided citations for many of the remarks that follow, the editors have decided not to include them, because a balanced historical account cannot be given of so many developments in such a short space.

1962 with a beam of highly energetic neutrinos originating from the decay in flight of an intense beam of pions. Thus, these neutrinos were produced in association with muons, not electrons. And in interacting in a heavy target, these neutrinos created only muons, not electrons. This led to assigning to the muon and its neutrino a new conserved quantum number that distinguished it from the electron and the electron's related neutrino.

The pion-muon-electron decay chain introduced two new coupling constants (i.e., interaction strengths, roughly analogous to the electronic charge), and when these turned out to be nearly equal to the nuclear β-decay constant and the μ-capture constant, this gave rise to the concept of the "universal Fermi interaction." Over the years, the universality of the theory has become even more evident, accommodating strange and charmed particle decays, parity violation, etc., and culminating in today's Weinberg-Salam electroweak gauge theory.

On the other hand, the crucial point of the mystery of the electromagnetic behavior of the muon as a Dirac particle was, and still is, its mass. The increasing precision of the measurements of this mass can be seen in Tables 8.1 and 8.2, which are derived from the papers of many authors deserving our admiration and gratitude. The conclusion drawn from this series of values, obtained between 1948 and today, is that the muon mass has a unique value, about 207 times the electron mass. At the present time, an interaction of the muon (other than the weak and electromagnetic) able to justify the great mass of such a "heavy electron" is unknown.

That is the reason why many attempts have been made since 1950 to detect any possible difference between the electromagnetic behavior of the muon and that of the electron. These have been fully justified by the confidence inspired by quantum electrodynamics (QED) in the interpretation of all the properties of the electron as a Dirac particle. This confidence, more than 20 years ago, led to experiments aiming to measure the so-called anomalous magnetic moment of the muon (or, briefly, the muon g factor). The g factor or gyromagnetic ratio is a measure of its magnetic moment $M = g(e/2mc)(\hbar/2)$; it is a dimensionless number.

For a Dirac particle of spin 1/2, independently of its mass, g should be 2 exactly; but the quantum fluctuations of the field around the particle induce an extra part of the magnetic moment. Then g is written: $g = 2(1 + a)$ where a is the "anomaly." Julian Schwinger showed that because of the emission and reabsorption of virtual photons, up to

Table 8.1. *Determinations of positive muon mass* m_{μ^+}

Method	Mass ratio m_{μ^+}/m_e	Researchers
Cosmic rays, photographic plate (method 1)	202 ± 8	Goldschmidt-Clermont *et al.* (1948)
Cosmic rays, photographic plate (method 1)	220 ± 26	Barbour (1949)
Cosmic rays, photographic plate (method 1)	217 ± 4	Franzinetti (1950)
Cosmic rays, Wilson cloud chamber (method 1)	215 ± 6	Fretter (1949), Retallack and Brode (1949), Brode (1949)
Proton synchrotron, photographic plate (method 1)	212 ± 6	Bradner *et al.* (1950)
Proton synchrotron, emulsion (method 1)	206.9 ± 0.4	Barkas *et al.* (1956)
Proton synchrotron, μ^+ stopped π^+ (method 2)	206.9 ± 0.2	Barkas *et al.* (1956), Cohen *et al.* (1957)
Muon magnetic moment from muon precession (μSR) and from muonium	206.76818 ± 54 (2.6 ppm)	
Zeeman effect	206.76860 ± 29 (1.4 ppm)	
Weighted average	206.76851 ± 25 (1.2 ppm)	Last two values

Source: V. W. Hughes and C. S. Wu (eds.), *Muon Physics, Vol. 1* (New York: Academic Press, 1975), p. 16.

the second order, again independently of the mass, $a = \alpha/2\pi$, where α is the fine-structure constant.[32] Then, up to the second order, the anomaly a is of the order of magnitude of 0.001. For the electrons and for the "heavy electrons," the value of a should be the same, and any difference is related to Feynman graphs of higher order.

Prior to the beginning of the $(g - 2)$ experiments at CERN around 1960, the value of g was obtained by measuring the Larmor spin precession frequency of the muon in a magnetic field. The muons were stopped in a nondepolarizing target, and the decay electrons emitted in one particular direction were detected. Thus the value of g was found to be within about 15% of that predicted by Schwinger.

The CERN experiments have instead been based on the idea that the muons produced by a beam of pions decaying in flight are longitu-

Table 8.2. *Determinations of negative muon mass* m_{μ^-}

Method	Mass ratio m_{μ^-}/m_e	Researchers
Cosmic rays, photographic plate (method 1)	202 ± 8	Goldschmidt-Clermont *et al.* (1948)
Cosmic rays, photographic plate (method 1)	220 ± 26	Franzinetti (1950)
Cosmic rays, Wilson cloud chamber (method 1)	215 ± 2	Fretter (1949); Retallack and Brode (1949), Brode (1949)
Proton synchrotron, π^- decay in flight (method 1)	206.5 ± 1.8	Lederman *et al.* (1951)
Muonic x rays, critical x-ray absorption, ^{31}P	206.76 ± 0.02 (100 ppm)	Wu and Wilets (1969)
Muonic x rays, Ge(Li) detector, intermediate-Z atoms	206.769 ± 0.02 (100 ppm)	Dixit *et al.* (1975), Tauscher *et al.* (1975), Engfer and Vuilleumier (1975)

Source: V. W. Hughes and C. S. Wu (eds.), *Muon Physics, Vol. 1* (New York: Academic Press, 1975), p. 19.

dinally polarized. Furthermore, in the subsequent decays the electrons reveal the direction of the muon spins, because they are preferentially emitted in the spin direction at the moment of decay. Hence a $(g - 2)$ muon experiment may be performed, trapping the longitudinally polarized muons in a uniform magnetic field, and then measuring the precession frequency of the spins. It has only to be added that it is necessary to use high-energy muons in order to lengthen their decay times, using the relativistic time dilation effect. The results reduced the error in the measure of $(g - 2)$ from the 15% previously mentioned to 0.4%; they were published from 1961 to 1965.

By now, refined experiments at CERN have given the situation expressed by the comparison of the following values:

$$a_{\mu^+} = \{1,165,910 \pm (11)\} \times 10^{-9},$$
$$a_{\mu^-} = \{1,165,937 \pm (12)\} \times 10^{-9}.$$

Besides this gratifying check of the CPT theorem [that a system in which one changes particles into corresponding antiparticles (C), inverts space (P), and reverses the direction of the flow of time (T) behaves identically as the original system], we have:

$$a_\mu(\text{exp}) - a_\mu(\text{th}) = (-13 \pm 29) \times 10^{-9},$$

providing one of the most sensitive tests of the muon's electrodynamics.[33] Therefore, for the muon, we may be sure that a_μ is a pure QED quantity, as it is for the corresponding anomaly of the electron.

Concluding this account of the research on muon physics carried out in the middle of this century, the situation might be summed up in a way not very different from the statement of Gell-Mann and Rosenbaum cited earlier. We know now that the muon is a particle of spin 1/2, and simpler than many others, because it is a lepton. Actually, we may suppose it to be properly described by the universal Fermi interaction and by quantum electrodynamics, an assumption that has been verified with increasing precision since the experiments on the $(g - 2)$ value previously mentioned. Today, measurements of the $(g - 2)$ value have reached an accuracy of the order of one part per million. Furthermore, at Fermilab, with a high intensity muon beam of several hundred GeV, it has been possible to study muon-hadron scattering in the same energy range as for electron scattering. Once again, no significant difference has been found between the behavior of the muon and that of the electrons, so that insofar as electromagnetic interaction is concerned, an electron and a muon behave identically. Moreover, in spite of the essential difference expressed by the two quantum numbers that distinguish the muon and its muonic neutrinos from the electron and the associated neutrinos, the two kinds of leptons are always characterized by an equal coupling strength. They have in common that they are strictly, but separately, conserved.

This degeneracy of muons with respect to electrons, called "universality," emphasizes the enigma of the mass difference between the two. So far, no answer has been found either through quantum electrodynamics or through the weak-interaction theory as it exists at present. I shall finish this chapter with a quotation from V. W. Hughes and T. Kinoshita:

> Although spontaneously broken gauge theories provide a promising framework within which masses of particles can be treated more rationally, it is not clear at present whether we can understand the differences of the muon and the electron along this line of reasoning. We may still be very far from the day when we will have a satisfactory answer to the question: "Why does the muon exist?"[34]

Notes

1 W. Bothe and W. Kolhörster, "Das Wesen der Höhenstrahlung," *Z. Physik 56* (1929), 751-77.

2 B. Rossi, "Some Results Arising from the Study of Cosmic Rays," in *International Conference on Physics, London, 1934, Vol. 1*, ed. by J. H. Awberry (London: Physical Society, 1935), pp. 233-47.

3 P. M. S. Blackett and G. P. S. Occhialini, "Some Photographs of the Tracks of Penetrating Radiation," *Proc. Roy. Soc. (London) A139* (1933), 699-726.

4 H. J. Bhabha and W. Heitler, "The Passage of Fast Electrons and the Theory of Cosmic Showers," *Proc. Roy. Soc. (London) A159* (1937), 432-58; J. F. Carlson and J. R. Oppenheimer, "On Multiplicative Showers," *Phys. Rev. 51* (1937), 220-31; H. Bethe and W. Heitler, "On the Stopping of Fast Particles and on the Creation of Positive Electrons," *Proc. Roy. Soc. (London) A146* (1934), 83-112.

5 W. Heitler, *The Quantum Theory of Radiation*, 2nd ed. (Oxford: Oxford University Press, 1947).

6 Luis Alvarez and Arthur H. Compton, "A Positively Charged Component of Cosmic Rays," *Phys. Rev. 43* (1933), 835-6; Thomas H. Johnson, "Coincidence Counter Studies of the Corpuscular Component of the Cosmic Radiation," *Phys. Rev. 45* (1934), 569-85; Bruno Rossi, "Directional Measurements on the Cosmic Rays Near the Geomagnetic Equator," *Phys. Rev. 45* (1934), 212-14; Sergio De Benedetti, "Absorption Measurements on the Cosmic Rays at 11°30′ Geomagnetic Latitude and 2370 Meters Elevation," *Phys. Rev. 45* (1934), 214-15; J. Clay and H. P. Berlage, "Variation der Ultrastrahlung mit der geographischen Breite und dem Erdmagnetismus," *Naturwiss. 20* (1932), 687; Arthur H. Compton, "Variation of the Cosmic Rays with Latitude," *Phys. Rev. 41* (1932), 111-13.

7 Carl Störmer, "Periodische Elecktronenbahnen im Felde eines Elementarmagneten und ihre Anwendung auf Brüches Modellversuche und auf Eschenhagens Elementarwellen des Erdmagnetismus," *Z. Astrophys. 1* (1932), 237-74; G. Lemaitre and M. S. Vallarta, "On Compton's Latitude Effect of Cosmic Radiation," *Phys. Rev. 43* (1933), 87-91.

8 Carl D. Anderson, "The Positive Electron," *Phys. Rev. 43* (1933), 491-4; Carl D. Anderson, "The Apparent Existence of Easily Deflectable Positives," *Science 76* (1932), 238-9.

9 Seth H. Neddermeyer and Carl D. Anderson, "Note on the Nature of Cosmic-Ray Particles," *Phys. Rev. 51* (1937), 884-6. Also, see Notes 10 and 11.

10 J. C. Street and E. C. Stevenson, "Penetrating Corpuscular Component of the Cosmic Radiation," *Phys. Rev. 51* (1937), 1005.

11 Y. Nishina, M. Takeuchi, and T. Ichimiya, "On the Nature of Cosmic-Ray Particles," *Phys. Rev. 52* (1937), 1198-9; Hideki Yukawa, "On the Interaction of Elementary Particles. I," *Proc. Phys.-Math. Soc. Japan 17* (1935), 48-57.

12 H. J. Bhabha, "Nuclear Forces, Heavy Electrons, and the β-Decay," *Nature 141* (1938), 117-18.

13 For a detailed discussion of the measurement of the decay mean life of mesons during this period, see Chapter 11 in this volume.

14 G. Bernardini, N. B. Cacciapuoti, B. Ferretti, O. Piccioni, and G. C. Wick, "The Genetic Relation Between the Electronic and Mesotronic Components of Cosmic Rays Near and Above Sea Level," *Phys. Rev. 58* (1940), 1017-26; M. Ageno, G. Bernardini, N. B. Cacciapuoti, B. Ferretti, and G. C. Wick, "The Anomalous Absorption of the Hard Component of Cosmic Rays in Air," *Phys. Rev. 57* (1940), 945-50.

15 Enrico Fermi, "The Ionization Loss of Energy in Gases and in Condensed Materials," *Phys. Rev. 57* (1940), 485-93.

16 For estimates of the lifetime by Yukawa and his collaborators, see Chapter 4 in this volume.

17 Franco Rasetti, "Evidence for the Radioactivity of Slow Mesotrons," *Phys. Rev. 59* (1941), 706-8; Franco Rasetti, "Disintegration of Slow Mesotrons," *Phys. Rev. 60* (1941), 198-204.

18 E. J. Williams and G. E. Roberts, "Evidence for Transformation of Mesotrons into Electrons," *Nature 145* (1940), 102-3.

19 R. D. Sard and E. J. Althaus, "Test of the Hypothesis that the Sea-Level Cosmic-Ray Meson Disintegrates into a Photon and an Electron," *Phys. Rev. 73* (1948), 1251; E. P. Hincks and B. Pontecorvo, "Search for Gamma-Radiation in the 2.2 Microsecond Meson Decay Process," *Phys. Rev. 73* (1948), 257-8; O. Piccioni, "Search for Photons from Meson-Capture," *Phys. Rev. 74* (1948), 1754-8.

20 J. Steinberger, "On the Range of the Electrons in Meson Decay," *Phys. Rev. 75* (1949), 1136-43.

21 W. Heisenberg, "Ueber den Bau der Atomkerne," *Z. Physik 77* (1932), 1-11; *78* (1932), 156-64; *80* (1933), 587-96. Part I and a portion of Part III are given in English translation in D. M. Brink, *Nuclear Forces* (Oxford: Oxford University Press, 1965), 144-60.

22 E. Fermi, "Tentativo di una teoria dell'emissione dei raggi 'beta'," *La Ricerca Scientifica 4(2)* (1933), 491-5; E. Fermi, "Tentativo di una teoria dei raggi β," *Nuovo Cimento 2* (1934), 1-19; E. Fermi, "Versuch einer Theorie der β-Strahlen. I," *Z. Physik 88* (1934), 161-71. The last paper is given in English translation in Charles Strachan, *The Theory of Beta Decay* (Elmsford, N.Y.: Pergamon Press, 1969), pp. 107-28. The German version is reprinted in Fermi's *Collected Papers, Vol. I,* ed. by E. Segré (Chicago: University of Chicago Press, 1962). The "exercise" remark is by F. Rasetti on p. 539 of the last work.

23 G. C. Wick, "Induced Radioactivity," *Rend. Accademia dei Lincei 19(6)* (1934), 319-24.

24 W. Heisenberg, "Radioactive Decay of the Meson," in *Cosmic Radiation,* ed. by W. Heisenberg (New York: Dover Publications, 1946), translated by T. H. Johnson from the original German (Berlin: Springer-Verlag, 1943).

25 S. Tomonaga and G. Araki, "Effect of the Nuclear Coulomb Field on the Capture of Slow Mesons," *Phys. Rev. 58* (1940), 90-1.

26 G. Bernardini, M. Conversi, E. Pancini, E. Scrocco, and G. C. Wick, "Researches on the Magnetic Deflection of the Hard Component of Cosmic Rays," *Phys. Rev. 68* (1945), 109-20; G. Bernardini, G. C. Wick, M. Conversi, and E. Pancini, "Positive Excess in Mesotron Spectrum," *Phys. Rev. 60* (1941), 535-6.

27 M. Conversi, E. Pancini, and O. Piccioni, "On the Disintegration of Negative Mesons," *Phys. Rev. 71* (1947), 209-10.

28 E. Fermi, E. Teller, and V. Weisskopf, "The Decay of Negative Mesotrons in Matter," *Phys. Rev. 71* (1947), 314-15.

29 C. M. G. Lattes, G. P. S. Occhialini, and C. F. Powell, "Observations on the Tracks of Slow Mesons in Photographic Emulsions," *Nature 160* (1947), 453-6 (part 1) and 486-92 (part 2).

30 Murray Gell-Mann and E. P. Rosenbaum, "Elementary Particles," *Scientific American* (July 1957), 72-88.

31 Robert B. Leighton, Carl D. Anderson, and Aaron J. Seriff, "The Energy Spectrum of the Decay Particles and the Mass and Spin of the Mesotron," *Phys. Rev. 75* (1949), 1432-7.

32 J. Schwinger, "On Quantum-Electrodynamics and the Magnetic Moment of the Electron," *Phys. Rev. 73* (1948), 416-17.
33 J. Bailey et al., "Final Report on the CERN Muon Storage Ring," *Nucl. Phys. B150* (1979), 1-74.
34 V. W. Hughes and C. S. Wu (eds.), *Muon Physics, Vol. 1* (New York: Academic Press, 1975), p. 14.

9 Some aspects of French physics in the 1930s

PIERRE V. AUGER

Born 1899, Paris, France; Docteur ès sciences, Université de Paris, 1926; atomic physics, cosmic rays; 12 Emile Faguet, Paris.

The decade from 1930 to 1940 witnessed several important discoveries in the field of atomic and nuclear physics. The most brilliant were those of Frédéric and Irène Joliot-Curie, who were the first to create radioactive isotopes using alpha radiation from polonium. Later, the study of neutron-induced nuclear transformation of uranium led Irène Joliot-Curie to discover the formation of bodies that were chemically analogous to lanthanum. Finally, after the discovery of uranium fission by Otto Hahn and Fritz Strassmann, Frédéric Joliot-Curie showed that the fission process was characterized by a large energy release and by the emission of more than two neutrons. With his collaborators, Hans Halban and Lew Kowarski, he concluded that the development of a chain reaction should be possible. The work of Joliot and his group was published in French journals and foreign journals and formed the basis for three patents, in collaboration with Francis Perrin, that contained the basic conditions for the development of energy-producing nuclear reactors, as well as for the construction of a highly powerful bomb.

As for my own work during this period, I would like to point out my research on neutrons emitted by polonium-beryllium sources, that is, emitted by beryllium submitted to bombardment by alpha particles from the polonium. I was able to measure their energy by studying the trajectories of protons knocked on by collision, these trajectories being observed in a cloud chamber filled with hydrogen and placed in a fairly

Ed. note: This chapter was translated from French by Mary K. Gaillard.

173

powerful magnetic field. The measurement of the curvature of the tracks allowed me to perform a statistical study of the energy of the incident neutrons. I was thus able to demonstrate the existence of relatively slow neutrons forming a truly continuous energy spectrum. An energy distribution of this type can occur only if the phenomenon yields at least three particles that share the available energy. In the case studied, these included the beryllium nucleus, having lost a neutron on bombardment by alpha radiation, the emitted neutron, and clearly the alpha particle itself, which must represent the third body. This shows that the alpha particle was not absorbed and that the neutron was simply knocked out of the beryllium nucleus. This was, in fact, the first observation (1932) of the phenomenon currently known under the name of spallation.

After this work, I turned to research on cosmic-ray physics. When I was working with Dmitry Skobeltzyn in 1929, we observed tracks of highly energetic particles coming from the upper atmosphere and traversing a Wilson cloud chamber placed in a magnetic field.[1] This provided direct proof of the corpuscular nature of cosmic radiation. I subsequently returned to this experiment in collaboration with Louis Leprince-Ringuet, and we measured the variation in intensity of cosmic radiation with latitude under the influence of the earth's magnetic field. I also observed, as early as 1934, the presence in cosmic radiation of highly penetrating particles in addition to electrons. In an article in *Nature*, I demonstrated the dual character of cosmic radiation based on experiments performed with cloud chambers and Geiger counters at high mountain altitude and also in underground laboratories.

It was in 1938 that I discovered a new phenomenon in the field of cosmic-ray physics, namely, extensive atmospheric showers. A new electronic device developed by my collaborator, Roland Maze, made it possible to obtain a very high degree of precision in coincidences between two or more Geiger counters.[2] The background from accidental coincidences was thereby reduced sufficiently that it was possible to detect phenomena that produced only a very small number of true coincidences per hour. Having placed two or three counters at horizontal distances of several meters, we observed coincidences due to simultaneously incident ionizing particles separated by such large distances as to exclude an interpretation in terms of local effects.

I conjectured that this result indicated a very important atmospheric phenomenon and with Maze placed one of the counters at the distance

Pierre Auger at the 10th anniversary of CERN in 1963 (credit: CERN photography department).

of 160 m in another building. Again, we obtained several coincidences per hour. The measurement of density of tracks in a shower per square meter and the evaluation of the particles' energies allowed me to show in 1938 that the primary cosmic radiation contained a component of particles with energies greater than 10^{15} eV, or 1 million GeV. We studied these extensive showers at high altitude with distances of 300 m between counters. After the war, studies by many workers led to numerous publications, and the maximum observed energy reached at least 10^{18} eV for showers covering a horizontal area of several hectares. I was able to show that these extensive showers contained a small proportion of highly penetrating particles, then called μ mesons, and thereby demonstrated their secondary origin. They are unstable and decay rapidly. I was able to measure their mean lifetime by a direct procedure with Roland Maze and Robert Chaminade.[3]

Notes

1 Pierre Auger and D. Skobeltzyn, "Sur la nature des rayons ultrapenetrants (rayons cosmique)," *Compt. Rend. 189* (1929), 55-7.

2 Pierre Auger and Roland Maze, "Extension et pouvoir pénétrant des grandes gerbes de rayons cosmiques," *Compt. Rend. 208* (1939), 1641-3.

3 Pierre Auger, Roland Maze, and Robert Chaminade, "Une démonstration directe

de la désintégration spontanée du méson," *Compt. Rend. 213* (1941), 381-3; Roland Maze and Robert Chaminade, "Une mesure directe de la vie moyenne du méson au repos," *Compt. Rend. 214* (1942), 266-9; M. Conversi and O. Piccioni, "Sulla disintegrazione dei mesoni lenti," *Nuovo Cimento 2* (1944), 71-87; M. Conversi and O. Piccioni, "On the Disintegration of Slow Mesons," *Phys. Rev. 70* (1946), 874-81.

10   The scientific activities of Leprince-Ringuet and his group on cosmic rays: 1933-1953

LOUIS LEPRINCE-RINGUET

Born 1901, Alès (Gard), France; Docteur ès sciences, Ecole Polytechnique; x rays, cosmic rays; Ecole Polytechnique, Paris.

A round-trip voyage by ship from Hamburg to Buenos Aires and back made it possible to elucidate with the help of electronic counters the effect of latitude, which had just been discovered by Clay using an ionization chamber. This effect was important because it allowed a decision as to whether or not the primary component of cosmic radiation was electrically charged. From 1932 on, it was the object of a great deal of research; in particular, a vast expedition over the entire surface of the earth was organized by Arthur Holly Compton. The research that we carried out in 1933 showed the importance of this effect at sea level on different components of radiation. It enabled us to verify the azimuthal effect (the predominance of radiation coming from the west), which is interpreted as being due to the predominance of positively charged particles in the primary radiation.[1]

The separation of different components of radiation into groups with well-defined properties and the knowledge of these properties led us to perform a high-altitude experiment at the international laboratory of the Jungfraujoch, 3,400 m above sea level, where we carried out several runs that determined the two principal components of cosmic radiation (the penetrating component and the soft component) and their absorption properties as a function of the atomic number of the element they traversed.[2]

We used the electromagnet of the Academy of Sciences at Bellevue to deviate the paths of high-energy particles observed in cosmic rays.

Ed. note: This chapter was translated from French by Mary K. Gaillard.

Starting in 1936, both at Bellevue and later at the Alpine stations, I worked on perfecting large cloud chambers controlled by electronic counters in order to determine the energy spectrum of cosmic rays and to determine their nature. I carried out this research both alone and in collaboration with J. Crussard, then with E. Nageotte, S. Gorodetzky, R. Richard-Foy, and M. Lheritier. In 1936, I was able to achieve energy measurements of cosmic rays up to 20 BeV, a limit that was, however, frequently exceeded by the hardest cosmic rays.

Several interesting properties were discovered during the course of the experiments at Bellevue. First, we found the predominance of positive particles in the totality of the penetrating radiation. Second, with J. Crussard, we found that there were particles of both charges and of relatively low energy (less than 300 MeV) that were nevertheless able to traverse a large thickness of lead (14 cm). These particles could not be protons, and it was extremely improbable that they were electrons. We therefore made the hypothesis that these could be new particles, but we were unable to confirm their existence, because at that time the properties of very energetic electrons were not well known. Nevertheless, these results were evidence in favor of the existence of the meson, which was identified by Anderson two years later. Finally, results on secondary cosmic rays and on certain nuclear effects were also obtained at Bellevue.[3]

From one of the pictures obtained at Bellevue in 1939 we were able (with Gorodetzky, Nageotte, and Richard-Foy) to determine the mass of a meson by applying the laws of mechanics (with relativistic corrections, of course, included), owing to a particularly favorable collision between a meson and an electron in the gas of the cloud chamber. This measurement gave a value of 240 ± 30 relative to the electron mass, which was taken as a unit mass.[4]

In 1937 a large Wilson cloud chamber with an iron-free magnetic coil for 8,000 A of current was installed at L'Argentière in the French Alps. This apparatus started functioning in 1938 and continued throughout the occupation. It enabled those physicists who stayed in France to continue their research on meson masses, on the energy spectrum, on the excess of positive particles, on their lifetimes, and on the nuclear effects of radiation. We had installed a small experimental center at the Lautaret pass (2,000 m altitude) near L'Argentière, and it was during this period that we began, thanks to a grant from the National Center for Scientific Research, the construction of a high mountain laboratory, 3,600 m of altitude, at the Aiguille du Midi near

Chamonix. One of the pictures showed the scattering of an incident particle of momentum 600 MeV/c from an electron, which allowed us to evaluate the mass of the incident particle, 990 electron masses ($\pm 12\%$), under the assumption of an elastic collision. This measurement should be considered as the earliest evidence in favor of the existence of a heavy meson.[5]

Starting in 1947, we studied nuclear phenomena produced by cosmic radiation, thanks to the method of thick and concentrated photographic emulsions, which had been developed at Bristol by C. F. Powell and G. P. S. Occhialini. We sent emulsion plates both to the Aiguille du Midi de Chamonix and to the hut at Vallot, and also into the upper atmosphere at an altitude of about 20,000 m, using sounding balloons provided by the national meteorology facility. With J. Heidmann, L. Jauneau, T. F. Hoang, and D. Morellet we studied thousands of events, including about 100 mesons events. In addition to plates exposed to cosmic radiation, we were able, thanks to the kindness of Professor E. Lawrence, E. Gardner, and C. M. G. Lattes, to examine plates that had been subjected to radiation at the large Berkeley cyclotron and that contained a large number of tracks of artificially produced mesons. We observed more than 500 meson-induced nuclear stars on these plates. The totality of all these phenomena allowed us to study, on the one hand, nuclear excitation induced by both heavy and light incident mesons and, on the other hand, large nuclear explosions.[6] This research was continued through 1953, with J. Crussard, D. Morellet, G. Kayas, A. Orkin-Lecourtois, and J. Tremblay, on heavy meson masses using the techniques of ionization, scattering, and path length.[7]

Between 1950 and 1955 the Ecole Polytechnique group, including Charles Peyrou, B. Gregory, A. Lagarrigue, R. Armenteros, F. Muller, and A. Astier, installed two large superimposed Wilson cloud chambers of 200 liters each at the Pic du Midi, at an altitude of 2,950 m. It seemed interesting to perform a large experiment that would combine the information on particle energy furnished by the curvatures of their trajectories in a magnetic field with the observation of their behavior as they traversed plates of matter. The chambers had large dimensions in order to increase the statistics of the experiment. The magnetic field was furnished by a large Helmholtz coil that provided a field of 6,000 G. A set of Geiger counters with special electronic devices triggered the chambers whenever a highly energetic nuclear phenomenon occurred above the apparatus. This apparatus ran regularly over a period of four years and provided 100,000 pictures. The scientific results obtained with it led to over a dozen publications, dealing mainly with the identification and the study of the properties of heavy mesons and hyperons, of which we were able to collect a sample of several thousand decays during our experiment. The apparatus was constructed principally by B. Gregory, A. Lagarrigue, and Charles Peyrou. Also participating in the experiments were A. Astier, R. Armenteros, and then, to some extent, W. B. Fretter of the University of California, R. R. Rau of Princeton University, J. Tinlot of the University of Rochester, and H. de Staebler of M.I.T.

The principal results were the following:

First, measurement of the mass of the K^+ meson and identification of its decay modes. The mass is determined by the curvature of the trajectory combined with the measurement of its path length when a particle is stopped in a plate chamber. In the early stages of our experiments, at a time when there was considerable uncertainty as to the mass of heavy mesons, we published a precise value for the K^+ meson mass: $920 \pm 30 \ m_e$. We sought to improve this value and were the first to show that the principal decay mode of the K^+ mesons was a two-body decay yielding a μ meson and a neutrino.

Second, anomalies in the behavior in the K^- meson. From the beginning of our experiment we had noticed the nearly complete absence of a negative counterpart for the K^+ meson. This result seemed to be in contradiction with charge symmetry. Our present understanding of these observations is that the K^- meson is the antiparticle of the K^+ meson and that considerably more energy is required for its production: One says that the K^- meson has the opposite strangeness to the K^+ meson. In 1953 we also published a spectacular event that displayed the production of a Λ^0 hyperon from a negative incident particle having an energy insufficient for the creation of a pair of strange particles.

Third, study of charged V decays. A large number of our pictures contained decays in the upper chamber (with the magnetic field) of the type known as charged V; that is, they appear in the form of a sharp angle on a fast track. The vertex of this angle is the point where the primary particle decays in flight, giving a charged secondary that is visible in the chamber and a neutral secondary that in some cases we were able to detect through its interactions in the plates of the lower chamber. In particular, we showed that an important fraction of these decays corresponded to Σ hyperons, which are more massive than protons. This identification was possible, in particular, by observing the interactions of the decay neutrons in the plate chamber. We were further able to identify a number of decay modes of charged kaons. Finally, one exceptional picture allowed us to identify the charged secondary from a very rare decay, the decay of the particle called Ξ^-, for which the decay mode is

$$\Xi^- \to \Lambda^0 + \pi^-.$$

The Λ^0 was observed in the chamber through its decay into a proton and a π^-.

The totality of this work enabled us to provide information on the lifetimes of different particles by observing the distribution in the chamber of the vertices of the decay angles.[8]

Fourth, the study of neutral decays (Λ^0 hyperons and θ^0 mesons). Our experimental apparatus was especially well adapted to the study of the decays of neutral particles. These appear in the form of an upside-down V. Thus one observes two particles of opposite signs produced in the decay of a neutral particle. The angle at the vertex and the curvatures of the trajectories can be measured. Both secondaries traverse the plates of the chamber and thereby provide information as to their nature. Around 1955, the sample of neutral decays that we accumulated was far superior to the total sample of events of the same type observed in all the other laboratories in the world. Specifically, we were able to provide a better value of the mass of the Λ^0, an important result, because it corrected certain anomalies in the interpretation of binding energies of excited nuclei (or hyperfragments). We were also able to provide a better value of the θ^0 mass and to establish definitively that the θ^0 meson decayed into two pions by observing the interaction of these two secondaries in the plate chamber. Most of these results were presented at the international conference on cosmic radiation in Bigorre in 1953.

From 1955 on, the entire Ecole Polytechnique laboratory oriented its research toward bubble chambers at the CERN particle accelerator, thus abandoning work on cosmic rays.[9]

Notes

1 P. Auger and L. Leprince-Ringuet, "Etude de la variation du rayonnement cosmique entre les latitudes 45 degrés Nord et 38 degrés Sud" (variation in intensity of cosmic rays between the latitudes 45°N and 38°S), *Compt. Rend. 197* (1933), 1242-4.
2 P. Auger and L. Leprince-Ringuet, "Analyze du rayonnement cosmique en haute altitude" (analysis of cosmic rays at high altitudes), *Compt. Rend. 199* (1934), 785-7;

P. Auger, L. Leprince-Ringuet, and P. Ehrenfest, "Absorption de la fraction molle du rayonnement corpusculaire cosmique" (absorption of the soft component of cosmic radiation), *Compt. Rend. 200* (1935), 1747-9.

3 L. Leprince-Ringuet and J. Crussard, "Étude des particules de grande énergie dans le champ magnétique de l'electro-aimant de Bellevue" (a study of high-energy particles in the magnetic field of the Bellevue electromagnet), *Compt. Rend. 201* (1935), 1184-7; *204* (1937), 240-2.

4 L. Leprince-Ringuet, E. Nageotte, G. Gorodetzky, and R. Richard-Foy, "Mesure de la masse d'un mésoton par choc élastique" (measurement of the mass of a mesoton by elastic scattering), *Compt. Rend. 211* (1940), 382-5; "Direkte Massenbestimmung eines Mesotrons mit Hilfe des elastischen Stosses," *Z. Physik 120* (1943), 588-97.

5 Louis Leprince-Ringuet and Michel Lhéritier, "Existence probable d'une particule de masse 990 m_e dans le rayonnement cosmique" (the probable existence of a particle of 990 m_e in the cosmic rays), *Compt. Rend. 219* (1944), 618-20; *J. Phys. Radium 7* (1946), 65-9.

6 L. Leprince-Ringuet, J. Heidmann, T. F. Hoang, L. Jauneau, and J. Strousma, "Observation of an Almost Complete Disintegration of a Silver Nucleus," *Compt. Rend. 225* (1947), 1144-6.

7 J. Crussard, C. Maffoux, D. Morellet, J. Trembley, and A. Orkin-Le Courtois, "Physique corpusculaire – observation d'un méson lourd du type K dans une plaque photographique nucléaire," *Compt. Rend. 234* (1952), 84-6; G. Kayas and D. Morellet, "Sur une méthode photometrique d'identification des particules de charge électronique dans les emulsions photographiques epaisses," *Compt. Rend. 234* (1952), 1359-61; J. Crussard, L. Leprince-Ringuet, D. Morellet, A. Orkin-Le Courtois, and J. Trembley, "Mésons X lents émis dans des 'etoiles cosmiques'," *Compt. Rend. 236* (1953), 872-4.

8 See the doctoral thesis of F. Muller. A. Hendel also wrote a doctoral thesis on the decay of τ mesons in the upper chamber.

9 The principal publications are as follows: "Production d'un V^0 par un primaire négatif de 1 BeV" (production of a V^0 by a negative primary of 1 BeV), *International Congress on Cosmic Radiation at Bagnères de Bigorre, 1953;* "Quelques résultats sur les V chargés" (some results on charged Vs), *International Congress on Cosmic Radiation at Bagnères de Bigorre , 1953;* "Mesures de masse de particules S par Moment-Parcours" (measurements of the mass of S-particles by the momentum-range method), *International Congress on Cosmic Radiation at Bagnères de Bigorre, 1953;* "Mass Measurements of Primaries of S Events by a Momentum Range Method," *Phys. Rev. 92* (1953), 1583-4; "Etude des mésons K chargés au moyen deux chambres de Wilson superposées" (a study of charged K mesons using two superimposed Wilson cloud chambers), *Nuovo Cimento 11* (1954), 292-309; "Résultats sur les particules V chargées" (results on charged V particles), *Nuovo Cimento (Suppl.) 12* (1954), 327-32; "Sur le signe de la particule K_μ" (on the sign of the K_μ particle), *Nuovo Cimento (Suppl.) 12* (1954), 324-6; "Further Discussion of the K_μ Decay Mode," *Nuovo Cimento 1* (1955), 915-41; "A V^0 Decay with an Electron Secondary," *Nuovo Cimento 4* (1956), 917-21; "τ-Mesons in the Momentum Cloud Chamber of the Ecole Polytechnique," *Nuovo Cimento (Suppl.) 4* (1956), 217-20.

11   The decay of "mesotrons" (1939-1943): experimental particle physics in the age of innocence[1]

BRUNO B. ROSSI

Born 1905, Venice, Italy; *laurea* Bologna, 1927; coincidence technique in cosmic-ray research; meson lifetime; γ-ray astronomy; Institute Professor, M.I.T. (emeritus).

I have chosen to talk about my work during the years 1939 through 1943 because, in some way, these years form a self-contained period of my personal life and of my scientific activity. It began with my arrival in the United States, as an exile from fascism. It ended with my shifting from peacetime work at Cornell University to wartime work at Los Alamos. Scientifically, it was almost all occupied by a research program on the spontaneous decay of μ mesons, or "mesotrons," as they were then called.

But let me go back a bit. The period immediately preceding my departure from Italy had not been particularly productive. I had spent an inordinate amount of time planning and supervising the construction of the new physics institute at the University of Padova. I had found it increasingly difficult to concentrate on physics, worried as I was about the threatening events in Europe and about the anti-Semitic decrees being issued in Italy in quick succession. Eventually, early in September 1938, I learned that I no longer was a citizen of my country and that my career as a teacher and scientist in Italy had come to an end.

It would be ridiculous to describe this situation as a tragedy. So many people in the world were the victims of incommensurably greater calamities. Still, it was a shock. Fortunately, I had friends abroad, and I hoped that with their help I would be able to make a fresh start in some other country. It was not going to be easy to leave Italy, not knowing if I would ever be able to go back. It was going to be even harder on my wife (we had been married for only a few months). But, setting aside her personal feelings, she urged me not to delay our departure more than was strictly necessary. While we were waiting for our passports, I wrote to Niels Bohr, who, most graciously, invited me to visit his institute. So Copenhagen was to be our first stop.

The two months we spent there were for us a most beneficial interlude, and those of you who have known Niels and Margrethe Bohr will not be surprised. The human kindness, the intellectual liveliness, the sane vision of human affairs that were the essence of the Copenhagen atmosphere went a long way toward clearing our minds and strengthening our confidence. I spent long hours at the institute, talking with people and catching up with my reading, gradually rekindling my enthusiasm for science.

While we were there, Bohr organized a conference that brought to Copenhagen many scientists, among them a number of cosmic-ray physicists. I strongly suspect that one of his motives was to give me the opportunity of meeting people who might be able to help me find a job. In any case, that is what happened, because, shortly thereafter, Patrick Blackett invited me to Manchester on a fellowship from the Society for the Protection of Science and Learning.

Manchester was very different from Copenhagen. Yet our six months there were another constructive period of our peregrinations. Manchester was a dismal city; but, starting from early spring, the surrounding moors were delightful, and we spent our weekends exploring them on bicycle. The people in Manchester – the Blacketts foremost – were most helpful and warm.

Blackett's laboratory was a stimulating place, with A. C. B. Lovell, Lajos Jánossy, George D. Rochester, J. G. Wilson, and other interesting people working there and, of course, with Blackett's inspiring presence. Most important for me was the opportunity of getting my hands dirty again by doing some experimental work. Among other things, I measured, with Jánossy, the absorption cross section of the γ-ray component of cosmic-rays in matter.[2] Of course, we did not make any world-shaking discovery; our result was just a fairly accurate check of the Bethe-Heitler theory. But it was a nice, clean experiment, rather elegant in its simplicity. From the point of view of the history of experimental techniques, it had a certain interest, because it was one of the first experiments making use of the anticoincidence method. (The circuit we designed for the recording of anticoincidences was a rather direct descendant of the coincidence circuit that I had developed 10 years earlier.)

In the meantime, Arthur Compton had written, inviting me to the University of Chicago on a fellowship from the Committee in Aid of Displaced Foreign Scholars. We hated to leave Europe, and we hated especially to leave England, which had been so good to us. But the situation in Europe looked very bleak; Blackett, too, was quite pessimistic and urged us to go. And so, in June 1939, we sailed from England, heading for the United States and Chicago.

This brings me to the main subject of this chapter, for it was in July 1939, shortly after our arrival in Chicago, that I became involved with the problem of the decay of μ mesons. It all began at Otsego Lake, in upper Michigan, where Arthur and Betty Compton had kindly invited us to recuperate from the stress of the preceding weeks. I welcomed the opportunity to discuss with Compton the possible directions of my work in Chicago. At that time, the instability of μ mesons was one of the hottest problems of cosmic-ray physics. As is well known, experiments had shown that the penetrating component of cosmic rays is composed of particles heavier than electrons, but lighter than protons. These particles had been given various names, the most commonly accepted being "mesotron." Independently, Hideki Yukawa had developed a theory of nuclear forces that predicted the existence of particles of mass intermediate between those of the electron and the proton. To account for the β decay, Yukawa had postulated that these particles were unstable, with a mean life of the order of microseconds, each decay process giving rise to an electron and a neutrino.

It was natural to identify the Yukawa particle with the cosmic-ray mesotron. This identification, of course, was wrong, and so was the lifetime assigned by Yukawa to his particle. But two wrong assumptions led to a conclusion that, as you know, turned out to be correct, namely, that cosmic-ray mesotrons were unstable with a mean life on the order of microseconds.

Attempts at detecting the electrons that were supposed to arise from the decay of mesotrons had been unsuccessful. On the other hand, the disintegration hypothesis appeared to receive some indirect support from certain experiments showing that mesotrons were more strongly absorbed by air than by condensed matter, when layers equivalent with respect to ordinary ionization losses were compared, and that, moreover, the anomalous absorption of air increased with decreasing density. It was argued that with a mean life of microseconds, a substantial number of mesotrons traveling through air would decay before ionization losses brought their energy below the detection limit, whereas practically no decay process would occur in the much shorter time required for mesotrons to be slowed down by condensed matter.

Many papers had been written, some by Werner K. Heisenberg and by Blackett, analyzing the possible effects of the anomalous attenuation of the mesotron beam due, supposedly, to their decay.[3]

While I was in Manchester I had made a critical analysis of the available experimental data on the atmospheric density effects purported to support the disintegration hypothesis (variation of the mesotron intensity with height, variation with zenith angle, variation with atmospheric temperature). I had concluded that none of these data, with the possible exception of some regarding the zenith-angle effect, could be taken as evidence for the instability of mesotrons.[4] The problem of the disintegration of mesotrons had been thoroughly discussed at an international cosmic-ray symposium held in Chicago immediately after my arrival. The conclusion had been that the experimental evidence for the decay of mesotrons could not yet be regarded as conclusive.

Further and better experiments were therefore needed. I thought that the most promising approach would be to make a direct and precise comparison between the attenuation of the hard component in air (where, as I mentioned, if mesotrons were unstable, both ionization losses and decay would contribute to the total attenuation) and its attenuation in some dense material (where the absorption would be due to ionization losses alone). This experiment would require accurate measurements at different levels, extending possibly to great altitudes.

Compton listened with interest as I was developing these ideas. Then he pointed out that the Mt. Evans region in Colorado was the ideal place for an experiment such as I had in mind. There was a road reaching the top of the mountain, more than 4,000 m above sea level. At the top, a small cabin had been built a few years before for the use of scientists working there. We could count on the support of Joyce Stearns, professor of physics at the University of Denver. I saw immediately that Compton's suggestion presented a unique opportunity, and I asked him if it might be possible to organize an expedition to Mt. Evans for the next summer. His answer was "Why not this summer?" I was taken aback. It was almost the middle of July. I had previously made a commitment to visit the summer school at Ann Arbor for about a week after leaving Otsego Lake. On the other hand, we would have to start from Chicago before the end of August in order to complete the measurements at Mt. Evans before the beginning of the snow. Thus, there was barely one month for building the experimental equipment and making the necessary logistic arrangements.

But, of course, I accepted the challenge, and back in Chicago I went immediately to work. Compton had arranged for Norman Hilberry and, later, for Barton Hoag to work with me. We built the Geiger-Müller counters, and I began to wire a coincidence circuit on a breadboard, as I used to do in Italy. Hilberry was horrified at such a primitive arrangement. He told me that in America, circuits were built on metal chassis and neatly mounted on racks. At first I was overwhelmed. How could we possibly build an elaborate instrumentation in the little time we had at our disposal, with me not having ever done anything of that sort? But then I relaxed and decided that I would continue to work at my breadboard circuit, while Hilberry would build an American-style instrument. It is no reflection on Hilberry's competence, but merely on his optimism, that his ambitious project could not be completed in time, and we had to leave for Colorado with my primitive circuit, which, however, proved perfectly adequate.

There was the problem of transportation. To buy or rent a truck was out of the question; at that time, little money was available to scientific laboratories. So Compton arranged to borrow an old bus that during the academic year was used by the Zoology Department for taking students on field trips. We set up our equipment in the bus, which thus became effectively our moving laboratory.

The experimental arrangement was very simple (Fig. 11.1). It consisted of three Geiger-Müller counters placed one above the other, separated by sufficient lead to filter out the soft component, and protected by lead walls against air showers. With this arrangement, the rate of threefold coincidences could be taken as a measure of the intensity of the hard component. A carbon absorber, made of graphite blocks, could be placed above the counter array. Because carbon has an atomic number close to that of air, we argued that ionization losses per gram per square centimeter of carbon and air would be practically the same.

We left Chicago on August 26. At the last moment, Compton had decided that we would need the help of a strong young man to move

The bus used in the 1939 experiment (credit: Bruno Rossi).

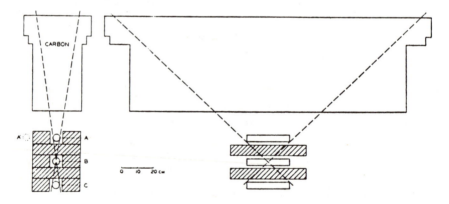

Figure 11.1. The 1939 experiment. From B. Rossi, N. Hilberry, and
J. B. Hoag in *Phys. Rev. 57* (1940), 462, Fig. 1.

the lead and the carbon, and he had persuaded a graduate student who
was a member of the Chicago swimming team to accompany us. His
name was Winston Bostick, and he proved to be a most useful and
agreeable addition to our little team (later on, he became a distinguished plasma physicist).

My wife and I, with Hoag, were riding on the bus, which, not being
a particularly fast vehicle, took three days to reach Denver. The memory of that trip through the unending midwestern plains, our first

Bruno Rossi at work inside the bus (credit: Bruno Rossi).

contact with the heart of America, was an unforgettable experience. We stopped a couple of days in Denver to take some measurements; then we proceeded to Echo Lake, at about 3,000 m above sea level, and stopped there for some more measurements. Finally, on September 3, we started for Mt. Evans.

The last leg of our trip was somewhat of an adventure. The paved road stopped at Echo Lake, and from Echo Lake to Mt. Evans there was, at that time, only an unpaved rocky road that was steep and narrow. Our bus, which until then had behaved quite well, was beginning to show signs of discomfort. We could not blame it; after all, it was not used to carrying tons of lead and graphite up to the highest mountains of the United States.

As a precaution, at Echo Lake we filled all available containers with water. Shortly after leaving Echo Lake, when the water in the radiator started boiling, my wife took charge of the problem. She recalls sitting on the hood, steadily pouring water in the radiator. Honestly, I cannot discount the possibility that through the years the details of this episode may have acquired a romantic tinge. In any case, she managed to husband the water supply so that eventually we made it to Summit Lake, halfway to the top of Mt. Evans. There we got a fresh supply of water, which enabled us to complete our journey.

For several days we drove up and down the mountain, alternating measurements at Mt. Evans and Echo Lake. By the middle of Septem-

Bruno Rossi and Mrs. Rossi on one of the twin peaks of Mt. Evans (credit: Bruno Rossi).

ber we were back in Denver, and toward the end of the month we returned to Chicago.

At each station we measured the counting rate with and without the graphite blocks above the counters. The results are shown in Fig. 11.2. The observed counting rates were plotted on a logarithmic scale against the total mass per square centimeter of air and carbon above the counters. The open circles refer to measurements taken without the carbon absorber; thus the solid curve connecting these points showed the dependence of the mesotron intensity on atmospheric depth. (I may note that no similar measurements of comparable accuracy had been performed previously.) The solid dots refer to measurements taken with the carbon absorber. Thus the dotted lines connecting the points which represent measurements performed at the various elevations with and without the carbon absorber showed the initial slopes of the logarithmic attenuation curves of mesotrons in carbon. These slopes were about one-half the corresponding slopes of the attenuation curves in air. This shows that, at all four elevations where we had performed our measurements, mesotrons undergo much stronger attenuation in a given layer of air than in a layer of carbon of the same mass per square centimeter. This was exactly what we had expected, assuming mesotrons were unstable, and I believe we were justified in

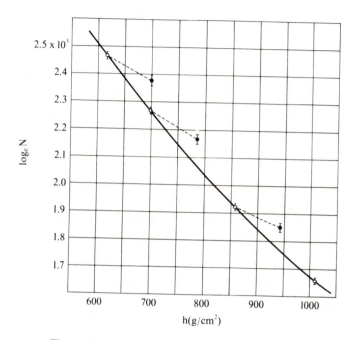

Figure 11.2. Results of the 1939 expedition, showing the decay of mesotrons in air. N is the intensity of cosmic ray mesons and h is the depth. From B. Rossi, N. Hilberry, and J. B. Hoag in *Phys. Rev. 57* (1940), 466, Fig. 3.

claiming that our observations had provided the first unambiguous evidence for the spontaneous decay of mesotrons.

It was also possible to use our results to estimate the mean life of mesotrons before decay. The difference between the slopes of the logarithmic attenuation curves in air and in carbon gave us the fractional attenuation due to decay of the mesotron flux in 1-g/cm² of air. From this quantity (and from the relationship between altitude, z, and barometric pressure, which we had also measured directly), we could immediately derive the mean free path before decay, that is, the quantity L defined by the equation

$$\frac{1}{N}\left(\frac{\delta N}{\delta z}\right)_{decay} = \frac{1}{L}. \tag{1}$$

Here N is the number of incident mesotrons, and δN is the number of those that decay traveling the distance δz. From our experiments we found that $L \approx 9.5$ km. Because of the relativistic dilation of time intervals, the mean life of a mesotron in motion with velocity v is $\gamma\tau$, where τ is the mean life at rest and $\gamma = (1 - v^2/c^2)^{-1/2}$. The corresponding mean free path before decay is

$$L = \gamma v\tau = p\tau/m, \tag{2}$$

where p is the momentum and m is the mass of the mesotron.

In the case of a non-monoenergetic mesotron beam, p should be replaced by an "effective" momentum, p_{eff}, *whose reciprocal equals the reciprocal of p* averaged over the momentum spectrum. For the computation of τ from our results, we estimated an effective momentum using the energy spectrum of mesotrons measured by Blackett at sea level; we obtained $p_{eff}c \approx 1.3 \times 10^9$ eV. We also assumed $mc^2 = 0.8 \times 10^8$ eV, which, at that time, was thought to be the most likely value for the rest energy of mesotrons. We then obtained for τ a value of about 2 µsec. With the present value of mc^2 (1.0566×10^8 eV) we would have obtained $\tau = 2.64$ µsec.

Shortly after the publication of our results,[5] we read accounts of experiments, similar in principle to our own, that had been performed in America and in Europe.[6] Among these were observations carried out at Chatillon and at Pian Rosà on the Matterhorn by my Italian friends Mario Ageno, Gilberto Bernardini, N. B. Cacciapuoti, Bruno Ferretti, and Gian Carlo Wick. All of these experiments were in qualitative agreement, showing evidence for anomalous absorption of mesotrons in air. However, there were fairly large quantitative differences in the results, easily explained by the low accuracy of the measurements.

Soon after our return from Colorado, I began thinking about the next step in my research program. As noted earlier, for the interpretation of our results we had assumed as valid the prediction of special relativity concerning the dilation of time intervals in a moving frame of reference. Of course, no one had any doubt on this score. Still, at that time, the relativistic dilation had not yet been verified experimentally. With the unstable mesotrons, nature had provided us with very fast moving clocks, and I thought that it would be nice to use them for verifying the relativistic formula. Also, I thought that this verification would provide crucial proof for the validity of the assumption that the difference in the apparent absorptions of mesotrons in air and in condensed materials was indeed due to the decay of mesotrons rather than to some other effect.*

---

* *Ed. note*: For some work of E. Fermi on another possible "anomalous absorption" effect, see Chapter 8 in this volume.

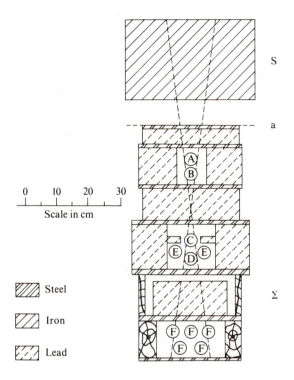

0 10 20 30

Scale in cm

S

a

Σ

Steel

Iron

Lead

Figure 11.3. The 1940 experiment. From B. Rossi and D. B. Hall in
Phys. Rev. 59 (1941), 224.

This was the motivation for an experiment that I undertook the
following summer in Colorado in collaboration with David Hall.[7] The
idea was to compare the attenuations in air and in condensed matter
of two groups of mesotrons: the first composed of mesotrons in a
narrow momentum interval, say between p_1 and p_2, the second composed of all mesotrons with momentum greater than p_2. Our experimental arrangement is shown in Fig. 11.3. I do not wish to go into
technical details, except for mentioning one point. As a measure of
the intensity of the mesotron group in the narrow momentum interval, we might have taken the difference between the mesotron intensities observed with two different absorbers over or between the
counters; this is, in fact, what other physicists engaged in similar
experiments were doing.[8] By this method, however, it would hardly
have been possible to reach the necessary statistical accuracy. So
what we did was to use an anticoincidence circuit to record directly

the mesotrons capable of traversing a given absorber and stopped by a given additional absorber.

In our experiment, the additional absorber was a 115 g/cm² thick layer of lead, Σ, that could be inserted between counters D and F, whereas the permanent absorber between the entrance surface of the instrument at *a* and the removable absorber Σ was equivalent to a 196 g/cm² thick layer of lead. Therefore, the difference between the rates of anticoincidences ABCD − F recorded with and without the absorber Σ in place could be taken as a measure for the rate of arrival on the instrument of mesotrons with residual ranges between 196 and 196 + 115 = 311 g/cm² of lead equivalent. On the other hand, the rate of simultaneously recorded coincidences between counters ABCD and F could be taken as a measure of the rate of arrival of mesotrons with residual range greater than 311 g/cm² of lead, the total equivalent lead thickness of the material above the lowest counters F. Counters E were used to guard against spurious effects simulating absorption of mesotrons in Σ.

Of course, the use of an anticoincidence arrangement for an experiment like ours looks quite trivial today, and hardly worth mentioning. But at that time the anticoincidence method was still quite new. Its application in this experiment illustrates a fact quite familiar to us experimentalists, namely, that often a seemingly minor change in the technique makes a crucial difference in the feasibility of an experiment. In this case, of course, the crucial factor was the large improvement of statistical accuracy brought about by the anticoincidence method.

The essential part of our experiment was a comparison between counting rates at Denver (1,616 m altitude), with no absorber above the instrument, and at Echo Lake (3,240 m altitude), with a 200 g/cm² iron absorber. According to the Bethe-Bloch theory, iron at 200 g/cm² is equivalent, with regard to ionization losses, to air at 147 g/cm², the mass of the air layer between the two stations. As expected, the counting rates of both coincidences and anticoincidences were higher at Echo Lake under the absorber than at Denver without absorber, the difference being clearly due to the decay of mesotrons on their way down from Echo Lake to Denver. As a convenient representation of the experimental results, we used again a "mean free path before decay," *L,* defined formally by the equation

$$\frac{1}{L} = \frac{1}{\Delta z} \ln \frac{N_D}{N_E},$$

(3)

where Δz $(-1,624\ m)$ is the vertical distance between Echo Lake and Denver and N_D and N_E are the counting rates in Denver and Echo Lake, respectively.

Our results are shown in Table 11.1. As mentioned earlier, the relativistic dilation of time intervals implies that the mean free path before decay should be proportional to the momentum. Our results, without yet proving this proportionality relation quantitatively, confirmed the theoretical predictions qualitatively by showing that L was considerably greater for the harder mesotron group ($R >$ g/cm^2 lead) than for the softer one (in the range 196-311 g/cm^2 lead). Moreover, because mesotrons in this latter group belong to a rather narrow interval of momenta, the effective momentum p_{eff} entering in the equation for the mean life,

$$\tau = mL/p_{\text{eff}}, \tag{4}$$

is largely independent of the poorly known mesotron spectrum. Of course, in the computation of the effective momentum, we had to take into account the momentum loss by ionization between Echo Lake and Denver. The result was $p_{\text{eff}}c = 5.0 \times 10^8$ eV. With this value of p_{eff}, with the value L shown in Table 11.1, and with the present value mc^2 $= 1.0566 \times 10^8$ eV for the mesotron rest mass, equation (4) yields τ ≈ 3.17 μsec.

As we returned from Colorado, in the fall of 1940, the Chicago chapter of our American experience was about to close. We had enjoyed our stay in Chicago. The friendly and intellectually stimulating atmosphere we found there had been largely responsible for our accepting America as our new home. We had no desire to leave, but my fellowship was only a temporary arrangement, and it was necessary to find a more permanent position. There were no faculty openings at the University of Chicago. However, a position was open in the Physics Department of Cornell University, and thanks to Hans Bethe, my name was considered for filling the vacancy. The previous April we had been invited to stop in Ithaca, on our way to the Washington meeting of the American Physical Society, where I gave a seminar; a short time afterward I was offered an associate professorship, which I accepted. So then, shortly before the beginning of the fall term, we packed our belongings and moved to Ithaca.

Among the graduate students at Cornell was a young man by the name of Kenneth Greisen. He became my first American student, and I could not have wished for a more intelligent, more enthusiastic, and

Table 11.1

	Residual range (Denver)		L (km)	$P_{\text{eff}}c$ (10^8 eV)	τ (10^{-6} sec) for mc^2 $= 1.0566 \times 10^8$ eV
	Minimum	Maximum			
Group 1	196 g/cm^2 Pb	311 g/cm^2 Pb	4.5 ± 0.6	5.0	3.17
Group 2	311 g/cm^2 Pb	∞	13.3 ± 0.9	–	–

more imaginative collaborator. As you know, in due time he became an outstanding scientist on his own merit. One result of our collaboration was a long review article.[9] In the first part of this article, we summarized the theoretical results concerning electromagnetic interactions of charged particles and photons. In the second part, we endeavored to coordinate and supplement the various results on shower theory scattered in the literature. Our aim was to provide cosmic-ray physicists with a set of formulae and diagrams, in a form directly applicable to the interpretation of their experimental data.

Greisen and I used the tool we had forged to investigate an important problem that had recently received a great deal of attention but had not been satisfactorily solved. This concerned the origin of the soft component of cosmic rays, which was thought to consist mainly of electrons, with a small admixture of slow mesotrons. The basic question was whether or not the electrons might be secondary products of mesotrons. We carried out a rigorous computation, at various altitudes, of the electron intensity arising from the decay of mesotrons, from collision processes of mesotrons with atmospheric gases, and from the subsequent cascade multiplication of these electrons. We then compared our results with measurements carried out by Greisen in Ithaca and in Colorado at altitudes of 259, 1,616, 3,240 and 4,300 m. The results were entirely unambiguous. They showed that the soft component increased with altitude much faster than the mesotron component. Whereas at low altitude the observed ratio of the soft component to the hard (mesotron) component was not much greater than the computed value, at 4,300 m it was about twice as large.[10]

The inescapable conclusion was that a sizable proportion of the electron component did not arise from secondary processes of mesotrons. In our closing remarks we favored the assumption that the additional electrons were the products of cascade processes of primary electrons. We were not clever enough to realize that our work had produced the first evidence for the existence of π^0 mesons!

In the meantime, I was still concerned with the related problems of the lifetime of mesotrons and the relativistic dilation of time intervals. I did not regard our previous mountain experiment as final, because only one of the two mesotron groups we had compared was formed by particles within a sufficiently narrow momentum interval to afford a reasonably reliable estimate of the effective momentum. To reach more conclusive results it was clearly desirable to compare the decay of two mesotron groups, both belonging to narrow momentum intervals.

Ken Greisen mounting the experiment at Echo Lake (credit: Bruno Rossi).

This was the aim of a further mountain experiment that Greisen and I carried out in the summer of 1941, in collaboration with three Colorado colleagues, J. C. Stearns and Darol K. Froman of the University of Denver and Phillipp Grant Koontz of Colorado State College.[11] The experimental arrangement was an obvious extension of that used the year before, and I shall not describe it here in detail.

We again compared measurements taken at Echo Lake and in Denver, compensating for the ionization losses in the air layer between the two stations by means of a suitable absorber. Of the two groups of mesotrons selected for our experiments, the first consisted of mesotrons reaching Denver with a residual range between 196 and 311 g/cm^2 of lead. For the second group, the lower and upper range limits had been increased by the addition of a 285 g/cm^2 iron absorber.

The mean free paths before decay derived from our experiment, the corresponding effective momenta, and the mean lives (computed for $mc^2 = 1.0566 \times 10^8$ eV) are shown in Table 11.2. It can be seen that in this experiment the softer group of mesotrons was practically identical with that selected in the 1940 experiment. It was reassuring to find that

Table 11.2

	Residual range (Denver)		L (km)	$P_{\text{eff}}c$ (10^8 eV)	τ (10^{-6} sec) for mc^2 = 1.0566×10^8 eV
	Minimum	Maximum			
Group 1	196 g/cm² Pb	311 g/cm² Fe	4.26 ± 0.46	5.15	2.96 ± 0.3
Group 2	196 g/cm² Pb	311 g/cm² Pb	8.5 ± 1.9	9.72	3.06 ± 0.7
	285 g/cm² Fe	285 g/cm² Fe			

the values of the mean free paths obtained in the two experiments were quite consistent with one another. Comparing the values of L and p_{eff} for the two mesotron groups, we see that, within the statistical accuracy, they are proportional to each other. This result provides the quantitative test of the relativistic dilation of time intervals that was the aim of our experiment.

Finally we endeavored to compute a "best" value for the mean life by combining the new data with those for the softer group of mesotrons obtained one year earlier (only data for mesotrons in narrow momentum intervals could be confidently used for this purpose). The result, adjusted to a value $mc^2 = 1.0566 \times 10^8$ eV for the rest energy of mesotrons, was $\tau \approx 2.96 \pm 0.2$ μsec.

After this experiment, Ken Greisen and I believed that no significant additional progress could be made by further measurements of the same kind. On the other hand, I was not yet satisfied with the measurement of the lifetime. The measurement of the decay of mesotrons in flight had given us the ratio τ/m of the mesotron mean life to its mass, but the mass of mesotrons, at that time, was not known with any accuracy. Moreover, despite all precautions, we could not discount the possibility of systematic errors in our measurements. Thus, although the order of magnitude of the mean life was by then well established, the uncertainty in its exact numerical value was probably greater than the 7% statistical error of our final result.

That was the motivation for undertaking a different kind of experiment that would give directly the mean life of mesotrons at rest, irrespective of their mass, and would also, hopefully, achieve a substantially greater accuracy. The principle of the experiment was the observation of the delayed emission of electrons arising from the decay of mesotrons brought to rest in an absorber.

A few experiments similar in principle to the one I had in mind had been reported recently. The only one that had given significant, if not quantitatively accurate, results was that of Franco Rasetti.[12] Using three circuits with different resolving times, Rasetti had measured the coincidence rates between pulses of counters recording the arrival of mesotrons on an absorber and pulses of counters recording the emission from the absorber of their decay electrons. From his observations, and assuming an exponential decay curve, he had obtained for the mean life of mesotrons at rest the value $\tau = 1.5 \pm 0.3$ μsec.

For the new experiment I secured the collaboration of Norris Nereson, another very capable graduate student at Cornell.[13] Our experi-

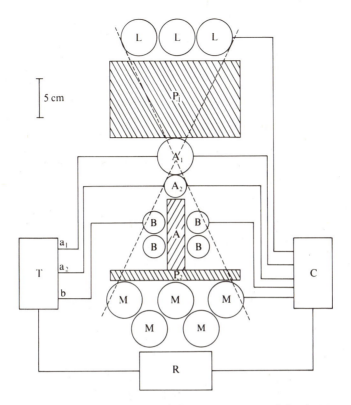

Figure 11.4. Experiment for the measurement of the decay curve of mesotrons at rest. From N. Nereson and B. Rossi in *Phys. Rev. 64* (1943), 199, Fig. 1.

mental arrangement is shown in Fig. 11.4. C is a circuit selecting coincidences between counters L, A_1, A_2, B − M. The resolving time was long compared with the mean life of mesotrons; therefore an anticoincidence was recorded whenever a mesotron came to rest in the absorber A after traversing counters L, A_1, A_2, and produced a decay electron that discharged one of the counters B. A small number of anticoincidences were also produced by spurious effects, but did not affect the final result.

The heart of the experiment was a "timing circuit" T, which measured the time intervals between the coincidences of counters A_1, A_2 (signaling the passage of mesotrons through these counters), and the subsequent pulses of counters B (signaling the emission of the decay electrons). The signals from the timing circuit (suitably lengthened) were recorded only when they were accompanied, within 1/100 of a second, by pulses from the anticoincidence circuit C, indicating that the mesotrons had stopped in the absorber A, and that their decay electrons had traversed counters B.

Our timing circuit had been developed by Nereson and myself specifically for the experiment I am describing.[14] Its basic principle was quite simple: the "early" signal started a current that began to charge a suitable condenser; the "late" signal stopped the current and discharged the condenser. The maximum voltage of the sawtooth pulse at the condenser was a function of the time elapsed between the arrivals of the two signals. The circuit was calibrated by pairs of artificial signals with variable and exactly known separations, the time scale being provided by a quartz-controlled oscillator. We estimated that 0.2 μsec was an upper limit for the calibration error.

Using various absorbers, we recorded a total of about 3,000 events with delays greater than 0.8 μsec. Events with shorter delays were disregarded, because a substantial number of them were due to spurious anticoincidences, in which case the observed delays were spontaneous delays of the counters (which we had measured and found to have a maximum value of just 0.8 μsec). The results were plotted on semilogarithmic plots (Fig. 11.5) with the abscissa as the time delay. An upper graph was the integral decay curve; each point gave the logarithm of the total number of delays greater than the corresponding abscissa. A lower graph was the corresponding differential decay curve, showing the logarithm of the number of observed delays in intervals of 0.4 μsec. Within the small statistical errors, both curves were straight lines, which means that the decay of mesotrons follows an exponential law. This, of course, was to be expected, although no decay curves of elementary particles had yet been measured. Nevertheless, that our experiment should have verified so nicely this expectation was reassuring and esthetically pleasing. From our decay curve we obtained the following value for the mean life: $\tau = 2.15 \pm 0.07$ μsec. It later turned out, to our satisfaction, that our value was remarkably close to the exact value of τ (2.198 $\pm$ 0.001 μsec) measured by means of artificially produced μ mesons.

I would like to open a parenthesis here. After the war I learned that while Nereson and I were working in the comfortable surroundings of Cornell University, two of my Italian colleagues, Marcello Conversi and Oreste Piccioni, in defiance of the harsh conditions prevailing in

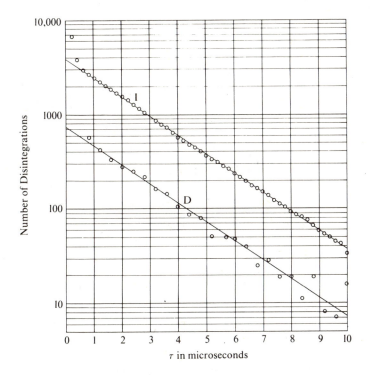

Figure 11.5. Integral (I) and differential (D) decay curves of mesotrons. From N. Nereson and B. Rossi in *Phys. Rev. 64* (1943), 200,
Fig. 3.

Rome under German occupation, had succeeded in carrying out a
carefully designed and very elaborate experiment aiming, like ours, at
a measurement of the mean life of mesotrons at rest.[15] They used a
method based on the observation of delayed coincidences; their result,
$\tau = 2.30 \pm 0.17$ μsec, was quite correct, within the limits set by the
statistical uncertainties.

Our experiment was concluded in the summer of 1943. By that
time we were in the midst of the war; for over a year, on a contract
with the radiation laboratory of M.I.T., I had been at work at
Cornell inventing and developing electronic circuits for radar application. Then came the call to Los Alamos. It was a time of hard
decisions, of strong and contrasting emotions. I remember finding
some measure of stabilizing comfort in the thought that the research
program that I had started four years earlier had been completed

and had helped establish some aspect of the physical reality; a modest accomplishment, to be sure, but destined to survive any human turmoil. (This is the strength or, depending on one's point of view, the weakness of the scientist.) Our work had produced the first unequivocal experimental evidence for the radioactive decay of mesotrons. It had achieved an accurate measurement of the mesotron mean life. As a by-product, it had verified the relativistic dilation of time intervals.

Today, thinking back to the work that produced these results and to the work in which other colleagues were engaged at that time, I am overtaken by a feeling of unreality. How is it possible that results bearing on fundamental problems of elementary particle physics could be achieved by experiments of an almost childish simplicity, costing a few thousand dollars, requiring only the help of one or two graduate students?

In the few decades that have elapsed since those days, the field of elementary particles has been taken over by the big accelerators. These machines have provided experimentalists with research tools of a power and sophistication undreamed of just a few years before. All of us oldtimers have witnessed this extraordinary technological development with the greatest admiration; yet, if we look deep into our souls, we find a lingering nostalgia for what, in want of a better expression, I may call the age of innocence of experimental particle physics.

Notes

1 This work was supported in part by the National Aeronautics and Space Administration under contract NASF-24441.
2 L. Jánossy and B. Rossi, "On the Photon Component of Cosmic Radiation and Its Absorption Coefficient," *Proc. Roy. Soc. (London) 175* (1940), 88-100.
3 H. Kulenkampff, *Verh. Deutsch. Physik. Ges. 19* (1938), 92; H. Euler and W. Heisenberg, "Theoretische Gesichtspunkte zur Deutung der kosmischen Strahlung," *Ergeb. Exakt. Naturw. 17* (1938), 1-69; P. M. S. Blackett, "Further Evidence for the Radioactive Decay of Mesotrons," *Nature 142* (1938), 992; B. Rossi, "Further Evidence for the Radioactive Decay of Mesotrons," *Nature 142* (1938), 993; P. M. S. Blackett, "On the Instability of the Barytron and the Temperature Effect of Cosmic Rays," *Phys. Rev. 54* (1938), 973-4.
4 B. Rossi, "The Disintegration of Mesotrons," *Rev. Mod. Phys. 11* (1939), 296-303.
5 B. Rossi, N. Hilberry, and J. B. Hoag, "The Disintegration of Mesotrons," *Phys. Rev. 56* (1939), 837-8; B. Rossi, N. Hilberry, and J. B. Hoag, "The Variation of the Hard Component of Cosmic Rays with Height and the Disintegration of Mesotrons," *Phys. Rev. 57* (1940), 461-9.
6 W. M. Nielsen, C. M. Ryerson, L. W. Nordheim, and K. Z. Morgan, "A Measurement of Mesotron Lifetime," *Phys. Rev. 57* (1940), 158; M. Ageno, G. Bernardini,

N. B. Cacciapuoti, B. Ferretti, and G. C. Wick, "The Anomalous Absorption of the Hard Component of Cosmic Rays in Air," *Phys. Rev. 57* (1940), 945-50; H. V. Neher and H. G. Stever, "The Mean Lifetime of the Mesotron from Electroscope Data," *Phys. Rev. 58* (1940), 766-70; A. Ehmert, "Ueber die Zerfallswahrscheinlichkeit des Mesons," *Z. Physik 115* (1940), 333-8.

7 B. Rossi and D. Hall, "Variation of the Rate of Decay of Mesotrons with Momentum," *Phys. Rev. 59* (1941), 223-8.

8 W. M. Nielsen et al., "A Measurement of Mesotron Lifetime" (Note 6).

9 B. Rossi and K. Greisen, "Cosmic-Ray Theory," *Rev. Mod. Phys. 13* (1941), 240-309.

10 B. Rossi and K. Greisen, "Origin of the Soft Component of Cosmic Rays," *Phys. Rev. 61* (1942), 121-8.

11 B. Rossi, K. Greisen, J. C. Stearns, D. K. Froman, and P. G. Koontz, "Further Measurements of the Mesotron Lifetime," *Phys. Rev. 61* (1942), 675-9.

12 F. Rasetti, "Disintegration of Slow Mesotrons," *Phys. Rev. 60* (1941), 198-204.

13 B. Rossi and N. Nereson, "Experimental Determination of the Disintegration Curve of Mesotrons," *Phys. Rev. 62* (1942), 417-22; N. Nereson and B. Rossi, "Further Measurements on the Disintegration Curve of Mesotrons," *Phys. Rev. 64* (1943), 199-201.

14 B. Rossi and N. Nereson, "Experimental Arrangement for the Measurement of Small Time Intervals between the Discharges of Geiger-Müller Counters," *Rev. Sci. Instr. 17* (1946), 65-71.

15 M. Conversi and O. Piccioni, "Misura Diretta della Vita Media dei Mesoni Frenati," *Nuovo Cimento 2* (1944), 40-70.

12   Particle physics in the 1930s: a view from Berkeley

ROBERT SERBER

Born 1909, Philadelphia; Ph.D., University of Wisconsin, 1934; quantum field theory, nuclear physics, particle accelerator theory, meson theory; associated in 1930s with Oppenheimer school; Columbia University (emeritus).

As seems allowable for a historical conference, this will be less a balanced account than a personal narrative. And, because I am a theorist, it will be an account more of ideas than of experiments. It is an account of the work of J. Robert Oppenheimer and his group on field theory and particle physics at Berkeley and Pasadena. At Pasadena there was the cosmic-ray work of Robert A. Millikan, Carl Anderson, Ira Bowen, and Henry Neher, and much of this narrative is related to that. There was also nuclear physics: under Charlie Lauritsen at Caltech, and at Berkeley under Ernest Lawrence. Oppenheimer's group did a good deal on nuclear physics and astrophysics, but I shall mention only the parts most closely connected to particle physics.

One thing that struck me in reconsidering the 1930s was the isolation of our California group. We were too poor to travel. Oppenheimer and Lawrence made a trip East occasionally and brought back news, and some visitors came: Paul A. M. Dirac, Niels Bohr, Enrico Fermi, M. S. Vallarta. Mostly we relied on the journals (there were no preprints in those days); so our work was often in parallel with work going on elsewhere.

Another thing that struck me was Oppenheimer's conviction that quantum electrodynamics was wrong, not only in its divergences but also in its finite predictions. The reason, of course, was that these predictions seemed in contradiction with the behavior of the cosmic rays, believed to be electrons and γ rays. Oppie at first disbelieved at mc^2 (the rest energy of the electron), then retreated to 137 mc^2, but could hardly write a paper without a lament. His view colored our work: A naive faith might have made us more resolute in trying to understand the real problems of quantum electrodynamics.

I shall begin earlier than my own arrival in Berkeley, with Oppenheimer's March 1930 paper showing that with the Dirac theory the line shift in atomic spectra was infinite.[1] To quote from the conclusion:

> We have treated these difficulties in some detail, because they show that the present theory will not be applicable to any problem where relativistic effects are important, where, that is, we cannot be guided throughout by the limiting case $c \rightarrow \infty$. The theory can thus not be applied to a discussion of the structure of the nuclei. It appears improbable that the difficulties discussed in this work will be soluble without an adequate theory of the masses of the electron and the proton; nor is it certain that such a theory will be possible on the basis of the special theory of relativity.

That was the tone. In the same issue of the *Physical Review* Oppenheimer discussed Dirac's suggestion, a couple of months old, identifying the antielectron as the proton.[2] Oppenheimer remarked that the same matrix elements of the current came into the calculation of Thomson scattering for particle and for antiparticle; so one could not plausibly get very different answers. He concluded that the proton must also be a Dirac particle, with its own antiparticle. Moreover, were the proton the electron's antiparticle, essentially the same matrix elements would lead to rapid electron-proton annihilation.

This paper was followed in April 1930 by another giving details of the calculation of the lifetime of a positron in matter.[3] It illustrates another Oppenheimer characteristic: His physics was good, but his arithmetic awful. The result was right, aside from a factor $(2\pi)^4$. Three months later Dirac gave a correct and complete treatment.[4] Dirac's paper still referred to "annihilation of electrons and protons."

Oppenheimer's carelessness was more serious in a May 1931 paper with Harvey Hall on the relativistic photoeffect.[5] Chung-yao Chao and also G. T. P. Tarrant had, in 1930, made absorption measurements of ThC'' γ rays in various elements, and they found a 25% excess by the time they reached Pb. But Oppenheimer and Hall said they got 25 times too much. The formula they gave was off by a factor 44/3. The error reinforced Oppenheimer's view that the theory was wrong in the MeV range.

We come now to a series of papers on the high-energy interactions of electrons and γ rays. In October 1931, Oppenheimer and J. F. Carlson used the Møller interaction to calculate the ionization loss of electrons in close collisions, but they did not put it together with the distant collisions to get a complete formula.[6] The "neutron" referred to in the title of the paper is Wolfgang Pauli's, which Oppenheimer heard him report on at the 1931 Ann Arbor summer school: a third element in addition to protons and electrons in nuclei, neutral, mass not much greater than the electron's, and magnetic moment comparable to the proton's. Oppenheimer and Carlson said the ionization produced by such a particle would approach a constant value at high energy, which they thought was in better agreement with cosmic-ray cloud chamber data than the electron's logarithmic increase, with energy, and so they suggested that Pauli's neutrons were the primary component of cosmic rays. They said: "No latitude effect shows primary cosmic rays are neutral." Although a latitude effect was first reported by Jacob Clay in 1927, it was not believed in California in 1931.

The suggestion that the primaries were not γ rays appears to have been taken by Millikan as heresy. A letter from Oppenheimer to Lawrence, written on January 3, 1932, on Oppenheimer's way home from the New Orleans meeting of the American Physical Society, said: "It was like you, Ernest, and very sweet, that you should whisper to me so comforting words about the Wednesday meeting. I was pretty much in need of them, feeling ashamed of my report, and distressed rather by Millikan's hostility and lack of scruple."[7]

A follow-up paper, in September 1932, gave the complete ionization loss formula and said that Hans Bethe had also derived the formula three months earlier.[8] This work was completed the following year by Felix Bloch, who gave the result for a Fermi-Thomas atom.[9] Oppenheimer and Carlson also gave up on "magnetic neutrons," which they said would lose a large fraction of their energy in a collision and would not produce cloud chamber tracks.

Sir James Chadwick discovered the real neutron in February 1932.[10] In March 1933 Anderson found the positron.[11] Evidence soon appeared also in nuclear physics, when Patrick M. S. Blackett and Giuseppe P. S. Occhialini suggested that the excess absorption of the ThC'' γ rays in Pb was due to pair production. Ernest Gray and Tarrant found 1/2 MeV radiation, but also 1 MeV, which was confusing.*

Oppenheimer and Milton Plessett, in July 1933, calculated pair production near threshold and at high energy.[12] They got a wrong factor, as usual, but the answer was not far off for ThC'' γ rays on Pb. They also pointed out the shower mechanism: no multiple production of pairs, but that fast electrons and positrons will themselves produce further pairs. But in the mass-absorption law for cosmic rays, serious deviations would be expected. "One is tempted to see in this discrepancy a failure of the theory when applied to radiation whose wavelength is of the order of e^2/mc^2 which marks the limit of applicability of classical electron theory." Thus doubts moved from $2mc^2$ to $137mc^2$.

Complete formulae for pair production and also bremsstrahlung were given later in 1933 by Walter Heitler and Fritz Sauter, and these were improved, in 1934, by Bethe and Heitler.[13] Also in 1934, Edwin McMillan in Berkeley measured the absorption in light and heavy elements of the 5.4MeV γ ray emitted in the proton-plus-fluorine reaction and showed that the lead absorption curve did indeed have a minimum.[14]

In a February 1934 paper, Oppenheimer and Wendell Furry struggled with the concept of a quantized Fermi field and found reasonable physical consequences.[15] They showed that the theory was symmetrical in electrons and positrons and could be formulated without reference to a filled negative sea. The problems associated with infinite subtractions were, of course, not eliminated. They discussed vacuum polarization and charge renormalization. They said that the remaining finite vacuum polarization effects would not be observable in atoms because they are not larger than the shifts "which arise from our ignorance of the reaction of the electron to its own radiation field"; however, it should be possible to see them in proton-proton scattering. The paper gave the correct formula for the modification of the Coulomb field at short distances. They also said that Lawrence had just returned from Brussels and had shown them Dirac's report to the Solvay congress on charge renormalization (which was published in 1934).[16]

In their paper, Oppenheimer and Furry commented that the pair-production cross section is much too large for the "known penetrating

---

* *Ed. note*: The 1 MeV radiation they reported was not one quantum annihilation radiation, but was due to improper experimental analysis. See H. A. Bethe, "On the Annihilation Radiation of Positrons," *Proc. Roy. Soc. (London) A150* (1935), 129-41.

power of cosmic rays." However, in 1934, C. F. von Weizsäcker, and independently Evan James Williams, showed that, viewed in the right coordinate system, bremsstrahlung and pair production involved only processes at energies of a few mc^2.[17] Oppenheimer, in a paper at the end of 1934, struggled to maintain his lack of faith.[18] He again pointed out the shower mechanism, which would lead to rapid degradation of high-energy particles. He concluded that the formulae were wrong or that there was some other and less absorbable component. This could not be the proton: Equal numbers of positives and negatives were seen, few slow protons and no corresponding antiprotons were seen, and the distribution of δ rays (i.e., recoil electrons) was wrong for protons. But, having argued himself into a correct conclusion (i.e., "some other and less absorbable component"), Oppenheimer refused to draw it; he made a tortured argument that if the correct theory were nonlinear, the presence of high-frequency components could damp low-frequency effects.

I came to Berkeley at this time, in September 1934. Edwin Uehling was working on the vacuum polarization for static fields, following the prescriptions of Dirac and Werner Heisenberg for eliminating singularities, and calculating the level shifts in the hydrogen atom. I extended the work to the general case of time-varying fields and studied the photon propagator.[19] In 1936 I tried some subtraction and regularization procedures, without success.[20] I also worked on two nuclear physics experiments that had significance for particle physics. One was an analysis of Milton White's p-p scattering experiment.[21] I have always believed that Milt did not get sufficient credit. The experiment was crude, but better than the accepted history suggests. A reason, perhaps, is an error of transposition I discovered when I looked up his paper in preparing for this talk. His paper stated that the depth of the p-p interaction, for e^2/mc^2 range, is V = 17.2 MeV, whereas the value I gave him was $V = 12.7$ MeV. This is not much further off on the high side than the value on the low side deduced by Gregory Breit, Edward U. Condon, and Richard D. Present the following year from the experiments of N. P. Hydenburg, Lawrence R. Hafstad, and Merle A. Tuve.[22] On looking back, I wondered why we did not compare the p-p and n-p interactions, until I realized that Milt's original publication preceded any measurement of the n-p cross section.

In 1936, William A. Fowler, L. A. Delsasso, and Lauritsen (this was Fowler's dissertation) measured the energies of light positron emitters and pointed out the mirror nucleus symmetry.[23] As a student of J. H.

Van Vleck, I knew all about Coulomb energy calculations, and I calculated the exchange contributions for Fermi-Thomas and shell models. However, Oppenheimer and Lauritsen decided that the data were not good enough to justify an elaborate theory, and so I did not publish. Fowler complained, at the Minneapolis symposium on the history of nuclear physics, that we stopped him from publishing the formula for the energy of a uniformly charged sphere. The statement actually made in the Delsasso, Fowler, and Lauritsen paper, that the experiment determined the mean value of e^2/r_{ij}, is less illuminating but more sophisticated. Somehow, the Delsasso, Fowler, and Lauritsen paper was overlooked, and Eugene Wigner rediscovered charge symmetry in 1937.

In June 1936, at the Seattle meeting, Oppenheimer gave results of shower calculations, and in February 1937 Oppenheimer and Carlson published their paper on showers.[24] They mentioned that Heitler and Homi J. Bhabha also got similar results, although the Heitler-Bhabha theory was incomplete in that it neglected ionization loss.[25] Oppenheimer and Carlson discussed the application to cosmic-ray showers and bursts and concluded that the phenomena could be accounted for, disagreeing with Heisenberg, who, using the Fermi β-decay interaction as an example, concluded that such theories give multiple production and supposed this responsible for the bursts.[26] They said that Anderson and Seth Neddermeyer had checked the behavior of shower particles to 400 MeV, and "evidence affords absolutely no indication of breakdown of theoretical formulae." So, finally admitting the theory right, they accepted the conclusion rejected two years earlier and said that there is "another cosmic-ray component, slowly absorbed, which is responsible for the continuation of the showers under thicknesses of absorber to which no electron or photon can itself penetrate. Some suggestions we think relevant to the solution of this problem will be discussed in another paper."

Before going to that, I should mention another California contribution at this time to the understanding of quantum electrodynamics. In 1937, Arnold Nordsieck, who got his Ph.D. under Oppenheimer, had a fellowship to work with Bloch at Stanford. Our relations with Bloch were close; in fact, our theoretical seminar was a joint Berkeley-Stanford affair. In 1937, Bloch and Nordsieck published their explanation of the infrared catastrophe.[27] While speaking of Bloch, I shall jump ahead a little: In 1939, Bloch, Oppenheimer, and Sidney M. Dancoff again attacked the divergence problems, studying the corrections of order α to electron scattering. A paper on this was published by Dancoff in 1939.[28] They had some of the right ideas, distinguishing between mass, vertex, and wave function divergences, and years later I heard Bloch lament to Oppenheimer that if Dancoff had not made a mistake in his calculations, "we would have seen the point at that time."

To return to cosmic rays: In May 1937, Anderson and Neddermeyer and, simultaneously, Jabez Street and E. C. Stevenson, announced the identification of a meson in the cosmic rays.[29] A month later, in June 1937, Oppenheimer and I made this promised suggestion: The particles discovered by Anderson and Neddermeyer and Street and Stevenson are those postulated by Hideki Yukawa to explain nuclear forces.[30] (Anderson and Neddermeyer were wiser; they suggested "higher mass states of ordinary electrons.") Yukawa's paper came out in 1935, but we know of no reference to it before our 1937 paper, and a very conscious purpose of our paper was to call attention to Yukawa's idea.[31]

It followed, we said, that the new particles are not primary cosmic rays but are produced by γ rays in nuclear collisions (and by pair production) in the upper atmosphere and are the penetrating component.* Also, the production of showers at sea level and below is due to knock-on electrons from the penetrating component. We calculated the production rate, which was about right. We were not unmindful of the paradox between nuclear interaction and penetrating component, but we could only say that we did not understand the nuclear interaction either, because Yukawa's suggestion did not describe the real nuclear forces.

Our original principal point, though, was that if Yukawa were right about the finite β-decay lifetime of the meson, this should show up in a deviation from the mass-absorption law between air and solid materials. But this did not get published. The reason was that Oppenheimer took our paper to Caltech and showed it to Millikan and returned much chastened; Millikan had objected strenuously, because his air-versus-water absorption measurements had already proved, he said, that no such effect existed. Oppenheimer said we should rewrite in a form less offensive to Millikan. Uncharacteristically, he left the rewriting to me, and after he had rejected about the fourth draft, I said in

---

* *Ed. note*: The actual phrase: "These particles need not then be primary cosmic rays, but may be ejected from nuclei by γ-rays (and formed by pair production) in the upper atmosphere."

Left to right: J. R. Oppenheimer, H. Yukawa, and S. Tomonaga at the Institute for Advanced Study, Princeton, November 1949 (credit: Ziro Maki and Michiji Konuma from Yukawa Hall Archival Library, Kyoto).

exasperation, "Let's cut it out completely." Oppenheimer blinked at me and said, "Do you really think so?" And out it came.

A year later, in 1938, Heisenberg and H. Euler made the same point, Bruno Rossi proved the air-solid absorption difference, and Blackett showed that it explained the temperature effect.[32] Millikan apparently regretted the position he had taken. In the Smith-Weiner book of letters is one from Oppenheimer to Millikan, dated January 1, 1939:[33]

> Thank you for your good note. . . . You are right about the radioactivity of the mesotron. I have been thinking of it for

two years now, and gave a seminar on it while Bohr was with us. The only evidence we had at the time came from Rossi's work in Eritrea, which he has just recently interpreted in this sense in a letter to *Nature*. We felt that the extension of your own earlier work on air and water absorption would provide a so much cleaner and less ambiguous test of the idea (which as you can guess rests on no very sure theoretical basis) that, perhaps mistakenly, we did not publish it, but just urged Bowen to get the job done.

Returning to our point in the 1937 paper about the secondary origin of the mesons: It raises the question of the nature of the primaries. After Clay's observation of the latitude effect, Carl Störmer, in 1930, applied his theory of the orbit of charged particles in a magnetic dipole field to the problem, and in 1933 G. Lemaître and Vallarta developed calculations including the shadowing effects of the earth.[34] In 1933, Arthur H. Compton also found a latitude effect, and in the following years Millikan and Neher made a world survey with ionization chambers and, with Bowen, made balloon measurements to very high altitudes.[35] With the advent of shower theory calculations of the transition curve, they felt in position to draw definitive conclusions. In a February 1938 paper they said that the primaries have been proved to be electrons and positrons and that the east-west effect shows that they are predominantly positrons.[36] They also said that photons produce a penetrating particle "which may through another collision, retransform its energy into electron-photon showers of the usual type." Apparently Millikan's objections to our paper went deeper than Oppenheimer had admitted to me, for to this, Millikan, Bowen, and Neher added a footnote: "We had discussed this suggestion at length at the Norman Bridge Laboratory as soon as, in February and March 1936, we had analyzed the results of the Madras flights made in October 1935. Oppenheimer and Serber of this group have already published a brief note embodying this suggestion."

Later in 1938 I wrote a paper sharpening the arguments of Millikan, Bowen, and Neher.[37] Hartland Snyder had just improved shower theory by using a Mellin transformation to solve the equations, but he still used approximate cross sections.[38] I used Snyder's method with the exact Bethe-Heitler cross sections and compared the results with the measurements of Millikan, Bowen, and Neher, as shown in Figure 12.1.

The difference curve gives the effect of a definite energy band of

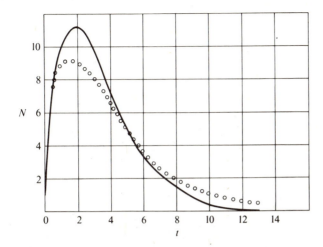

Figure 12.1. Multiplication curve for an electron of 11 BeV averaged over all directions of incidence. The circles give the San Antonio-Madras difference curve of Bowen, Millikan, and Neher. From R. Serber in *Phys. Rev. 54* (1938), 319, Fig. 1.

charged primaries with a mean energy of 11 BeV. The normalization, made on the assumption that all the energy eventually shows up as ionization, is absolute: $N = 1$ represents one incident primary (this normalization makes the curves have the same area). One sees the penetrating component; the argument that it is not primary is this: It gives about $N = 1/2$ at $t = 11$, and if the number of incident electrons were consequently reduced by half, we should only have half the multiplication, well below the observed curve.

Figure 12.1 appeared so convincing that it probably delayed recognition of the truth, that the primaries are protons. This was argued by Thomas H. Johnson and J. Griffiths Barry in 1939 on the basis of the altitude dependence of the east-west effect and was proved, in 1941, by Marcel Schein, William P. Jesse, and Ernesto Wollan, who showed that there is a hard component increasing to the highest altitudes.[39]

Returning to the 1937 paper, to the fact Yukawa's original theory did not give realistic forces. In December 1937, Yukawa and Shoichi Sakata published a paper on scalar meson theory.[40] Yukawa's 1935 paper was a vector theory, that is, an analogy with the electrostatic potential. The first attempt to get realistic forces, I think, was a paper I gave at the Stanford meeting of the American Physical Society, also in December

1937.[41] I supposed both scalar and vector mesons: the scalar to give central forces; the vector, with tensor coupling to the nucleons, to give the right spin dependence and also to give the anomalous nuclear magnetic moments. I said that the theory was satisfactory on the latter two points, but I had trouble with the sign of the central force and, more particularly, with the charge independence. In April 1938, a paper by Yukawa, Sakata, and Mituo Taketani gave the vector theory with both vector and tensor couplings, and in May 1938, N. Kemmer gave all these and also pseudoscalar theory.[42] Yukawa and Sakata and Kemmer also suggested the neutral meson to get charge independence.

It is remarkable that all of us derived tensor forces (Kemmer said there appears to be nothing in nuclear theory against them), but no one thought of the effect on the deuteron. Jerome Kellogg, Norman Ramsey, Isidor Rabi, and Jerrold Zacharias found the quadrupole moment only in 1939.[43] I asked Rabi about it, and he was surprised he had not known of these papers at the time, and even more surprised that no one had told him in the ensuing 40 years.

On charge independence: At the same time, Oppenheimer and I found strong nuclear evidence. In a February 1938 paper, we discussed the reaction $p + {}^{11}B$ leading to a narrow level in ${}^{12}C$ that decays by emission of both 16-MeV γ rays and long-range α rays.[44] We showed that this must be due to a strong isotopic spin selection rule and said that the ${}^{12}C$ level was $T = 1$ and would have an analogue state in ${}^{12}B$ (we thought the ground state, but actually it is the first excited state). The ideas came from Professor Breit; we found the first applications. Our paper forgotten, analogue states were rediscovered 14 years later by Robert K. Adair.

At the end of 1939, Oppenheimer, Snyder, and I returned to the question of the showers produced by the penetrating component.[45] We showed that for spin-1/2 mesons, knock-on electrons account for showers up to 20 BeV and bremsstrahlung in Pb for bursts to 250 BeV, although we mistakenly thought the latter had the wrong Z dependence. But by this time we were convinced the mesons were vectors, which gave too much bremsstrahlung. Following criteria given by Heisenberg, we questioned the validity of the Born approximation of bremsstrahlung for spin 1. A more careful treatment, in 1941, by Oppenheimer's students Robert F. Christy and Shuichi Kusaka got good agreement with Schein and Piara S. Gill's bursts in Pb for meson spin 0 or 1/2, but it proved one could not reduce the theoretical estimate for spin one sufficiently.[46] And Herbert C. Corben and Julian Schwinger

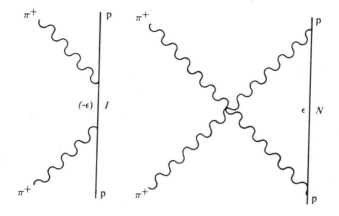

Figure 12.2. Feynman diagrams for the scattering of a positive π meson (π^+) by a proton (p). N denotes a neutron (of energy ϵ, and I denotes a positive doubly charged nucleon isobar (of energy $-\epsilon$). Drawn by author.

showed that nonminimal electromagnetic coupling was even worse.[47] A following note by Oppenheimer concluded that the only remaining possibility was a pseudoscalar meson, a correct conclusion on misleading evidence (as Rabi said, "Who ordered the μ meson?").[48] Oppenheimer also said that a pseudoscalar would give the right sign of the tensor force; that part of the argument was correct.

Back to the questions raised by our 1937 paper: how the nuclear mesons could also be penetrating led to considerations relevant to particle theory, although not to the question addressed.

Yukawa and Sakata, in their paper on scalar mesons, concluded that there was no difficulty; the mesons would have a small scattering cross section. This was an error; they thought it was like the Klein-Nishina case. There are two Feynman diagrams for the latter, similar to those shown in Figure 12.2. In the limit of nucleon mass going to infinity, the energy denominators are $\pm \epsilon$, and the diagrams cancel. This will also be right for neutral mesons, but for charged mesons only one diagram is allowed because of charge conservation; in the case shown, the other will require a nucleon of charge 2. Missing this point, Yukawa and Sakata gave a cross section three orders of magnitude too small. This situation led Bhabha, in 1940, to suggest nucleon charge isobars, and Heitler and S. T. Ma suggested both charge and spin isobars (e.g., a doubly charged isobar will restore the first diagram of Figure 12.2).[49]

Another attack was by Heisenberg in 1939, who considered as an example the neutral pseudoscalar theory and argued that the inertial reaction due to self-interaction of the spin of the nucleon with the meson field would have a damping effect and reduce the cross section.[50] He used a nonrelativistic classical theory, with extended source for the nucleon, and indeed found a small cross section, proportional to the square of the source radius. I thought Heisenberg's model might do some good for nuclear forces, relieving the $1/r^3$ singularity of the tensor forces, and I began work on that problem.

In 1940, Gregor Wentzel gave his strong-coupling theory, nonrelativistic for the nucleon, like that of Heisenberg, but a quantum mechanical treatment.[51] This, indeed, as he showed, led to nucleon isobars. Oppenheimer and Schwinger, in 1941, considered charged scalar and neutral pseudoscalar strong-coupling theories, got isobar energies, and calculated scattering cross sections.[52] These were still large in the scalar case, but for the pseudoscalar case they confirmed Heisenberg's result. At the same time, Dancoff and I, at Illinois, also did the same theories and calculated the nuclear forces for the pseudoscalar.[53] We found that as soon as the tensor force became larger than the isobar, splitting it destroyed its tensor character, and it became a spin-independent central force, but still with a $1/r^3$ behavior; so we were even worse off than when we started.

In the fall of 1941, Dancoff went to the Institute for Advanced Studies to work with Pauli. Pauli became interested in what we had been doing and put Dancoff to work on the symmetrical pseudoscalar theory. They found that it predicted the 3,3 resonance, later to become famous.[54]

Finally, I would emphasize that progress in physics depends more often on the development of experimental techniques than on theoretical ideas. So perhaps the most useful thing I did for particle physics was my work in 1939 and 1940 with Donald W. Kerst on the betatron, which laid the foundation for the design of future particle accelerators.

I leave my narrative dangling here: At that point we all went off to war.

Notes

1 J. R. Oppenheimer, "Note on the Theory of the Interaction of Field and Matter," *Phys. Rev. 35* (1930), 461-77.
2 J. R. Oppenheimer, "On the Theory of Electrons and Protons," *Phys. Rev. 35* (1930), 562-3.

3 J. R. Oppenheimer, "Two Notes on the Probability of Radiative Transitions," *Phys. Rev. 35* (1930), 939-47.

4 P. A. M. Dirac, "On the Annihilation of Electrons and Protons," *Proc. Cambridge Phil. Soc. 26* (1930), 361-75.

5 Harvey Hall, "Relativistic Theory of the Photoelectric Effect. Part I. Theory of the K-Absorption of X-Rays"; Harvey Hall and J. R. Oppenheimer, "Part II. Photoelectric Absorption of Ultragamma Radiation," *Phys. Rev. 38* (1931), 57-79.

6 J. F. Carlson and J. R. Oppenheimer, "On the Range of Fast Electrons and Neutrons," *Phys. Rev. 38* (1931), 1787-8. *Ed. note*: For a discussion of this paper and its relationship to Pauli's suggestion, see Laurie M. Brown, "The Idea of the Neutrino," *Phys. Today 31* (1978), 23-8.

7 Robert Oppenheimer, *Letters and Recollections*, ed. by Alice Kimball Smith and Charles Weiner (Cambridge, Mass.: Harvard University Press, 1980), p. 147.

8 J. F. Carlson and J. R. Oppenheimer, "The Impacts of Fast Electrons and Magnetic Neutrons," *Phys. Rev. 41* (1932), 763-92; H. Bethe, "Bremsformel für Elektronen relativistischer Geschwindigkeit," *Z. Physik 76* (1932), 293-9.

9 F. Bloch, "Bremsvermögen von Atomen mit mehreren Elektronen," *Z. Physik 81* (1933), 363-76.

10 J. Chadwick, "Possible Existence of a Neutron," *Nature 129* (1932), 312.

11 Carl D. Anderson, "The Positive Electron," *Phys. Rev. 43* (1933), 491-4.

12 J. R. Oppenheimer and M. S. Plesset, "On the Production of the Positive Electron," *Phys. Rev. 44* (1933), 53-5.

13 W. Heitler and F. Sauter, "Stopping of Fast Particles with Emission of Radiation and the Birth of Positive Electrons," *Nature 132* (1933), 892; H. Bethe and W. Heitler, "On the Stopping of Fast Particles and on the Creation of Positive Electrons," *Proc. Roy. Soc. (London) A146* (1934), 83-112.

14 Edwin McMillan, "Some Gamma-Rays Accompanying Artificial Nuclear Disintegrations," *Phys. Rev. 46* (1934), 868-73.

15 W. H. Furry and J. R. Oppenheimer, "On the Theory of the Electron and Positron," *Phys. Rev. 45* (1934), 245-62.

16 P. A. M. Dirac, "Discussion of the Infinite Distribution of Electrons in the Theory of the Positron," *Proc. Cambridge Phil. Soc. 30* (1934), 150-63.

17 C. F. von Weizsäcker, "Ausstrahlung bei Stössen sehr schneller Elektronen," *Z. Physik 88* (1934), 612-25; E. J. Williams, "Nature of the High Energy Particles of Penetrating Radiation and Status of Ionization and Radiation Formulae," *Phys. Rev. 45* (1934), 729-30.

18 J. R. Oppenheimer, "Are the Formulae for the Absorption of High Energy Radiations Valid?" *Phys. Rev. 47* (1935), 44-52.

19 E. A. Uehling, "Polarization Effects in the Positron Theory," *Phys. Rev. 48* (1935), 55-63; R. Serber, "Linear Modifications in the Maxwell Field Theory," *Phys. Rev. 48* (1935), 49-54.

20 Robert Serber, "A Note on Positron Theory and Proper Energies," *Phys. Rev. 49* (1936), 545-50.

21 Milton G. White, "Collisions of High Energy Protons in Hydrogen," *Phys. Rev. 47* (1935), 573-4.

22 G. Breit, E. U. Condon, and R. D. Present, "Theory of Scattering of Protons by Protons," *Phys. Rev. 50* (1936), 825-45; M. A. Tuve, N. P. Heydenberg, and L. R. Hafstad, "The Scattering of Protons by Protons," *Phys. Rev. 50* (1936), 806-25.

23 W. A. Fowler, L. A. Delsasso, and C. C. Lauritsen, "Radioactive Elements of Low Atomic Number," *Phys. Rev. 49* (1936), 561-74.

24 J. R. Oppenheimer, "On the Elementary Interpretation of Showers and Bursts,"

Phys. Rev. 50 (1936), 389; J. F. Carlson and J. R. Oppenheimer, "On Multiplicative Showers," *Phys. Rev. 51* (1937), 220-31.

25 H. J. Bhabha and W. Heitler, "Passage of Fast Electrons through Matter," *Nature 138* (1936), 401; H. J. Bhabha and W. Heitler, "The Passage of Fast Electrons and the Theory of Cosmic Showers," *Proc. Roy. Soc. (London) A159* (1937), 432-58.

26 W. Heisenberg, "Zur Theorie der 'Schauer' in der Höhenstrahlung," *Z. Physik 101* (1936), 533-40.

27 F. Bloch and A. Nordsieck, "Note on the Radiation Field of the Electron," *Phys. Rev. 52* (1937), 54-9.

28 S. M. Dancoff, "On Radiative Corrections for Electron Scattering," *Phys. Rev. 55* (1939), 959-63.

29 Seth H. Neddermeyer and Carl D. Anderson, "Note on the Nature of Cosmic-Ray Particles," *Phys. Rev. 51* (1937), 884-6; J. C. Street and E. C. Stevenson, "Penetrating Corpuscular Component of the Cosmic Radiation," *Phys. Rev. 51* (1937), 1005.

30 Hideki Yukawa, "On the Interaction of Elementary Particles. I," *Proc. Phys.-Math. Soc. Japan 17* (1935), 48-57.

31 J. R. Oppenheimer and R. Serber, "Note on the Nature of Cosmic-Ray Particles," *Phys. Rev. 51* (1937), 113.

32 H. Euler and W. Heisenberg, "Theoretische Geschichtspunkte zur Deutung der kosmischen Strahlung," *Ergeb. Exakt. Naturw. 17* (1938), 1-69; Bruno Rossi, "The Disintegration of Mesotrons," *Rev. Mod. Phys. 11* (1939), 296-303; P. M. S. Blackett, "On the Instability of the Barytron and the Temperature Effect of Cosmic Rays," *Phys. Rev. 54* (1938), 973-4.

33 See Robert Oppenheimer, *Letters* (Note 7), p. 206.

34 J. Clay, "Penetrating Radiation," *Proc. K. Akad. Amsterdam 30* (1927), 1115-27; Carl Störmer, "Periodische Elektronenbahnen im Felde eines Elementarmagneten und ihre Anwendung auf Brüches Modellversuche und auf Eschenhagens Elementarwellen des Erdmagnetismus," *Z. Astrophys. 1* (1930), 237-74; G. Lemaître and M. S. Vallarta, "On Compton's Latitude Effect of Cosmic Radiation," *Phys. Rev. 43* (1933), 87-91.

35 Arthur H. Compton, "A Geographic Study of Cosmic Rays," *Phys. Rev. 43* (1933), 387-403.

36 I. S. Bowen, R. A. Millikan, and H. Victor Neher, "New Evidence as to the Nature of the Incoming Cosmic Rays, Their Absorbability in the Atmosphere, and the Secondary Character of the Penetrating Rays Found in Such Abundance at Sea Level and Below," *Phys. Rev. 53* (1938), 217-23.

37 R. Serber, "Transition Effects of Cosmic Rays in the Atmosphere," *Phys. Rev. 54* (1938), 317-20.

38 H. Snyder, "Transition Effects of Cosmic Rays in the Atmosphere," *Phys. Rev. 53* (1938), 960-5.

39 Thomas H. Johnson and J. Griffiths Barry, "The East-West Symmetry of the Cosmic Radiation at Very High Elevations Near the Equator and Evidence that Protons Consitute the Primary Particles of the Hard Component," *Phys. Rev. 56* (1939), 219-26; Marcel Schein, William P. Jesse, and E. O. Wollan, "The Nature of the Primary Cosmic Radiation and the Origin of the Mesotron," *Phys. Rev. 59* (1941), 615.

40 Shoichi Sakata and Hideki Yukawa, "On the Interaction of Elementary Particles. II,," *Proc. Phys.-Math. Soc. Japan 19* (1937), 1084-93.

41 R. Serber, "On the Dynaton Theory of Nuclear Forces," *Phys. Rev. 53* (1938), 211.

42 Hideki Yukawa, Shoichi Sakata, and Mitsuo Taketani, "On the Interaction of Elementary Particles. III," *Proc. Phys.-Math. Soc. Japan 20* (1938), 319-40. N. Kem-

mer, "Quantum Theory of Einstein-Bose Particles and Nuclear Interaction," *Proc. Roy. Soc. (London) A166* (1938), 127-53.

43 J. M. B. Kellogg, I. I. Rabi, N. F. Ramsey, Jr., and J. R. Zacharias, "An Electrical Quadrupole Moment of the Deuteron," *Phys. Rev. 55* (1939), 318-19.

44 J. R. Oppenheimer and R. Serber, "Note on Boron Plus Proton Reactions," *Phys. Rev. 53* (1938), 636-8.

45 J. R. Oppenheimer, H. Snyder, and R. Serber, "The Production of Soft Secondaries by Mesotrons," *Phys. Rev. 57* (1940), 75-81.

46 R. F. Christy and S. Kusaka, "The Interaction of γ-Rays with Mesotrons," *Phys. Rev. 59* (1941), 405-14; R. F. Christy and S. Kusaka, "Burst Production by Mesotrons," *Phys. Rev. 59* (1941), 414-21; Marcel Schein and Piara S. Gill, "Burst Frequency as a Function of Energy," *Rev. Mod. Phys. 11* (1939), 267-76.

47 H. C. Corben and Julian Schwinger, "The Electromagnetic Properties of Mesotrons," *Phys. Rev. 58* (1940), 953-68.

48 J. R. Oppenheimer, "On the Spin of the Mesotron," *Phys. Rev. 59* (1941), 462.

49 H. J. Bhabha, "On Elementary Heavy Particles with Any Integral Charge," *Proc. Indian Acad. Sci. 11* (1940), 347-68; W. Heitler and S. T. Ma, "Inner Excited States of the Proton and Neutron," *Proc. Roy. Soc. (London) A176* (1940), 368-97.

50 W. Heisenberg, "Zur Theorie der explosionsartigen Schauer in der kosmischen Strahlung. II," *Z. Physik 113* (1939), 61-86.

51 Gregor Wentzel, "Zum Problem des statischen Mesonfeldes," *Helv. Phys. Acta 13* (1940), 269-308; G. Wentzel, "Zur Hypothese der höheren Proton-Isobaren," *Helv. Phys. Acta 14* (1941), 3-12.

52 J. R. Oppenheimer and Julian Schwinger, "On the Interaction of Mesotrons and Nuclei," *Phys. Rev. 60* (1941), 150-2.

53 S. M. Dancoff and R. Serber, "Nuclear Forces in the Strong Coupling Theory," *Phys. Rev. 61* (1942), 394; R. Serber and S. M. Dancoff, "Strong Coupling Mesotron Theory of Nuclear Forces," *Phys. Rev. 63* (1943), 143-61.

54 W. Pauli and S. M. Dancoff, "The Pseudoscalar Meson Field with Strong Coupling," *Phys. Rev. 62* (1942), 85-108.

13 The observation of the leptonic nature of the "mesotron" by Conversi, Pancini, and Piccioni[1]

ORESTE PICCIONI

Born 1915, Siena, Italy; *laurea*, University of Rome, 1938; elementary particle physics experimentalist; University of California at San Diego.

The pioneers: Montgomery and associates and Rasetti

In the spring of 1938, a few months before his definite transfer to the United States, Enrico Fermi bestowed on me the *laurea in fisica, 110/110 con lode*. Considering my knowledge at that time, I clearly had been given the *lode* on credit. I felt the debt had perhaps been paid off when I heard in 1946 that Fermi had given a seminar on the experiment that Marcello Conversi, Ettore Pancini, and I had done, joking that he would not "dare to pronounce those names."[2]

The experiment, the last of a series of four on stopped muons, showed that the negative "mesotrons" stopped in carbon were not absorbed, but decayed normally, like the positive ones. In the interpretation given by Fermi at the seminar, in which he also cited a theoretical paper of Sin-itiro Tomonaga and Toshima Araki, our experiment inevitably disproved the belief that the mesotron was the particle predicted by Hideki Yukawa as the mediator of nuclear forces and of nuclear β radioactivity.[3] As a consequence, the mesotron was a particle that represented a new family in our theories, and it was rebaptized "muon," with its family name changed from meson to lepton, because of its unwillingness to be absorbed. For 30 years the muon reigned as the only known heavy lepton, until the τ appeared in 1976.

My introduction to the mesotron had occurred in 1940 in the Italian mountains near the Matterhorn, when I was helping Gilberto Bernar-

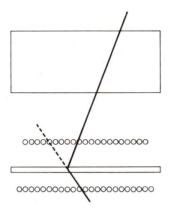

Figure 13.1. Schematic layout of the experiment of Montgomery and associates.[5] Circles represent Geiger-Müller counters. The large box is a lead filter; the small box is the muon-stopping lead plate. A muon trajectory is shown, decaying into an electron and a neutrino (dotted line). From C. G. Montgomery et al. in *Phys. Rev. 56* (1939), 636, Fig. 1.

dini, Bruno Ferretti, and others, with medium enthusiasm except for skiing. It was only when I saw the 1941 paper of Franco Rasetti that my enthusiasm rose to a high level.[4] Luckily, Conversi joined me, and we studied that experiment together, as well as the one of the Montgomeries, Ramsey, and Cowie, who must be considered the very first pioneers on the subject, despite their negative result.[5]

The layout of their counters (Figure 13.1) consisted, with rudimentary simplicity, of only two trays of counters connected to a true "delayed coincidence" circuit, very simply conceived. They studied the rate of spuriously delayed counter pulses but could not reduce them beyond the level of 6% of all nonstopping muons. Consequently, they expected a signal-to-noise ratio of 23 true decays to 140 spurious ones (it was actually less than 10 to 140). These authors did not find muon decays, but they showed so clearly, and discussed quantitatively, the problem of the spurious delays that the authors of the next effort could easily see that a tray of veto counters at the bottom would be an adequate remedy. Thus Rasetti made the muon beam more collimated, so that it avoided the electron counters and only a few veto counters would be needed to eliminate the spurious delays. The decay electrons were detected at the sides of the stopping material (Figure 13.2).

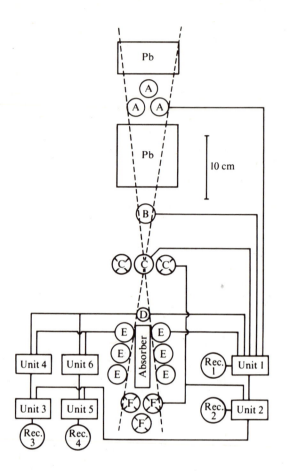

Figure 13.2. Rasetti's experiment.[4] Units 1, 4, and 6 register FC coincidences with resolving times t_2, t_3, and t_4 equal to 15, 2, and 1 μsec, respectively. From F. Rasetti in *Phys. Rev. 60* (1941), 199, Fig. 1.

He avoided the task of building a true delayed coincidence circuit by counting simultaneously the rates n_4, n_3, and n_2 of three prompt coincidence circuits with resolving times, respectively, $t_4 = 1$, $t_3 = 2$, and $t_2 = 15$ μsec, so that $n_3 - n_4$ gave the number of coincidences with a delay between 2 and 15 μsec. There was no information on whether the muon or the electron had arrived first. With only two time channels, the experiment could not prove the exponential decay, but making that assumption, the measured mean life was 1.5 ± 0.3 μsec. (I think the detailed data actually show 1.5 ± 0.2, because $n_2 - n_3$ and $n_2 - n_4$ are not independent; see the work of Rasetti.[4])

Remembering that the spurious delays observed by Montgomery and associates had a typical value of microseconds, one could not assume that Rasetti, who wrote very cautiously on this point, had indeed observed muon decays. From the correct value of the mean life, it follows that he had probably more than 40% spurious counts in the first channel. At that time, however, Rasetti's experiment appeared encouraging, and it illustrated for future authors a better geometrical scheme (though of much lower counting rate) than that of Montgomery and associates.

Our electronic development

Conversi and I believed strongly that in both of the previous experiments inadequate electronics had been used. To measure and work with time differences like 10^{-6} sec, the intrinsic time resolution of the electronics should have been, in our opinion, less than 10^{-7} sec, and the pulses should have rectangular, not triangular, shapes. We introduced for our own use the concept of the rise time, and we kept our eyes on it, rejecting once and for all the popular style of making do with slow rises at the price of critical adjustments of voltages. We blamed that old style for the trouble of Montgomery and associates, and we believed that if we could build an adequate electronic apparatus, we could afford to go back to their geometrical layout, which gave a much higher counting rate than the "defensive" assembly of Rasetti.

We thus needed small values R for the resistors on the plates of our vacuum tubes to charge rapidly the "parasitic" capacitances C and to obtain small rise times RC. To have sufficient pulse height, we then needed large current variations in the resistors, which could be obtained with the large transconductance tubes used in the developing field of television. We started experimenting with the RCA and Philips television tubes, being especially fascinated by the secondary emission tube EE50 made by Philips.

In our institute, electronics had at best a second-class status, for instance with respect to the time-honored trade of glassblowing, which I sincerely hated. As a consequence, our being passionately involved in that art made us separate and independent from our professors. When Rome was bombed, we took refuge in a classroom half below ground level at the Liceo Virgilio, close to the war-immune Vatican and miles from the university. We took our apparatus, a few boxes of resistors, a scope, and soldering irons.

There we heard on the radio the strangely worded announcement of the armistice. I was an army officer and thus was desired by the German Army. I left Rome with others in an attempt to cross the front line and was arrested by the Germans. After a month, during which the activity had stopped, I returned to the Virgilio to work and to hide with Conversi and our friend Mezzetti. There my unforgettable friend Franco Lepri, a bright electronics expert, would come for long consultations, and members of the underground would bring us gloomy news like that of the massacre in Fosse Ardeatine. Edoardo Amaldi paid a few visits to "our laboratory." Bernardini had left Rome. Thus, times were not pleasant, and our work was the only pleasure we had.

We continued, designing from scratch every element of our electronic apparatus. We grew fond of a circuit made with two vacuum tubes, the second reacting on the first one to reinforce its input pulse.[6] Thus the fluctuations of the Geiger-Müller counters were no longer felt so that artificial periodic pulses could be used to study and to calibrate the time response of the whole apparatus. Although we designed this circuit without any help from books, we discovered later that a related circuit was described in the literature to produce a continuous oscillation and was called a multivibrator. The proper name, as used today, is univibrator.

The most important use of the multivibrator was to introduce between the muon counters and the electron counters a reliably constant delay that we could vary easily by changing a little plug-in unit containing two resistors and one capacitor. We could change the delays frequently enough to average all unwanted fluctuations. We also used the multivibrator to measure, as a routine check, the gradual decrease of the coincidence rate versus increasing delays, from zero to 1.5 μsec, which was the most sensitive test of the ensemble. It also gave us the actual delay of one branch of counters with respect to another, a parameter that was essential to our second experiment, testing the Tomonaga effect.

Another circuit we developed was a twofold coincidence based on the Philips tube EE50.[7] The secondary emission electrodes of that tube and the plate were equivalent to a switch, and the circuit worked as two switches in series, with a better rise time than that offered by the classic Rossi circuit.

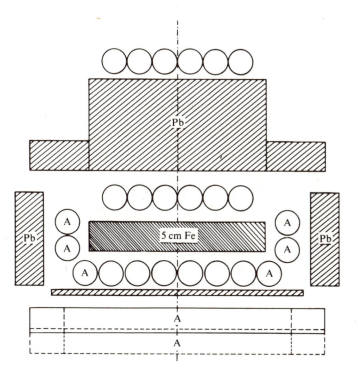

Figure 13.3. Layout of Geiger-Müller counters and absorbers for our experiments 1 and 2.[8,11] From top, the first and second sets of six counters detect an incoming muon. If the muon stops in the absorber (5 cm of iron in the figure) and emits an electron at a later time, a delayed coincidence is counted. The *A* counters veto a count if the muon crosses them. The thin lead slab prevents the decay electrons from vetoing themselves. From M. Conversi and O. Piccioni in *Phys. Rev. 70* (1946), 861, Fig. 1.

Seeing the decay of the mesotron

In hindsight, some of our religious electronic concerns may have been excessive, but when we finally looked at our decay electron rate versus delay (Figures 13.3 and 13.4), we did not regret any part of our labor.[8] The goal of proving the exponential decay had been achieved. The good alignment of the four points left no doubt that our measured mean life of 2.3 μsec ± 6.5% (6% off the present value 2.197) was correct, despite being out of the range of the experiments of Rasetti (1.5 ± 0.2) and of Pierre Auger, Roland Maze, and Robert

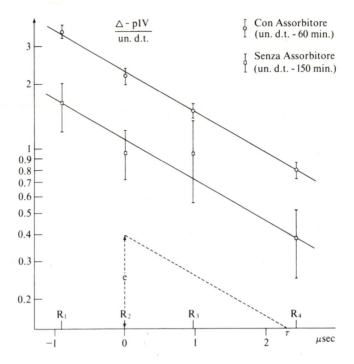

Figure 13.4. Exponential decay of muons with 5-cm Fe absorber (top points) and without it (bottom points). From M. Conversi and O. Piccioni, in *Nuovo Cimento IX* (1944), 67, Fig. 20.

Chaminade (1 ± 0.3). We had received a report of the French work during our preparations for the experiment.[9]

It gave us a warm feeling to realize that we had been the first to prove that the mesotron decayed, and decayed with a mean life and energy of the decay product quite proper for a Yukawa particle, and that practically all of the penetrating cosmic-ray particles had that property.

Actually, we were only the first within that part of the world with which we were in communication. The Nazis had given us an advantage of some two years by chasing Bruno Rossi from Padua and keeping him away from his researches, but despite that, at Cornell, he saw the meson decay about a year before we did.[10] Our results, which were delayed in being published in *Nuovo Cimento* until April 1944 because of the war, appeared in the *Physical Review* in 1946. Understandably, they were largely ignored. The French group of Chaminade, André

Fréon, and Maze reported in 1944 a measured mean life of 2.2 ± 0.2 μsec, with an experiment also done during the hard times of the war.[11]

Confirmation of the Tomonaga theory

Our second experiment[12] was to check the prediction of Tomonaga and Araki that negative muons would be absorbed by nuclei before they decayed. Because by that time previous work had well established that 55% of the mesotrons were positive and 45% negative, we planned to show that only 55% of the stopped particles decayed. We had to estimate the fraction of decay electrons emerging from the stopping plate. Fortunately, our wide Montgomery-type geometrical arrangement allowed the use of a plate of horizontal dimensions much larger than its thickness, for which a reliable calculation could be done, but we needed the value E_{el} of the energy of the decay electrons, for which no direct measurements existed. It was common procedure to assume that the muon decayed into two bodies and that all electrons had the same energy, 40 to 50 MeV. Indeed, experiments at various altitudes, including those of Bernardini and associates of the Rome group, indicated such a value.

However, to minimize the effect of such an uncertainty, we used an iron stopping plate only 0.6 cm thick, so that the electron absorption would be small. Moreover, our geometry allowed a comparison with our previous data taken with 5 cm of iron, from which we derived the (effective) "penetration" of the electrons in iron. We obtained 2.5 cm, which indicated a comfortably small loss for electrons generated in the 0.6-cm plate. The result was that the fraction of the mesons decaying was $\eta = 0.49 \pm 0.07$. Taking into account possible systematic errors, we had $\eta < 0.56 \pm 0.08$. Tomonaga's prediction that negative muons were captured was thus confirmed. Our precaution of measuring the average penetration reduced the effect of our ignorance of the electron energy spectrum. Rasetti had found $\eta = 0.42 \pm 0.15$ using aluminum, and Maze and Chaminade, also using aluminum, had found $\eta = 1$. Their inconsistency was no surprise, however, because neither of those pioneer works had displayed the exponential decay or used sufficient precautions for such a difficult measurement.

Adding the lenses

Reflecting on our results, we considered the obvious possibility of using the Rossi-Puccianti lenses (Figure 13.5). In fact, a curious (but not unheard of) type of question arose concerning the "ownership of

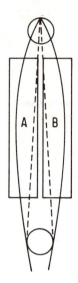

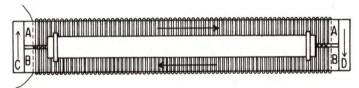

Figure 13.5. Rossi-Puccianti lenses.[13] The side view (at top) shows the trajectory of a "wanted" charge sign (solid line) and of an "unwanted" one. The top view shows the direction of the magnetic field and the position of the top Geiger-Müller counter. From B. Rossi, in *Nature 128* (1931), 300.

an idea." Back in 1940, Bernardini had used the Rossi lenses (described in 1931, 12 years before our work) to study the positive-negative ratio of muons and other subjects.[13] Conversi had been his first and major collaborator, and everyone at the institute was very familiar with those simple devices and with their ability to separate positive from negative particles to some extent.[14] Because Tomonaga's effect also consisted in a difference between negative and positive particles, the qualitative notion of using the lenses to check the theory was rather obvious. Apparently, much before the time when we had any reason to consider the lenses, Pancini had expressed to someone the simple notion of using them. Conversi and I held that given the obviousness of that suggestion, we should feel free to use our apparatus and add the lenses without investigating the priority of anyone. However, when the whole apparatus with the lenses was ready to count, and Pancini came back to Rome, after a harsh time in northern Italy, we invited him to join us, burying the question of "who had said it first." We gained an excellent collaborator.

The adventurous idea of using solid magnetized iron to study the high-energy cosmic particles was first described (in 1929) by Dmitry Skobeltzyn, despite the fact that at that time such particles were supposed to be electrons.[15] The message was read by Mott-Smith, in the United States, who built an iron magnet and tried it. Independently of those authors (as asserted by Mott-Smith), Rossi also tried at first simple magnetized iron pieces like Mott-Smith. But then, with the collaboration of Professor Luigi Puccianti of Pisa (the professor of Fermi, Bernardini, and myself), he originated in 1931 the very clever compact design of Figure 13.5. The magnetic field circulated as shown in Figure 13.5, so that the iron bars worked as shown, focusing for one sign (solid-line trajectories) and defocusing for the opposite sign (dotted lines). Rossi used the lenses to determine the sign of the cosmic particles, but because of the approximate equality of the positive and negative components (55 to 45), the result was not conclusive. The question whether **B** or **H** acted on the muon was debated and was gradually resolved in favor of **B**.[16]

Placing two lenses on top of each other, Bernardini and Conversi confirmed the charge sign selection expected according to calculations made by Gian Carlo Wick.[17] However, the use of the lenses to select mesons that emerged with an energy low enough to be stopped in our stopper was new, and not at all guaranteed to succeed because of the greater effect of Coulomb scattering. We tried exactly the same lens design used by Rossi, which had also been used by Bernardini without changes (Figure 13.6). The reduction in counting rate posed some new problems that were successfully met by our electronics instrument. The result was an hourly rate of 0.33 ± 0.04 converging positive muons and 0.08 ± 0.02 for negative.[18] A rate of 0.27 for negatives was expected if negative muons decayed like the positive ones. The rejection rate of the lenses, taking into account that negative muons are all absorbed in iron, was thus only $0.33/.08 = 4$. However, our second experiment and its conclusions were clearly confirmed.

Why carbon?

Once we were sure of the capture of negative muons, we kept asking Professors Wick and Ferretti what would be emitted as a result

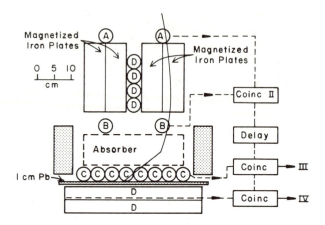

Figure 13.6. Layout of our experiments 3 and 4.[18,20] The D counters provide a veto of the count. From M. Conversi, E. Pancini, and O. Pancccioni in *Phys. Rev. 71* (1947), 209, Fig. 1.

of the capture. After refusing to answer for some time, eventually they ventured the notion that photons of energy comparable to the muon mass might be emitted.

We immediately started plans to observe such photons and promptly concluded that we should have a stopping material as transparent to photons as possible, which determined the choice of carbon. The experiment (which I did later at M.I.T. and found no photons, Figure 13.7) should have been done without lenses, there being no point in rejecting positive mesons at the price of the intolerable reduction of the counting rate that the lenses imposed.[19] However, as a matter of routine prudence and to check the stopping power of carbon in comparison to iron, we introduced a carbon absorber in the same apparatus where we had done the Tomonaga experiment with iron. The Tomonaga effect disappeared. We felt sure that something was wrong with our instruments, but we failed to come up with any explanation. We increased the current, magnetizing the lenses until the windings almost burned, and put back the 5-cm iron stopper, for which the decay rate for negative versus positive particles was then 0.03 ± 0.025 versus 0.67 ± 0.065. For practical purposes, the lenses allowed no wrong-sign particles to be detected. For negative focusing, the addition of 4 cm of carbon to the 5 cm of iron caused the decay rate to increase from 0.03 to 0.27, showing that indeed all negative muons decayed in carbon, because for positive focusing the 4-cm carbon alone gave 0.36

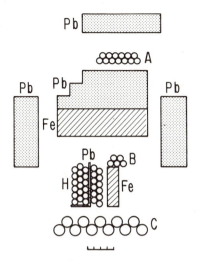

Figure 13.7. Schematic of the apparatus I used at M.I.T. to search for photons originating from the decay or the absorption of muons. When an event "*AB* (not *C*)" occurs, the 30 counters H are pulsed like a wire chamber; then the wires scan mechanically, and a picture is taken. The thin lead plate is there to convert the photons that originated in the iron stopper. From O. Piccioni in *Phys. Rev. 74* (1948), 1755, Fig. 1.

± 0.045, just as expected because of the positive excess (0.36/0.27 = 1.33 ± 0.18 versus 0.55/0.45 = 1.22). As a check, we added a 1.2-cm iron plate (the equivalent of 4 cm of carbon) to the 5-cm iron plate. For negative focusing, the decay rate was still negligible (Table 13.1).[20]

We owed the cleanness of our result both to the perfect functioning of our electronic apparatus and to having achieved a rejection ratio better than anyone could have hoped for with that simple device.

I want to emphasize, however, that the evidence for the noncapture of negative muons could also have been reached, although with less aesthetic attraction, if we had proceeded in the line of our experiment 2, measuring the decay rates with various thicknesses of carbon and above all comparing with aluminum once we had become aware of the large yield from carbon. The Z^4 dependence is, in fact, such a powerful tool that T. Sigurgeirsson and A. Yamakawa at Princeton, with a very simple setup that gave a modest counting rate, confirmed our result in only eight days, comparing beryllium with sulfur.[21] The same evidence was offered by an experiment that, to close the loop initiated in Rome,

Table 13.1 *Results of measurements on β-decay rates for positive and negative mesons*

Sign	Absorber	III	IV	Hours	M/100 hr
a) +	5 cm Fe	213	106	155.00′	67 ± 6.5
b) −	5 cm Fe	172	158	206.00′	3 (± 2.5)[a]
c) −	None	71	69	107.45′	−1
d) +	4 cm C	170	101	179.20′	36 ± 4.5
e) −	4 cm C + 5 cm Fe	218	146	243.00′	27 ± 3.5
f) −	6.2 cm Fe	128	120	240.00′	0 (+ 4 − ?)[a]

[a] I estimated the errors in preparing this chapter.

I did in the mountains of Colorado, at 11,000 ft, after joining the staff of Brookhaven National Laboratory. The experiment (Figure 13.8) involved the same basic general configuration as that of our Rome apparatus, but it distinguished particles coming unaccompanied ("normal" penetrating particles) from those that were locally produced and discharged at least two counters in the tray A.[22] I measured the ratio of the decay rate for carbon to that for sulfur, for normal as well as for locally produced mesons. This ratio was 1.85 ± 0.05 for the normal particles and 1.11 ± 0.1 for the local production. Both types of events displayed the same exponential decay of 2.2 μsec mean life (Figure 13.9). Evidently the decay rate for the locally produced particles, strongly presumed to include both negative and positive charges, did not increase when carbon instead of sulfur was used as the stopping material. This fulfilled the expectation for Yukawa mesons. Clearly, the locally produced mesons were the same as the pions discovered at Bristol; thus the experiment also showed that the muons from pion decay were indeed the same objects as the normal penetrating cosmic particles, which could then be thought of as all originating from pions, as anticipated by Robert Marshak.[23]

As mentioned earlier, while visiting M.I.T. in the Rossi group, I searched for photons either from capture or from decays. The suggestion of photons had been renewed by B. Pontecorvo.[24] The experiment (Figure 13.9) was performed with an early version of a (slow) wire chamber and found no photons from either process.

In summary, the cycle that for me started in Rome and finished in Colorado first indicated the correctness of the Yukawa theory by proving the decay of mesotrons, continued confirming the theory with the two experiments on the capture in iron, indicted the theory because of

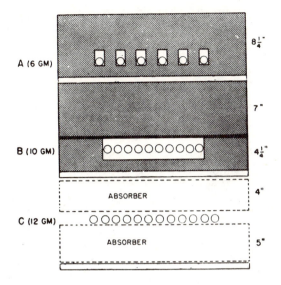

Figure 13.8. Schematic of the apparatus I used at Berthoud Pass, Colorado.[22] A "normal" incoming muon will discharge only one counter of the A set. A local penetrating shower will discharge two or more As. In either case, a pulse in C, delayed with respect to AB, is required. From O. Piccioni in *Phys. Rev. 77* (1950), 2, Fig. 1.

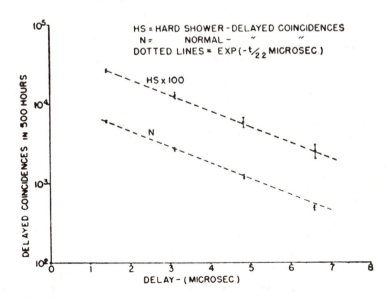

Figure 13.9. Delay distribution for the apparatus of Figure 13.8. From O. Piccioni in *Phys. Rev. 77* (1950), 4, Fig. 2.

the noncapture in carbon, then confirmed the Bristol work in favor of the theory. The resultant knowledge from this cycle was that the Yukawa meson was the mediator of the nuclear forces but not of the β decay. The universal Fermi interaction thus emerged from the quantitative comparison of the β nuclear decay with the decay and capture of muons.[24] The muon was recognized as a heavy lepton, and its mass and decay properties made it a wonderful probe for fundamental researches like parity violation and the study of the gyromagnetic ratio *g*.

The Z^4 dependence was studied in the beautiful works by Rossi and G. E. Valley using a cloud chamber instead of the Rossi lenses[25] and by Harold K. Ticho using the lenses.[26]

From then on, accelerators and committees took over, offering experimental opportunities unavailable with cosmic rays and depriving us of the ski slopes of Plateau Rosard and Berthoud Pass, and perhaps decreasing our candor also in other respects.

Acknowledgments
I thank Ms. P. Fisher and Dr. W. Mehlhop for their help in writing this manuscript.

Notes

1 Funds for this work were provided by the U.S. Department of Energy.
2 E. Fermi, E. Teller, and V. Weisskopf, "The Decay of Negative Mesotrons in Matter," *Phys. Rev. 71* (1947), 314-15; E. Fermi and E. Teller, "The Capture of Negative Mesotrons in Matter," *Phys. Rev. 72* (1947), 399-408.
3 S. Tomonaga and G. Araki, "Effect of the Nuclear Coulomb Field on the Capture of Slow Mesons," *Phys. Rev. 58* (1940), 90-1; H. Yukawa, S. Sakata, and M. Taketani, "Interaction of Elementary Particles. Part III," *Proc. Phys.-Math. Soc. Japan 20* (1938), 319-40.
4 Franco Rasetti, "Mean Life of Slow Mesotrons," *Phys. Rev. 59* (1941), 613; Franco Rasetti, "Disintegration of Slow Mesotrons," *Phys. Rev. 60* (1941), 198-204.
5 C. G. Montgomery, W. E. Ramsey, D. B. Cowie, and D. D. Montgomery, "Slow Mesons in the Cosmic Radiation," *Phys. Rev. 56* (1939), 635-9.
6 M. Conversi and O. Piccioni, "Sulle registrazione di coincidenza a piccoli tempi di separazione," *Nuovo Cimento 1* (1943), 279-90.
7 O. Piccioni, "Un nuovo circuito di registrazione a coincidenze," *Nuovo Cimento 1* (1943), 56-70.
8 M. Conversi and O. Piccioni, "Misura diretta della vita media dei mesoni frenati," *Nuovo Cimento 2* (1944), 40-70; M. Conversi and O. Piccioni, "On the Mean Life of Slow Mesons," *Phys. Rev. 70* (1946), 859-73.
9 Pierre Auger, Roland Maze, and Robert Chaminade, "Une démonstration directe de la désintégration spontanée du méson," *Compt. Rend. 213* (1941), 381-3; Roland Maze and Robert Chaminade, "Une mesure directe de la vie moyenne du méson au repos," *Compts. Rend. 214* (1942), 266-9.

10   Bruno Rossi and Norris Nereson, "Experimental Determination of the Disintegration Curve of Mesotrons," *Phys. Rev. 62* (1942), 417-22; Norris Nereson and Bruno Rossi, "Further Measurements on the Disintegration Curve of Mesotrons," *Phys. Rev. 64* (1943), 199-201.

11   Robert Chaminade, André Fréon, and Roland Maze, "Nouvelle mesure directe de la vie moyenne du méson à repos," *Compt. Rend. 218* (1944), 402-4.

12   M. Conversi and O. Piccioni, "Sulla disintegrazione dei mesoni lenti," *Nuovo Cimento 2* (1944), 71-87; M. Conversi and O. Piccioni, "On the Disintegration of Slow Mesons," *Phys. Rev. 70* (1946), 874-81.

13   Bruno Rossi, "Magnetic Experiments on the Cosmic Rays," *Nature 128* (1931), 300-1.

14   G. Bernardini and M. Conversi, "The Deviation of Cosmic Rays in a Ferromagnetic Body," *Ricerca Sci. 11* (1940), 840-8; M. Conversi and E. Scrocco, "Ricerche sulla componente dura della radiazione penetrante eseguite per mezzo di nuclei di ferro magnetizzati," *Nuovo Cimento 1* (1943), 372-413.

15   D. Skobeltzyn, "Ueber eine neue Art sehr schneller β-Strahlen," *Z. Physik 54* (1929), 686-702.

16   C. F. von Weizsäcker, "Durchgang schneller Korpuskularstrahlen durch ein Ferromagnetikum," *Ann. Physik 17* (1933), 869-96.

17   G. Bernardini, M. Conversi, E. Pancini, E. Scrocco, and G. C. Wick, "Researches on the Magnetic Deflection of the Hard Component of Cosmic Rays," *Phys. Rev. 68* (1945), 109-20.

18   M. Conversi, E. Pancini, and O. Piccioni, "On the Decay Process of Positive and Negative Mesons," *Phys. Rev. 68* (1945), 232.

19   O. Piccioni, "Search for Photons from Meson-Capture," *Phys. Rev. 74* (1948), 1754-8.

20   M. Conversi, E. Pancini, and O. Piccioni, "On the Disintegration of Negative Mesons," *Phys. Rev. 71* (1947), 209-10.

21   T. Sigurgeirsson and A. Yamakawa, "Decay of Mesons Stopped in Light Materials," *Phys. Rev. 71* (1947), 319-20.

22   O. Piccioni, "Local Production of Mesons at 11,300 Feet," *Phys. Rev. 77* (1950), 1-5.

23   R. E. Marshak and H. A. Bethe, "On the Two-Meson Hypothesis," *Phys. Rev. 72* (1947), 506-9.

24   B. Pontecorvo, "Nuclear Capture of Mesons and the Meson Decay," *Phys. Rev. 72* (1947), 246-7.

25   G. E. Valley and B. Rossi, "On the Mean Life of Negative Mesons," *Phys. Rev. 73* (1948), 177-8.

26   Harold K. Ticho, "The Capture Probability of Negative Mesotrons," *Phys. Rev. 74* (1948), 1337-47.

Appendix

The recollections of my colleague Conversi (see Chapter 14) might well give the reader the impression that our experiment was to an appreciable extent the outgrowth of the experiments of Bernardini, Conversi, Pancini, Scrocco, and Wick[1] (BCPSW), because of both their technical results and a specific suggestion by Pancini. Instead, our experiments started under the stimulus of the physics of meson decay, quite independently of their work. Incidentally, my advisor was Professor Edoardo Amaldi.

The leptonic discovery (our last experiment) was the consequence of our efforts to find answers to successive questions about the end of the life of the "mesotron," rather than the consequence of "combining the two techniques." Clearly, our work required many techniques and tools. The lenses were one of such tools, and they had been in the inventory of experimental physics for 13 years; a very gracious tool, but not at all indispensable. We did not use it in our first proof of the meson capture in iron; thus we could have done without it for the work with carbon, which was to be a search for photons from capture. It is also notable that for the same research as ours, Rossi, in 1945, used a cloud chamber instead of his own invention, presumably because he doubted its adequacy.

In fact, Bernardini and associates had used the lenses (strictly a replica of Rossi's model) only for high-energy muons, detecting particles with an energy above a certain value and with an array of two lenses. Neither their data nor their computations could establish that one lens would work well for low-energy mesons stopping in our absorber (particularly because of the different effect of multiple scattering) and that it would give us a high rejection of particles of the wrong sign (a parameter not acutely needed by BCPSW). However, we did not need a certificate of viability: The notion that the effective field deflecting the particles, H_e, was roughly equal to B was sufficient for us to try.* There was no risk, no "technique" to learn, in fact no need of "previous personal involvement" in order to try that simple and beautiful device built by Rossi in 1931.[2] The viability was proved by the agreement of our experiment III with our experiment II. For our final experiment IV, Table 13.1 shows that the unexpectedly high rejection value and the articulated comparison of several measurements, rather than the BCPSW data, provided the much needed confidence in our result.

Rossi and Puccianti, with fine intuition, simply assumed H_e equal to B, and Rossi constructed a model that reached a high value of B with a small gap, an achievement not a priori obvious. He could not confirm the wrong expectation that the rays were only negative electrons, and that started the wrong speculation that H_e might be closer to H than to

---

* *Ed. note*: H_e being near B means an effective magnetic field thousands of times larger than H, the magnetic field in air.

B. By 1937-9, however, the electron hypothesis was substituted by the correct one on negative and positive mesons with a small positive excess. The result of Rossi became then clear, and the H_e = H idea lost any credibility.

Moreover, between 1931 and 1940 a series of theoretical and experimental works had appeared.[3,4] All those works, especially that of Alvarez, converged on the result that H_e was equal to B. Bernardini and associates wrote that the hypothesis of H_e being "considerably smaller than B" had been dismissed by Weizsäcker in 1933.[1] Rossi, in 1964, described the lenses and did not refer to the experiments of BCPSW as a proof that H_e was equal to B.[4] Nonetheless, it is certain that the experiments of BCPSW were a nice confirmation that H_e was not much smaller than B (their data did not offer a stronger conclusion) and that they made all of us aware of Rossi's device in a much more vivid way than we could have obtained by reading the journals: The lenses had become a household item.

This takes me to another point in Conversi's text (see Chapter 14). In such circumstances, no professional debt of acknowledgment could have resulted toward whoever had said first that the lenses could have been added to our apparatus. It would be too easy to walk over to a laboratory and take credit for giving obvious and unsolicited advice. Actually, I do not even know to whom Pancini talked, although I know that he did not talk to me nor to Conversi, nor did we ever see as much as a sketch on the classical "back of an envelope," and I cannot recollect whether or not I had the same obvious thought before I heard of Pancini's suggestion, in the form of a simple wish, just as Pancini. I do recollect that I interpreted that Pancini had suggested the use of two lenses in a vertical array, which geometry would have resulted in a prohibitively low intensity.

However, in 1944, because our friend had been kept away from research longer than us, and because of the expectation that he would not join us for our experiment III, we wrote at the end of our paper of experiment II a sweeping acknowledgment that "Pancini had suggested" the use of the lenses, naturally omitting the circumstances just described. We did not notice that such an omission gave to our words a meaning that we did not have in mind. Significantly, we did not write the credit for the "suggestion" in the actual publication of experiment III. Nevertheless, our acknowledgment has sometimes been given the widest of meanings to the point of quoting Pancini as suggesting a capture experiment with a carbon absorber. Pancini, of course, contributed most valuably to our experiments III and IV with his intelligent experimental intuition and his careful work. However, our carbon experiment was by no means an "extension to other elements" of experiment III. The Tomonaga theory absolutely excluded a Z dependence, and no one had such a thought in mind.

Thus, in my recollection, there was no reason to choose carbon, other than that of searching for photons (which inevitably required carbon), as I wrote in my report. I add that after our experiment with iron, Ferretti discussed with me the possibility of photon emission.[5] I can confidently dismiss the second reason, namely the fact that French authors had indicated "approximate agreement" with absence of capture.[6] Because our experiment had given us the value of the mean life with evidence of good accuracy, we knew from their wrong mean life that we could not rely on their conclusion about the capture. In fact, we know now that such a conclusion was wrong because aluminum does capture almost completely for experiments with a minimum delay of 2 μsec, such as that of Maze and associates. Moreover, Rasetti's conclusion indicated complete capture in aluminum, in disagreement with the French authors. If, nonetheless, we wanted to check the results of Chaminade and associates, we surely would have used aluminum (thinking of some experimental error), never carbon. No one at that time imagined any "reasonable" Z dependence of the Tomonaga capture, with which to relate the behavior of carbon with that of aluminum.

Significantly, Conversi recalls that we insisted in not believing our result for carbon, rather than considering it a confirmation of the French work. Having so corrected Conversi's personal recollection, I want to restate that our French colleagues who worked with great style in such adverse circumstances certainly contributed to our knowledge and determination to pursue our research.

Going back to the subject of the lenses, I emphasize again that our experiment with carbon could well have been done without magnets, as we had done with iron, by comparing data with two or more absorber plates. The Princeton work reported earlier in Chapter 13 reinforces the statement.

I am glad, however, that we used the lenses, not only because they added an aesthetic touch to the experiment but even more because with the name of Rossi now associated with our discovery I feel less uneasy when I reflect that were it not for Nazism, undoubtedly Rossi would have discovered the leptonic nature of the mesotron much before we did.

Notes

1 G. Bernardini, M. Conversi, E. Pancini, E. Scrocco, and G. C. Wick, "Researches on the Magnetic Deflection of the Hard Component of Cosmic Rays," *Phys. Rev. 68* (1945), 109-20; G. Bernardini and M. Conversi, *Ricerca Sci. 11* (1940), 858; G. Bernardini, G. C. Wick, M. Conversi, and E. Pancini, "Positive Excess in Mesotron Spectrum," *Phys. Rev. 60* (1941), 535-6; *Ricerca Sci. 12* (1941), 1227.

2 Bruno Rossi, "Magnetic Experiments on the Cosmic Rays," *Nature 128* (1931), 300-1.

3 C. F. von Weizsäcker, "Durchgang schneller Korpuskularstrahlen durch ein Ferromagnetikum," *Ann. Physik 17* (1933), 869-96; Luis Alvarez, "On the Interior Magnetic Field in Iron," *Phys. Rev. 45* (1934), 225-6; W. F. G. Swann and W. E. Danforth, Jr., "Deflection of Cosmic-Ray Secondaries in Magnetized Iron," *Phys. Rev. 45* (1934), 565; W. E. Danforth and W. F. G. Swann, "The Deflection of Cosmic-Ray Charged Particles in Passing Through Magnetized Iron," *Phys. Rev. 49* (1936), 582-91; W. F. G. Swann, "A Theoretical Discussion of High Energy Charged Particles in Passing Through Magnetized Iron," *Phys. Rev. 49* (1936), 574-82.

4 Bruno Benedetti Rossi, *High Energy Particles* (New York: Prentice-Hall, 1952).

5 B. Ferretti, "The Absorption of Slow Mesons by an Atomic Nucleus," in *International Conference on Fundamental Particles and Low Temperature, Cambridge, 22-27 July, 1946, Vol. 1* (London: Physical Society, 1947), pp. 75-7.

6 Robert Chaminade, André Fréon, and Roland Maze, "Nouvelle mesure directe de la vie moyenne du méson au repos," *Compt. Rend. 218* (1944), 402-4.

14 The period that led to the 1946 discovery of the leptonic nature of the "mesotron"

MARCELLO CONVERSI

Born 1917, Tivoli, Italy; *laurea*, University of Rome, 1940; elementary particle physics experimentalist; University of Rome, Italy.

The discovery of the leptonic nature of the cosmic ray meson (the "mesotron," in the language of the time) was the culmination, reached in 1946, by Ettore Pancini, Oreste Piccioni, and myself, of a series of experiments based on two techniques developed in Rome early in the 1940s.

Some aspects of the struggle that led to this discovery (in today's language, a second charged lepton) have been vividly recalled by Piccioni in Chapter 13. I shall try to bring out other aspects through personal recollections, which inevitably have some autobiographic character and deal in part with the physics environment in Rome in that period.

My career as a physicist started with World War II, at a time when my intellectual interests were divided between physics and music, with perhaps some temperamental preference for the latter. On the afternoon of June 10, 1940, I was dictating my university thesis to a typist when, from a loudspeaker in the typist's office, I heard Mussolini announcing the declaration of war on France and Great Britain. One month later I received my doctorate (*laurea*) at the University of Rome, after discussing my thesis, which had been prepared at the Physics Institute under the supervision of Bruno Ferretti. The drawback of a strong amblyopia to my left eye gave me the advantage of being rejected for military service and avoiding participation in the war on the side of the German Nazis against the free world. So I soon was able to go back to work.

As a consequence of the evolution of the political situation, most of the Italian physicists who had been working on frontier problems had left Italy during the 1930s. In particular, of the celebrated Roman school, only Edoardo Amaldi was still at the Physics Institute in 1940. After the premature disappearance of Ettore Majorana, all the other members of the "old group" (Bruno Pontecorvo, Franco Rasetti, Emilio Segré, et al.) had left Italy, including the most prominent person, Enrico Fermi, who left at the end of 1938. Amaldi (who had a prominent role in the revival of physics after the war, first in Italy and later in Europe through the creation of CERN) took a few essential steps to keep alive what was left of the Roman school. At his initiative, Gian Carlo Wick came to Rome late in 1939 to hold the chair left by Fermi. Furthermore, arrangements were made with Gilberto Bernardini, who was then a professor at the University of Bologna, to come to Rome for a few days each week in an attempt to concentrate there a sort of national effort in the field of cosmic rays, to which Bruno Rossi had made fundamental contributions before leaving Italy. Bernardini's intellectual charm, which included enthusiasm and also a cultural refinement outside of physics, had a great influence on the small community of the younger generation of physicists; this included, in addition to Ferretti, a few young experimentalists, Mario Ageno, Carlo Ballario, Bernardo N. Cacciapuoti, Pancini, Piccioni, and Eolo Scrocco, all either holding, or near to holding, positions as assistant professors at the University of Rome.

Early in September of 1940, Scrocco (my oldest friend, who later became professor of theoretical chemistry) put me in touch with Bernardini, who invited me to work with him on a new attempt to see the effect of magnetic deflection on the high-rigidity cosmic rays. The nearly equal numbers and energy distributions of positive and negative cosmic-ray particles at sea level had hindered the observation of such an effect in all past attempts,[1-3] in which it had not been clear whether or not the magnetic induction, **B**, or the magnetizing field, **H**, acted on the particles traversing a magnetized iron bar.[4] Bernardini suggested altering the natural charge distribution of the particles by means of a "magnetic lens" of a type first developed by Rossi and then placing a second lens just below the first to analyze the altered distribution.[2] As is well known, these magnetic lenses are basically made up of two iron bars with opposite magnetization that are the essential parts of a closed magnetic circuit. The system behaves as a converging magnetic lens for charged particles of a given sign that traverse the iron bars from the top and as a diverging magnetic lens for particles of the opposite sign.

I found the proposal appealing, and we started to work on it immediately with great enthusiasm. With the help of a technician, and with the participation of Pancini for a short time before he left for military service, we quickly developed the two magnetic lenses and the three Geiger-Müller counters (2 cm in diameter, 50 cm in length) to be placed above, in the middle of, and below the lens doublet. By counting the threefold electronic coincidences among the pulses of the three counters, with the magnetizing current in the bottom lens alternately "parallel" and "antiparallel" with respect to that in the top lens, we soon recorded a large difference in the two counting rates.[5] This clearly demonstrated the viability of the method for further applications to cosmic-ray physics.

In fact, after a detailed quantitative treatment, made essentially by Wick, the method was applied successfully to measure the numbers and energy spectra[6-8] of the positive and negative mesons capable of traversing the lenses and, in particular, to determine the ratio between the lifetime and mass of the meson. This turned out to be about 2.1×10^{-8} sec-MeV-cm^{-2}, very close to its present value.[7]

However, by far the most important application of the magnetic lenses, essentially in their primitive version (i.e., without using the lens doublet as in the previous applications), came a few years later when this technique was combined with that of "delayed coincidences" developed in 1942–3 by Piccioni and myself, as will be explained later.

Piccioni, with a few more years of experience than I, had a deep understanding of and keen enthusiasm for electronics, and most of the development that followed was due to his great competence and ingenuity in the field. In particular, he invented a new type of fast coincidence circuit based on the use of secondary emission tubes that was an essential part of the electronic apparatus later developed and used for the direct determination of the meson lifetime, the experiment Piccioni and I already had in mind late in 1941 when we decided to work together.[9]

Piccioni has vividly described (Chapter 13) the steps we had to take to arrive at this determination: We had essentially to develop fast counting systems and rapid time resolution techniques similar to those developed at the same time, but quite independently, in the United States, with which there was no possibility of communication after December 1941.[10]

Of course, we knew of the successful direct measurement of the meson lifetime as reported by Franco Rasetti, whose work had been in a sense the starting point for ours.[11] However, Rasetti's result did not include demonstration of the exponential character of the meson decay curve, because it was based on the measure of the logarithmic decrement observed for only two time intervals.

Piccioni and I had developed an electronic system of delayed coincidences that allowed us to record the counting rates for different values of a delay introduced artificially between the signal corresponding to the arrival of a meson stopped in an iron absorber and the signal corresponding to the emitted decay electron. All this work was done at a time when the Physics Institute was being kept alive by a small number of people, because many of the staff were in the war. I remember Amaldi lecturing early in the morning (6:30 A.M.) to allow young physicists in the military service, such as Pancini and Piccioni, to attend his lectures on nuclear physics.

Our electronic system was nearly completed when, on July 19, 1943, Rome was bombed for the first time by the Americans. About 80 bombs fell within the perimeter of the university campus, one of them just outside the window of our laboratory a few minutes after I had left, having moved the electronic system far away from the window.

The following day we decided to take all our apparatus out of the university campus, which is near the railroad freight station San Lorenzo. I found a site in the basement of the Liceo Virgilio, near the Vatican City, that appeared convenient and was presumed not to be one of the goals of future incursions. There we installed our apparatus and started a series of checks on it. But we were soon interrupted again as a consequence of the occupation of Rome by the Nazis, immediately following the armistice (September 8, 1943). In fact, Piccioni, with a few other army officers, tried to cross the front line in an attempt to join the Allied troops coming toward Rome from southern Italy, but they were captured by the Germans. Fortunately, they were released after only a few weeks. I, on the other hand, became involved in the start of a political underground movement that later developed into the so-called Resistance; however, I stayed in Rome, where my mother was seriously ill (she died three weeks later). Thus our scientific activity was completely interrupted for about one month.

The period of the Nazi occupation was the darkest and most dramatic period covered by these personal recollections. When we resumed our work at the Liceo Virgilio, we had to hide from Germans who made frequent raids in the city to conscript young males for military service. We also took a serious risk by keeping with us for some time weapons and a radio transmitter, which came from the underground antifascist movement. The transmitter came from the Partito d'Azione (via Amaldi) in bad condition; it was repaired by our friend Franco Lepri, who was a great expert in radio and electronics techniques. We had occasional visits from a few friends and one or two students interested in our work. But contacts with the Physics Institute (far away on another side of the city, and attended by only a few students in those times) were only sporadic.

Even though it was difficult to progress with our experiment under these conditions, we managed to do so, and early in 1944 Piccioni and I had conclusively demonstrated that the cosmic-ray particles that stopped in our 5-cm-thick iron absorber did indeed undergo spontaneous decay, exhibiting an exponential decay curve characterized by a lifetime only slightly in excess of the 2.2-μsec value known today.[12] Even though we were (obviously) not aware of the fact that we had measured the Fermi coupling constant, we were very proud of our result and were convinced we were the first to record the exponential decay curve of a free particle. Because of the total lack of communication with the United States, we did not know that in fact a similar result had already been obtained there in the beautiful experiment of Rossi and Norris Nereson.[13]

A second experiment was carried out immediately afterward, with essentially the same apparatus, to test the theoretical predictions on the behavior of mesons at rest.[14] The theory of Sin-itiro Tomonaga and Toshima Araki was based on the assumption, then universally accepted, that the constituent of the hard component of the cosmic radiation was identical with the particle introduced by Hideki Yukawa to explain the short range of nuclear forces.[15] Accordingly, of the mesons at rest in any material, only those that are positively charged should undergo spontaneous decay, whereas the negative ones should be absorbed, via atomic and then nuclear capture, in a time much shorter than the meson lifetime. Consistent with these predictions, we found that indeed about half of the mesons that were stopped in a 0.6-cm-thick iron absorber underwent spontaneous decay.

A more direct and conclusive test required separately stopping positive or negative mesons by the technique of the magnetic lenses, in which I had been personally involved since 1940. Piccioni and I discussed this possibility at length in terms of counting rates and rejection

Experiment of Conversi, Pancini, and Piccioni. At the left is a dem
onstration model of the experiment; at the right are the original
magnetic lenses used in the experiment to concentrate the positive
or negative mesons separately on the "stopper" (iron or carbon).
The experiment was made possible by the use of "fast electronic
delayed coincidences," an advanced technique developed by Conversi and Piccioni during the years of World War II (credit: Renato
Cialdea).

efficiency for the mesons of the unwanted sign, and we concluded that
the experiment was feasible. Early in 1943, however, Pancini had already suggested, in quite general terms, that for this purpose the
method of the magnetic lenses be combined with that of the delayed
coincidences, the latter being developed then by Piccioni and myself.
Later, Pancini was in the war, and after the German occupation, he
joined the partisans in northern Italy. There was no hope of having
him with us immediately, so Piccioni and I decided to go on with the
new experiment, with the understanding that Pancini's name might
appear in the presentation of possible future results deriving from it.

Preparation of the experiment started when we were still in the
Liceo Virgilio, but it was completed in the Institute of Physics, where

we moved soon after the Allied troops liberated Rome (June 5, 1944). Data taking started in the spring of 1945, and Pancini was able to participate in the runs, because he came back to Rome soon after the liberation of northern Italy (April 25, 1945).

The results of this experiment confirmed, beyond any doubt, that whereas the positive mesons that were stopped in iron underwent spontaneous decay, essentially all the negative ones did undergo nuclear capture, well in line with the theoretical predictions of Tomonaga and Araki.[16]

The next step was to extend the test of these predictions by replacing the iron absorber with one of low-Z material. In my personal recollection, the choice of a low-Z material (which turned out to be carbon) was determined for two reasons: the indication, reported by Rasetti[17] (see also the work of Roland Maze and Robert Chaminade[18]) that more than half of the mesons stopped in an absorber of aluminum undergo spontaneous decay; the hope of seeing, via electron-positron pair formation, energetic γ rays possibly resulting from the capture of negative mesons at rest.

Piccioni, in particular, insisted on the importance of this latter point (and later carried out an important experiment to search for photons from muon capture, obtaining, however, a negative result), especially after a discussion he had with Ferretti and Wick.[19] And Piccioni himself found the graphite cylindrical rods 5 cm in diameter (the only shape available on the market in those days) that replaced the iron absorber in the new experiment.[20]

When we found that delayed coincidences were recorded at essentially the same rates (about 0.3 per hour) for negative and positive mesons focused on the graphite absorber, we at first believed that something was wrong with our apparatus. But we failed to find any instrumental explanation of the simple fact that delayed coincidences were counted with graphite in both situations, even when we increased the magnetizing current up to the maximum value that the coils could stand, whereas their counting rate disappeared when the iron absorber alone was placed back and negative particles were focused on it.

So this was a new effect, at variance with the conclusions of Tomonaga and Araki, who had foreseen, on sound theoretical grounds, that negative mesons had to be captured even in light nuclei in a much shorter time than their lifetime. Apparently the mesotron could not be identified with the Yukawa particle!

As promptly demonstrated by Fermi, Edward Teller, and Victor F. Weisskopf,[21] the discrepancy with the theoretical predictions could be expressed in terms of a factor 10^{12}. Fermi and Teller also ruled out, through detailed calculations, the remote possibility that for some unknown reason the overall time of slowing down and atomic capture of a negative meson was much longer in carbon than in iron.[22]

The striking result of our experiment prompted the bold idea, put forward by Robert E. Marshak and Hans Bethe before the great discovery of the charged pion and the $\pi \rightarrow \mu$ decay at Bristol, that the mesons observed in the cosmic radiation near sea level were secondary weak-interacting particles, arising from the decay of mesons created by the primary cosmic rays in the upper layers of the atmosphere.[23] It also prompted the derivation of the Z^4 Wheeler law governing muon capture and led to the idea of a universal Fermi interaction, which evolved in the following years, after the discovery of the Cabibbo angle.[24] It was, finally, the start of the new field of "mesic atoms." But perhaps its greatest impact on the evolution of particle physics was its contribution to the development of the lepton concept and the opening of the new field of lepton physics, only recently enriched by the discovery of yet another charged heavy lepton.[25]

Notes

1 D. Skobeltzyn, "Ueber eine neue Art sehr schneller β-Strahlen," *Z. Physik 54* (1929), 686-702.

2 Bruno Rossi, "Magnetic Experiments on the Cosmic Rays," *Nature 128* (1931), 300-1.

3 L. M. Mott-Smith and G. L. Locher, "A New Experiment Bearing on Cosmic-Ray Phenomena," *Phys. Rev. 38* (1931), 1399-408.

4 C. F. von Weizsäcker, "Durchgang schneller Korpuskularstrahlen durch ein Ferromagnetikum," *Ann. Physik 17* (1933), 869-96.

5 G. Bernardini and M. Conversi, "Sulla deflessione dei corpuscoli cosmici in un nucleo di ferro magnetizzato," *Ricerca Sci. 11* (1940), 840-8.

6 G. Bernardini, M. Conversi, E. Pancini, and G. C. Wick, "Sull eccesso positivo della radiazione cosmica," *Ricerca Sci. 12* (1941), 1227-43; G. Bernardini, M. Conversi, E. Pancini, and G. C. Wick, "Positive Excess in Mesotron Spectrum," *Phys. Rev. 60* (1941), 535-6.

7 M. Conversi and E. Scrocco, "Ricerche sulla componente dura della radiazione penetrante esequite per mezzo di nuclei di ferro magnetizzati," *Nuovo Cimento 1* (1943), 372-413.

8 G. Bernardini, M. Conversi, E. Pancini, E. Scrocco, and G. C. Wick, "Researches on the Magnetic Deflection of the Hard Component of Cosmic Rays," *Phys. Rev. 68* (1945), 109-20.

9 O. Piccioni, "Un nuovo circuito di registrazione a coincidenze," *Nuovo Cimento 1* (1943), 56-70.

10 M. Conversi and O. Piccioni, "Un circuito di conteggio a demoltiplicazione di 16 con tubi a vuoto," *Nuovo Cimento 1* (1943), 12-24; M. Conversi and O. Piccioni, "Sulle registrazione di coincidenza a piccoli tempi di separazione," *Nuovo Cimento 1* (1943), 279-90.

11 Franco Rasetti, "Mean Life of Slow Mesotrons," *Phys. Rev. 59* (1941), 613; Franco Rasetti, "Disintegration of Slow Mesotrons," *Phys. Rev. 60* (1941), 198-204.

12 M. Conversi and O. Piccioni, "Misura diretta della vita media dei mesoni frenati," *Nuovo Cimento 2* (1944), 40-70; M. Conversi and O. Piccioni, "On the Mean Life of Slow Mesons," *Phys. Rev. 70* (1946), 859-73.

13 Bruno Rossi and Norris Nereson, "Experimental Determination of the Disintegration Curve of Mesotrons," *Phys. Rev. 62* (1942), 417-22; Norris Nereson and Bruno Rossi, "Further Measurements on the Disintegration Curve of Mesotrons," *Phys. Rev. 64* (1943), 199-201.

14 M. Conversi and O. Piccioni, "Sulla disintegrazione dei mesoni lenti," *Nuovo Cimento 2* (1944), 71-87; M. Conversi and O. Piccioni, "On the Disintegration of Slow Mesons," *Phys. Rev. 70* (1946) 874-81.

15 S. Tomonaga and G. Araki, "Effect of the Nuclear Coulomb Field on the Capture of Slow Mesons," *Phys. Rev. 58* (1940), 90-1.

16 M. Conversi, E. Pancini, and O. Piccioni, "On the Decay Process of Positive and Negative Mesons," *Phys. Rev. 68* (1945), 232.

17 Franco Rasetti, "Mean Life of Slow Mesotrons" (Note 11).

18 Roland Maze and Robert Chaminade, "Une mesure directe de la vie moyenne du méson au repos," *Compt. Rend. 214* (1942), 266-9.

19 O. Piccioni, "Search for Photons from Meson-Capture," *Phys. Rev. 74* (1948), 1754-8.

20 M. Conversi, E. Pancini, and O. Piccioni, "On the Disintegration of Negative Mesons," *Phys. Rev. 71* (1947), 209-10.

21 E. Fermi, E. Teller, and V. Weisskopf, "The Decay of Negative Mesotrons in Matter," *Phys. Rev. 71* (1947), 314-15.

22 E. Fermi and E. Teller, "The Capture of Negative Mesotrons in Matter," *Phys. Rev. 72* (1947), 399-408.

23 R. E. Marshak and H. A. Bethe, "On the Two-Meson Hypothesis," *Phys. Rev. 72* (1947), 506-9; C. M. G. Lattes, G. P. S. Occhialini, and C. F. Powell, "Observations on the Tracks of Slow Mesons in Photographic Emulsions. 1. Existence of Mesons of Different Mass," *Nature 160* (1947), 453-6.

24 John A. Wheeler, "Mechanism of Capture of Slow Mesons," *Phys. Rev. 71* (1947), 320-1; B. Pontecorvo, "Nuclear Capture of Mesons and the Meson Decay," *Phys. Rev. 72* (1947), 246-7; O. Klein, "Mesons and Nucleons," *Nature 161* (1948), 897-9; G. Puppi, "Sui mesoni dei raggi cosmici," *Nuovo Cimento 5* (1948), 587-8 (in Italian); Nicola Cabibbo, "Unitary Symmetry and Leptonic Decays," *Phys. Rev. Letters 10* (1963), 531-3.

25 M. L. Perl et al., "Evidence for Anomalous Lepton Production in e$^+$-e$^-$ Annhilation," *Phys. Rev. Letters 35* (1975), 1489-92.

15 On the discovery of the neutral kaons[1]

ROBERT W. THOMPSON

Born 1919, Minneapolis; Ph.D., M.I.T., 1948; Los Alamos, 1943-5; University of Indiana, 1948-53; University of Chicago from 1953; experimentalist in cosmic rays and elementary particle physics.

The first observations of the phenomenon of forked tracks were made at Manchester, England, by George D. Rochester and Clifford C. Butler in the course of a magnetic cloud chamber investigation of cosmic-ray-induced penetrating showers under lead, with the original electromagnet of Patrick M. S. Blackett.[2] They pointed out that their two events (one neutral and one charged) could be explained in terms of the spontaneous decay of a new type of heavy unstable particle. Their observations were confirmed and their arguments placed on relatively secure statistical grounds by the Caltech cloud chamber group, who made some of their observations at mountain altitude.[3] At the suggestion of Carl D. Anderson and Blackett, the generic term V particle was adopted to refer to the unstable parent particles, which on decay are responsible for the phenomenon of forked tracks.

The Manchester group reassembled their apparatus at Pic-du-Midi in the Pyrenees at nearly 3,000 m, obtaining a number of new examples of V tracks.[4] Contemporaneously, the Indiana group independently reached similar, but more quantitative, conclusions.[5] Among nine examples of V^0 decay at sea level, with a 12-in. magnetic Wilson chamber, the latter authors found three examples in which the positive decay fragment, from ionization and momentum, was found to have a mass near to that of a proton. In two of the examples the negative decay fragment could be shown to be light and therefore was inferred to be a muon or pion.

251

The use of the Q value (i.e., the kinetic energy of the decay products with the decaying particle at rest) was introduced for the first time in connection with the decay of the new heavy unstable particles. For the decay scheme

$$V^0 \rightarrow p + \pi-, \tag{1}$$

the two examples yielded Q values of 31 ± 5 and 34 ± 10 MeV, remarkably close to the presently accepted value $Q = 36.9$ MeV for the particle now known as the Λ^0. These were probably the first identified Λ^0 particles. The group later published the value $Q = 37$.[6] The Q Value of event 50 (of the 1953 article), assuming two-body decay into two pions, was 231 ± 49 MeV, which is in agreement with the currently accepted value. It is thus probable that this event was the first observed example of what was called the θ^0 meson, but in modern notation is denoted K_S^0.

Rather detailed descriptions of the magnet coils, the magnet yoke, and the thermal jacket have been given elsewhere.[7] One of the principal features of the apparatus was that the cloud chamber was thermally insulated from the thermal jacket by Micart spacers. When a "slow expansion" was made for purposes of "cleaning" the chamber, it was observed that the envelope of the uppermost droplets was flat, corresponding to the flat upper surface of the chamber. Farther down, the envelope assumed a slow "wave motion," with amplitude midway down the chamber of only about 0.8 cm. This was a gratifying sight to witness, because it meant that gaseous distortions were largely eliminated because of proper design of the thermal jacket.

In approximately 600 hr of continuous operation, with various triggering systems, 21,100 stereoscopic photographs were taken. All unusual events found on scanning these photographs were catalogued for further examination:

Total photographs, 21,100
Penetrating showers, 4,183
V^0 events, 230
Angular deflections in gas, 105
$V^\pm$ events, 38
Events in gas of chamber, 47
K^+ decay, 1

The evidence for the existence and the properties of the θ^0 mesons is perhaps best explained in terms of the "Thompson surface," the surface of constant Q value; this is a full-fledged three-dimensional representation of the data. The computed values are plotted in a three-dimensional surface, stereoscopic photographs of which are shown in Figure 15.1. It can be seen, by viewing the figure in stereo, that with the exception of four clearly anomalous points the main group delineates a Thompson surface. The vertical semiaxis indicates a center-of-mass momentum of the fragments of a little more than 200 MeV-cm^{-1}; the average is very nearly equal to zero, indicating equal or nearly equal decay momenta, and the eccentricity of the limiting plane indicates a parent mass of just under 1,000 m_e.

The scatter of the points about the theoretical surface is nicely consistent with the internally estimated errors. Thus the data indicate in a natural way the presence of a predominantly two-body decay ($K_S^0 \rightarrow \pi^+ + \pi^-$) plus four anomalous cases ($K_L^0 \rightarrow \pi^+ + \pi^- + n^0$?). However, the existence of anomalous events that we considered to be definitely incompatible with the Q surface for $K_S^0 \rightarrow \pi^+ + \pi^-$ for the main group required a more detailed and quantitative discussion of two-body versus multibody decay. If, in addition to the two charged decay fragments, one or more neutral particles (hence unobserved in the cloud chamber) are produced in the decay, it can be shown that the corresponding representative points scatter widely, but lie within the surface for the decay scheme $K_S^0 \rightarrow \pi^+ + \pi^-$.

The weighted mean of the Q values for $\theta^0 \rightarrow \pi^+ + \pi^- + 214$ MeV was

$$Q(\pi^+, \pi^-) = 213.9 \pm 2.8 \text{ MeV.}$$

No improvement in this value was obtained until the advent of Luis Alvarez's liquid-hydrogen bubble chamber at Berkeley. Our Q value is in excellent agreement with that currently accepted.

The anomalous events that lie within the two-body surface of constant Q value are probably examples of the three-body decay of the neutral K meson, known as the τ mode (first observed in nuclear emulsion by the Bristol group as early as 1949). The fact that the masses of the θ ($m_{\theta^0} = 966 \pm 10$ m_e) and the τ ($m_\tau = 966 \pm 4$ m_e) were very nearly equal suggested that the θ was identical with the τ. Thus one had

$$\theta^0 \begin{cases} \nearrow \pi^+ + \pi^- \\ \searrow \pi^+ + \pi^- + \pi^0. \end{cases}$$

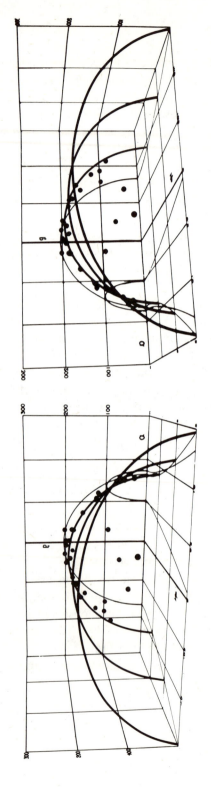

Figure 15.1. The Thompson surface, or surface of constant *Q*-value. This figure may be viewed in stereo with the aid of a plane mirror. From R. W. Thompson, J. Burwell, and R. W. Hugget in *Nuovo Cimento (suppl.) 4* (1956), 305, Fig. 8 and R. W. Thompson in *Progress in Cosmic Ray Physics V. III* (Amsterdam: North-Holland, 1956), 299, Fig. 21.

Figure 15.2. A cosmic-ray example of the reaction $\pi^- + p \rightarrow \theta^0 + \Lambda^0$. From R. W. Thompson et al. in *Phys. Rev. 95* (1954), 1577, Fig. 1 and R. W. Thompson in *Progress in Cosmic Ray Physics V. III* (Amsterdam: North-Holland 1956), 328, Plate X.

However, one particle could not decay into two different states of parity, or so it was generally believed. The celebrated plot of Richard H. Dalitz was important in providing an ingenious method of analyzing the emulsion data.[8] This, the so-called θ-τ puzzle, led T. D. Lee and C. N. Yang to their Nobel Prize for discovery of the nonconservation of parity.

One of the most important results from the Brookhaven cosmotron, shortly after its completion, was the clear evidence obtained by R. P. Shutt and his collaborators for the production of neutral heavy unstable particles in association.[9] Confirmation of the results of these workers was obtained by the Indiana group.[10] In addition to the confirmation of the phenomenon in an event produced in the natural cosmic radiation, the Indiana event provided important clarification as to the nature of the heavy unstable particles observed by Shutt and associates. The event (R-530), shown in Figure 15.2, was one of the very few in that epoch of cloud chamber work in which the measurements were sufficiently accurate to permit complete vector analysis of the event. This vector diagram is shown in Figure 15.3. We confirmed that the heavier of the two neutral decay fragments was indeed a Λ^0. However, the other neutral particle appeared to be a normal θ^0 meson. The measurements made with our so-called rectangular magnet chamber

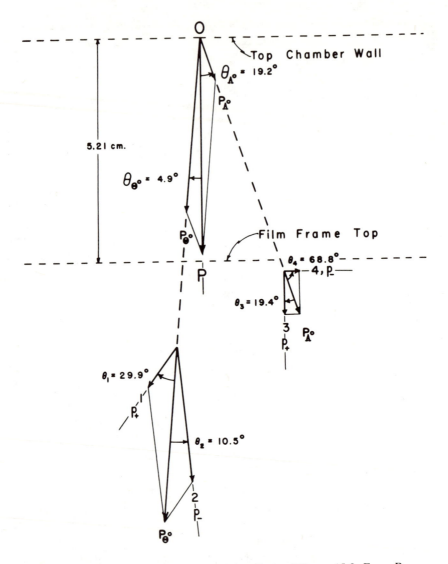

Figure 15.3. Complete vectorial analysis of Figure 15.2. From R. W. Thompson et al. in *Phys. Rev.* **95** (1954), 1579, Fig. 5.

were sufficiently accurate to show that the incident particle itself was a pion. Thus we deduced that we were observing the reaction

$$\pi^- + p \rightarrow \theta^0 + \Lambda^0.$$

Angular-momentum conservation then required that the θ^0 have integral spin, providing the first experimental evidence that the θ^0 meson is a boson.

The Indiana Q-value measurements have rather remarkably stood the test of time. It should be pointed out that the currently accepted Q values were determined with the aid of accelerators, where beams are defined in respect to nature, direction, and momentum, and that relatively enormous research funds were needed to obtain relatively minor Q-value measurement improvements. It is a question of making measurements of sufficient accuracy to resolve the data, rather than the desire to measure for the sake of accuracy alone.

Notes

1 This work was supported by a grant from the U.S. Office of Ordnance Research and by a grant from the Frederick Gardner Cottrell Foundation.
2 G. D. Rochester and C. C. Butler, "Evidence for the Existence of New Unstable Elementary Particles," *Nature 160* (1947), 855-7; P. M. S. Blackett, "The Measurement of the Energy of Cosmic Rays. I-The Electromagnet and the Cloud Chamber," *Proc. Roy. Soc. (London) A154* (1936), 564-73.
3 A. J. Seriff, R. B. Leighton, C. Hsiao, E. W. Cowan, and C. D. Anderson, "Cloud-Chamber Observations of the New Unstable Cosmic-Ray Particles," *Phys. Rev. 78* (1950), 290-1.
4 R. Armenteros, K. H. Barker, C. C. Butler, A. Cachon, and A. H. Chapman, "Decay of V-particles," *Nature 167* (1951), 501-3.
5 R. W. Thompson, Hans O. Cohn, and R. S. Flum, "Cloud-Chamber Observations of the Neutral V-Particle Disintegration," *Phys. Rev. 83* (1951), 175.
6 R. W. Thompson, A. V. Buskirk, L. R. Etter, C. J. Karzmark, and R. H. Rediker, "The Disintegration of V^0 Particles," *Phys. Rev. 90* (1953), 329-30.
7 R. W. Thompson, J. R. Burwell, and R. W. Huggett, "The θ^0-Meson," *Nuovo Cimento (Suppl.) 4* (1956), 286-318.
8 R. H. Dalitz, "Decay of τ Mesons of Known Charge," *Phys. Rev. 94* (1954), 1046-51.
9 W. B. Fowler, R. P. Shutt, A. M. Thorndike, and W. L. Whittemore, "Production of Heavy Unstable Particles by Negative Pions," *Phys. Rev. 93* (1954), 861-7.
10 R. W. Thompson, J. R. Burwell, J. R. Huggett, and C. J. Karzmark, "Evidence for Double Production of V^0 Particles," *Phys. Rev. 95* (1954), 1576-9.

Part IV

Discussion and commentary

16   First round-table discussion

Roger H. Stuewer (chairman), Robert W. Seidel, Donald F. Moyer,
Victor F. Weisskopf, Gilberto Bernardini, Silvan S. Schweber, Paul
A. M. Dirac, and Herbert L. Anderson

M. Barnóthy (Forró): There are a few names that should not
be forgotten if we wish to place developments in the study of cosmic
radiation in a historical perspective. Obviously the first and most important name to mention is that of Victor Hess. I do not know how
many are aware that the controversy between Hess and Millikan was a
very bitter one. I know it, because when eventually, due to the good
offices of Erich Regener and Werner Kolhörster, Hess got the Nobel
Prize for the discovery of cosmic radiation, he wrote me a 16-page
handwritten letter in which he explained in detail how Millikan, until
about 1926, disbelieved his results and the existence of the *Höhenstrahlung*, but thereafter claimed to have discovered the radiation (for
which Millikan aptly coined the name cosmic radiation). I am not
implying that Millikan's careful measurements made between 1923 and
1926 in Lake Muir and Lake Arrowhead did not contribute significantly to our knowledge. Hess's letter, which would be of historical
interest, had to be burned, unfortunately, before we escaped from
Hungary in 1948. I published the content of this letter in 1937 in The
Journal of the Hungarian Society for Natural Sciences (*Természet
Tudományi Kozlöny*).[1]

The second important discovery was that of Walther Bothe and
Kolhörster in 1929; they established the corpuscular nature of the radiation by measuring coincidences between two Geiger counters with a
4-cm gold absorber between the counters. This experiment inspired us
in 1929 to build the first cosmic-ray telescope, a zenith instrument with
which we tried to establish an east-west asymmetry of the radiation

and to see whether the radiation does not originate in some special astronomical objects such as the sun or novae. By 1933 we already had several large zenith instruments. With these instruments, placed in the tower of the Eötvös Institute and in a coal mine near Budapest at 300 and 1,000 m water equivalent (m.w.e.) depth, during the years 1933-43, we established with multiple correlation how the intensity of the radiation depends on changes in the barometric pressure, air temperature, and geomagnetic field intensity. The monitoring of the intensity over many years according to sidereal time enabled us to observe a periodicity linked with the rotation of the Milky Way, proving thereby that a part of the radiation has an extragalactic origin. With a horizontal telescope of 1° zenith-angle opening, and 3 m of lead absorber between and above the counters, we studied the interaction of high-energy (>50 GeV) decay products of μ mesons (neutrinos) with lead nuclei. These investigations, together with the absorption curve of the radiation in lead at 1,000 m.w.e., the increase of the intensity at large zenith angles, and the observed positive temperature effect at great depth, led us to the inference that strong interactions must take place between neutrinos of very high energy created in μ-meson decays and nucleons.

Tribute should be paid to the great contributions of Regener and his group (Ehmert, Pfotzer) in Stuttgart. They extended our knowledge of cosmic radiation over a wide range of intensity. In the years 1936-9 they made careful balloon measurements of the intensity of the radiation up to 27 km above sea level and studied its absorption in Lake Constance, down to 250 m depth.

J. Clay in Holland was the first to discover in 1930 the geomagnetic latitude effect of the radiation. His and Gemert's observation in 1939 of the intensity of the radiation at nine different levels in a coal mine, varying from 10 m.w.e. to 1,380 m.w.e., showed that the power spectrum describing the decrease of the intensity with increasing depth (energy of the rays) changes its slope abruptly at 427 m.w.e. This finding is concordant with our results and inferences on the nature and behavior of very high energy decay products of μ mesons.

Seidel: I would also like to respond briefly that this dispute between Hess, Kolhörster, and Millikan broke into print in 1928 in a published letter from Hess and his collaborators, repudiating Millikan's claim to have been the discoverer of what was then variously called the Millikan rays and the cosmic rays.[2] There is also a Millikan article or two in response to that article; so there are, in addition to this letter that has been destroyed, historical publications and manuscripts relating to this particular dispute.[3] It has not altogether been forgotten.

Moyer: I think that Professor Dirac (Chapter 2) introduced a theme that has been repeated in the other chapters: the dramatic change in the direction of work in response to unexpected results. He mentioned his change in interest from Bohr orbits to Heisenberg's quantum condition, a change in direction that occurred when he read Schrödinger's theory, a change in direction when the Jordan-Wigner second quantization for fermions took him beyond classical analogies, his interest in negative energy, and, after the vindication of the positron, his work on the vacuum polarization. A number of other dramatic changes of direction were discussed by Professor Bernardini: recognition of the penetrating rays, the showers, the positron, the muon, the lifetime problem. We've been talking a lot about the Conversi-Pancini-Piccioni paper, and also Weisskopf talked about the major change of direction when he was a graduate student because of the three Dirac papers and the discovery of the positron, meaning one had to do quantum field theories with many electrons. He also mentioned the cosmic-ray showers and renormalization theory. Anderson, as soon as he began getting pictures in the cloud chamber, was seeing many more things than he expected: plus and minus particles and penetrating radiation. These are exactly the things that a historian is interested in, these dramatic changes. The historian is interested in studying and narrating the actions of the scientists in these situations, and of course explaining the reasons signified by the actions if possible. We would like to distinguish between reasons based on material reality (the discovery of the positron is certainly an example) and reasons based on local conditions. In Professor Anderson's case, it was quite clear. Millikan insisted that he build the cloud chamber. So the reason he did that was what I call a local condition. I read a long list that no one remembers, but these are the types of things that I would like lots more detailed information about. What actions occurred in these changes of direction? And what parts of these responses were responses to material reality and to local conditions?

Weisskopf: I think your examples are not all commensurate.

Moyer: Right. They weren't intended to be.

Weisskopf: For example, the fact that Millikan liked cloud chambers is a very local phenomenon. I'm not even sure whether

Anderson would have used the cloud chamber, whether Millikan liked it or not. And if he hadn't, then somebody else would have – Blackett, or somebody in this country. Also, you may have overestimated, and in some cases underestimated, the revolutionary character of the changes. The most revolutionary of all events was the discovery of quantum mechanics in the twenties. Maybe some people will disagree with me. Heisenberg's new way of looking at things, which turned out to be not so different from Schrödinger's, was a real revolutionary new idea that had an enormous consequence on thinking. The second revolutionary idea was the influence of relativity on quantum mechanics, as exemplified by Dirac's equation and many other things, like the Klein-Gordon equation. I don't know whether going to the many-particle idea from the one-particle idea was such a revolutionary step. I think that it came as a consequence of the other steps. I would not call this a prime effect; it was a natural development of quantum mechanics. It had to go in this direction, because it was obvious, when Heisenberg methods were applied to radiation theory, that we deal with systems of infinite degrees of freedom; methods had to be found sooner or later of the kind that Wigner and Jordan and others introduced (second quantization). Perhaps if Fermi had solved the problem, we all would have understood it a little earlier, but that's not the point. I believe the essential thing is the impact of quantum mechanics and relativity on the treatment of systems with many degrees of freedom. That means the combination of Einstein, Heisenberg, Dirac, and Niels Bohr. Niels Bohr has not been mentioned enough in our discussions. I think that is a very big mistake. Niels Bohr is one of those people who did not work so much by himself, with the exception, of course, of the Bohr model of 1913. (I am now speaking of the quantum mechanics period, of the twenties.) The influence of Niels Bohr on the developments of the twenties was absolutely enormous. It's pretty hard for people to find it out from the published material. It would be very important if some of the people who were there would get together with historians and write it down. There will be an excellent occasion for this, because in five years we will celebrate the 100th birthday of Niels Bohr.

Stuewer: H. B. G. Casimir recently made a remark that people went to Copenhagen to think, to philosophize, rather than to calculate, as in the case of, say, Munich or Göttingen during this period. Would you agree with that, Dr. Weisskopf?

Weisskopf: Definitely. Absolutely and definitely. In Munich and Göttingen you learned to calculate. I couldn't express it better. In

Copenhagen you learned to think. I must say that in Zurich with Pauli I learned both.

Bernardini: I would like to examine the transition that happened (and I refer in particular to Hess, who has been mentioned rightly now) between what had been for years cosmic rays mainly as geophysics and the subsequent stage when they became the physics of the particles that compose cosmic rays, with the precise aim of identifying the kinds of particles. I wanted to mention this transition because, in these last few years, it went also in the opposite direction: Elementary particle physics, the physics that we now know, is exerting its great influence on astrophysics. For instance, we understand now the cosmological structure of extragalactic bodies far more than 30 or 40 years ago, as a consequence of knowledge about the elementary particles.

Schweber: I would like to continue on the initial thread that was raised, and perhaps limit myself to the field-theoretic aspects of the thirties and forties. Even a cursory look at the field theory of the thirties and forties, particularly of the post-World War II period, reveals a marked contrast, which is, talking broadly, the revolutionary stand of most of the field theorists of the thirties as compared with the conservative stand of the post-World War II generation. What I have in mind are Heisenberg's theories of fundamental length, Dirac's work in which he considered Hilbert spaces with negative norms, Bohr's constant refrain "This is not crazy enough!" as compared with the solution advanced by Feynman, Schwinger, Tomonaga, and Dyson, which was at its core basically conservative. The latter asked to take seriously the received formulation of quantum mechanics and relativity and to explore the content of the synthesis. Yet, as you, Professor Dirac, have stressed (Chapter 2), your own revolutionary epistemological stand was marked by a conservative ontological stand, meaning an unwillingness to accept new particles. Not only you. If you look at Pauli, he brings forth theories of a neutrino, and yet he's not willing to take a stand and publish it.

So my question to you is this: How do you explain this dichotomy? What are the sources that maintain the revolutionary theoretical stand, particularly in your theories, your willingness to adopt new and revolutionary theories, yet make you take a conservative stand with respect to new particles? The other thing that came to mind as I listened to your presentation (since one reason for a meeting is to draw insights from the lessons you have learned) is that I hear you saying "You must be bold. You must have courage. You must take a stand. You must be revolutionary." You deplored Schrödinger's lack of courage in not publishing his initial relativistic equation, the second-order equation. Yet you benefited from his willingness to wait and get the nonrelativistic equation, the first-order equation. It allowed you to get the general transformation more simply. There clearly would have been a delay, in terms of trying to assimilate the second-order kind of approach. So my question is this: Is it a universal stand that one should take? Should one always be bold? Should one publish even though it may mean for the community at times not being able to assimilate what you are saying?

Dirac: I don't know if one can give general rules about it. One can just consider the various examples in the past, and, as I mentioned, there is the example of the spin of the electron. People were very willing to agree to authorities. Pauli and Lorentz were strongly opposed to the spin. It needed young men like Goudsmit and Uhlenbeck to push forward their idea. Kronig himself wasn't strong enough to push forward against Pauli. Goudsmit and Uhlenbeck only succeeded because of the impetuosity of Ehrenfest. I don't think one can give any rules about whether one should be bold or not. One doesn't want to stick one's neck out and afterward be shown to have put forward a stupid idea. But still, what may be thought of first as a stupid idea may turn out to have some truth in it.

People were mentioning the big changes in ideas, and so far as my life was concerned, there was the one tremendous change brought about by Heisenberg's introduction of noncommutative algebra for the dynamical variables. That was completely unexpected, completely away from everything that one had learned, and it was quite a wrench to accept it at all. And then, once one did accept it, one was on quite a new footing. One was in a new world. The development of this new world led to wonderful results. The whole related development of quantum theory sprang from this one fundamental change.

Weisskopf: I would like to make one small remark concerning Pauli. I think it is incorrect that Pauli refrained from publishing the neutrino because he was afraid of publishing it. Pauli never liked to publish. Indeed, he always said that "I had this idea also, but I am so glad that somebody else worked it out so that I can do something else." So I believe that Pauli was very happy that Fermi worked out the neutrino hypothesis. The lack of publication of the neutrino was not a lack of courage.

Dirac: I never understood Pauli's objection to the spin. Did you? It was sufficiently strong to crush Kronig absolutely.

Weisskopf: Whatever Pauli says is sufficiently strong to crush anybody. But I would say that I don't understand it either. The only way that I can explain it would be that Pauli did not like the idea of expanding the classical concept of angular momentum that far away from its classical applicability.

Dirac: You mean the half-quantum.

Weisskopf: Yes. The half-quantum number is so far away from what one really means by angular momentum, shown by the well-known misunderstandings of the Einstein-Podolski-Rosen difficulty, if it is formulated in terms of spin measurements on a pair of fermions in a singlet state. I think the idea of extrapolating the angular momentum to a quantum number one half was somewhat against his taste; that's all I can say.

Dirac: The need had already been shown previously in the spectroscopic analysis of the Zeeman effect.

Weisskopf : Which he did, actually.

Dirac: He did that himself. Well then he must have seen the need for half-quantum numbers.

Weisskopf: Yes, but what he objected to was to associate it with angular momentum.

Bernardini: I should make a little personal premise. I had the honor and pleasure to be almost a friend of Pauli, mainly because he liked wine, and I went a few times with him near Frascati, and then around Florence, to drink good wine. Anyhow, on one of these occasions, I remember that I spoke with him (I was rather a young fellow, of course) about the neutrino. And his reaction, as far as I remember, was just in the sense now indicated with the angular momentum. He was a man of geniality, but he had such a solid structure in his way of thinking that he didn't like revolutions. At that time he didn't like the idea, pursued in that period, to abandon the principle of conservation of energy, and then he brought out the idea of the neutrino. This, I think, was a great motivation for him; but, of course, he was so wise and so enlightened that he didn't have the opportunity to defend in other ways these personal convictions. I may only remind Viki [Weisskopf] (because in this sense we have additional elements of community) that Pauli once made a statement that probably Viki remembers; it came at the end of a lecture concerning general physics, metaphysics, etc. It is the following: "We are inclined to believe that what is discovered in physics related to conservation of energy, momentum, etc., has to be right and correct because [and this is a statement that impressed me] the human mind and the objects that we see and that we discover belong to the same perennial order." In other words, it was an expression of faith in this congeniality, harmony, that has to be between a man who is trying to investigate nature and the answer that nature is supposed to give.

Stuewer: The question of the aesthetics of science, or the aesthetics of physics in particular, is a very provocative one. I wonder if either of you, Dr. Weisskopf or Dr. Dirac, would like to comment on whether Pauli's aesthetics agreed with yours?

Dirac: I didn't know him that intimately.

Weisskopf: It could not disagree with mine, because I got everything from him.

Dirac: Pauli very often bet on the wrong horse when a new idea was introduced. He was very much opposed to the positron idea to begin with, for about three months, or was it six months?

Weisskopf: Yes. He was also opposed, as I said, to your idea of filling the vacuum.

H. Anderson: We've been discussing in this symposium a good deal about how theorists do theory and then experimentalists do experiments, and they get together somehow, but that the experimentalists are not very much inspired by the theory, and the theorists somehow manage to anticipate some of the developments in working out some of their ideas. But I want to mention another kind of theorist, because of my familiarity with Fermi's work. He was the kind of theorist who didn't work that way at all. If you examine all the theoretical works of Fermi, and there were some very distinguished ones, as you know, they were always done with the idea of explaining some experimental fact. If you read Fermi's paper on the β-ray theory, which was a foundation of field theory and came about as a result of his coming to understand the idea of the second quantization, you'll find that he developed the theory and immediately applied it to explain the β-ray spectra. At the end of almost every paper in which he made a theoretical development, there was always a calculation to show that the theory gave some agreement with the experiment. This, in a sense, was a protective thing. You were not as likely to be wrong if you developed a theory that agreed with experiment. Thus, Fermi's method was never to publish unless he had made that test.

Schweber: I would like to follow up and direct a question to Professor Weisskopf. The generation in which you were trained had the philosophical bias that was given to it by Heisenberg, which was to look at observable quantities, and Dirac keeps on stressing that point. If you look at the early papers of Oppenheimer on field theory, this necessity for looking at observable quantities is pretty clear. In fact, let me read you a little excerpt from that paper.[4] After he has calculated the self-energy of the electron and found it divergent logarithmically, he says that "the energy level of the normal state is thus infinitely displaced by the interaction of the particles with the field. The question which we now have to consider is whether or not the energy differences between two states are displaced by a finite or infinite amount." The question that of course comes to mind is this: After you had heard from Furry and Carlson that the divergence was logarithmic and had remembered Oppenheimer, you would have gotten a finite answer for the difference in the energy levels. The question that really arises. . . .

Weisskopf: No. Indeed, the answer to this question is that the differences are also infinite. Because, after all, everything is proportional to the Rydberg, and the Rydberg is proportional to the mass of the electron; if the mass of the electron is infinite, then everything is infinite. And so this cannot be solved that simply. In other words, in order to get finite shifts, one has to (excuse the word) renormalize the theory and extricate that infinity, or, in other words, put the actual electron mass in it. Wigner and I ran into this difficulty when we calculated the line width. We also found an infinite shift, but we disregarded it. We had the excuse that it may have been caused by the fact that our calculations were nonrelativistic. But Oppenheimer should have known the answer to this at that time.

Schweber: What I was really driving at is this: Would an experiment have made a difference? In other words, would Pasternack's indication of deviations at that time have given you a stimulus to take the notion seriously? What really is in the back of my mind is that you took it seriously in relation to charge renormalization. Subtraction was clearly also in the air in dealing with vacuum self-energies. People were doing it; so the notion of renormalization was not very foreign. You yourself keep on asking "Could the generation of the thirties overcome the divergence difficulties of QED?" And you keep on saying yes. There are enough instances in particular cases like charge renormalization and vacuum self-energy that you were not very far away.

Weisskopf: Bob Serber could testify to this better than I. My answer to your question would be definitely yes. If there had been a Lamb-Retherford experiment at that time, probably a lot of us, I'm sure Euler and myself, would have sat down and tried to calculate it, and I don't see why we couldn't have calculated it pretty much in the same way as French and I, or Kroll and Lamb, did later. We probably would not have used the elegant methods developed by Feynman, Dyson, etc., but we probably would have gotten the correct result, as we did in 1948. It is interesting that one of the reasons why I think we did not do it was that there was no experiment, or practically no experiment. Another reason was the tremendous attraction of nuclear physics at that time. A new field of physics opened up, and you don't like to sit on the sidelines and calculate a vague Pasternack effect when other people are about to explain nuclear resonances. I would be very interested in Bob Serber's opinion.

Serber: I think one of the reasons we didn't do it was the conviction that Oppenheimer always spread that the theory was wrong, even in its finite parts. A more naive approach might have led to a more successful effort. The effort wasn't really made. After the vacuum polarization work, I did try more general subtraction physics and regularization, but without success. We gave it up after a short time. And then Felix Bloch and Oppenheimer got together a little later and straightened out some of the problems that were important. They separated mass and vertex and wave function renormalization, and years later I heard Bloch lament to Oppenheimer that if Sid Dancoff hadn't made a little mistake in the calculation, they would have seen the point right then. So it was close; but as Viki said, people were also diverted by nuclear physics and what was going on in astrophysics.

Seidel: I would like to travel from the mundane world of theory to the ethereal realm of cosmic-ray experiments in the 1930s and ask a question that does relate in a sense to local conditions, and particularly to the Italians working in cosmic-ray experiments in the 1930s. There was reference in the chapter of Carl Anderson (Chapter 7) to the cosmic-ray work of Robert Millikan, which was not in any real theoretical sense connected with the development of quantum electrodynamics, although he did derive certain consequences as to the absorbability of electrons in the atmosphere from the Dirac theory and other theories. On the other hand, Millikan was tremendously successful in the 1930s in building cosmic rays as a field of study in the United States, chiefly through a very strategic use of his entrepreneurial skills to publicize his work in cosmic rays, which resulted in them being named after him for a time, and also in the dispute with Hess and his collaborators. He was successful in winning support for the study of cosmic rays in the United States, through the Carnegie Institution of Washington, that went not only to Caltech but also to the University of Chicago for Compton's work and to a number of other individuals and institutions. Therefore, in this period of the history of science, we see a large-scale organization of scientific effort that was controlled to a certain extent by what was called the Advisory Committee on Cosmic Rays at the Carnegie Institution of Washington, which granted money for this purpose throughout the United States. There was a financial and experimental network set up at that time to support cosmic-ray studies. I was wondering what, if any, analogue one would find with the Italian situation. How was their work supported? What sorts of informal networks of communication existed, and in particular, what was their reaction to the work done by the Americans? Was there any sense of national rivalry?

Bernardini: Bruno Rossi may check if what I am saying now is correct. At that time, Italy was very much dependent on the power of some very distinguished personalities. And in Florence we had the lucky situation to have a director of the Institute of Physics, Professor Garbasso, who was in his time a fairly good theoretician, particularly in relation to Lorentz's works, etc., who was, since the very beginning, supporting the work that had been done by Rossi. I joined them three or four years later. Because he had also a respectable position, he was now and then Minister of Education, etc., he found the money to support particularly this kind of activity on cosmic rays that was gradually developing in Florence under the guidance of Bruno Rossi. And I may add that there was also involved, I think, an organization, almost brand new, started at that time. Its name was Consiglio Nazionale delle Ricerche, and Garbasso had some links and influence also in the money that was given there, and this was the start of the possibility for us to make some (for that time and for Italy) sizable expenditures. However, I would like to insist on this difference, which I mentioned before, between what was going on in the United States and in Italy. One aspect was related to the structure, to the characteristics of those particles, that is, to the research that had been done concerning the influence of the magnetic field of the earth on the cosmic rays.

Rossi: I don't think of much to add to what Gilberto [Bernardini] said. It's quite correct, and it is true that through Antonio Garbasso we got the support we needed. Garbasso was at that time many things. He was director of the institute that he had built. He was the mayor of Florence. That was very important. I remember a year that they wanted to cut off the electricity from the institute, because the institute did not have enough money to pay for it, but because Garbasso was the mayor, they didn't do it. We were without heat; I remember that we were wearing very heavy coats. Anyway, he was the mayor of Florence, he was a senator, and (I think that Gilberto is right) I think he was Minister of Education for some time. He had great merits. He was very enthusiastic, not about his own work, which he had renounced, but about the work of the younger people around him. Gilberto mentioned that Garbasso had done some quite creditable work before World War I, in which he had taken part. After World War I, he got involved in politics and renounced his own scientific activities, but he still kept a very strong interest in physics. He was still coming to the institute three times a week to deliver his lectures, and whenever any one of us had an idea that he thought was good, he very enthusiastically supported it, and I think that's a great merit.

Piccioni: I've recognized that we're on the Italian scene at the moment. I want to say that our experiment on the capture and noncapture of μ mesons was not done under those happy circumstances of having abundant funds to buy what we needed. It was even hard to find at all the needed materials. It was done during the darkness of World War II. With the help of Amaldi we could buy some RCA tubes on the black market. Of course, in 1943 the important goal was to avoid being "drafted" by the Germans. We succeeded almost 100%. At one time the Germans caught me, but, fortunately they did not like me. They asked me if I was a partisan. I said no, and they said, "We prefer the Polish partisans because they know how to die." What price glory! They sent me back to Rome, and I continued my work. Immediately after the war, funds were still scarce. However, the cost of our experiment was very little.

Stuewer: Could you tell us the source of funds that enabled you to buy the equipment on the black market?

Piccioni: Let me inject that my salary during most of the preparation was paid by the Italian Army, because I was an officer, and I was stealing 50% of my time for physics. About the black market, Amaldi had some ways of getting some residuals of a few lire, and there was a good friend of mine, Lepri, who was at the same time very smart in electronics, and quite an expert in the black market. However, we certainly did not have the funds that were available in, let's say, 1941, when with the guidance of Gilberto we went to the mountains to do cosmic rays and stayed quite elegantly at one of the best resort hotels of Cervinia.

Bernardini: I may only add that in Rome there existed an old man, Corbino, who directed the institute and who was very much the equivalent of Garbasso. He called Enrico Fermi from Florence, and Fermi went to Rome just because Corbino was able to establish a special chair to have Fermi there. And he was also a man, as Bruno [Rossi] was saying, who was Minister of Education and many other things. It is this kind of boss that has always existed in Italy, and probably in all parts of the world, but in Italy certainly. He strongly supported everything good that was going on. What I may say is the following (and Bruno, who was there, may testify): Since the very beginning, when Bruno started to go to Rome to visit Fermi, Fermi was extremely interested in cosmic-ray research. I remember that when the war arrived, and Conversi, Pancini, and Piccioni had started their experiment, I was more or less there, going back and forth, because at that time I was already in Bologna. They had the courage to move their equipment to the Liceo, in order to avoid the consequences of the bombardment, etc. This was rather dramatic, but not particularly for the lack of means, because the means were more or less what we were used to having at that time in places like Rome and Florence.

Stuewer: Could you perhaps recapture for us the effect on the Rome group of the departure of Fermi in 1938? How did that influence you, directly or indirectly? Or is that too difficult a question?

Bernardini: First of all, Bruno and I were together in Florence; then we were divided (he was already in Padua, I believe). So we were separated from Rome, and our visits to Rome were essentially when we had something that we considered both interesting and disquieting. Only then we went to speak with Fermi. But the main point for me was essentially to maintain these contacts. We started to do these experiments on the magnetic lenses, and we went on; but for these preliminary experiments with Marcello Conversi I don't remember that we had special troubles. On the contrary. We had been supported by the preliminary results and then by the theoretical work of Gian Carlo Wick, who is two years younger than I and who was a student of Fermi. He helped us, Marcello Conversi and myself and Pancini, to believe that what we were finding, particularly the effect of the concentration of these particles, was really due to the fact that the fields inside were strong enough to deflect these particles, as Professor Pucci-anti suggested to me. Considering your remark, I would say that I remember the war (but probably this is completely different from Pic-cioni) as a dramatic event, but for other reasons, not for the physics we were able to do.

Thompson: First of all, I should like to speak (in a humorous vein) on the Italian group. We all concur that the famous experiment of Conversi, Pancini, and Piccioni, establishing that the μ meson was not the Yukawa meson, and the famous experiment of Rasetti, etc., are great contributions to the history of physics. What I want to call your attention to is this: It was not entirely an Italian affair, because there was a certain already world-famous physicist, Bruno Rossi, who emigrated to the United States and took up a position at Cornell. There, with Norris Nereson, he determined the lifetime of the μ to an accuracy in unbelievable agreement with the currently accepted value. This with pre-World War II tubes and without benefit of a cathode-ray oscilloscope. It is not surprising that J. R. Oppenheimer, then director of the atom bomb laboratory at Los Alamos, brought Rossi there as an electronics expert! While there, Rossi headed a group devoted to the study of particle detectors. One of the results of this group was my invention, entitled "shallow (multiple wire) proportional counter."

At the conclusion of World War II, Rossi took up a position as professor of physics at the Massachusetts Institute of Technology. Once settled at M.I.T., Rossi built a school of cosmic-ray physics such as had never been seen before. For graduate students, candidates for the Ph.D., he had Matt Sands, H. S. Bridge, Robert Hulsizer, R. W. Willi-ams, R. W. Thompson, etc. During my sojourn there, he had an ex-tremely able group of collaborators, including at various times such persons as Robert B. Sard, Wayne Hazen, George Valley, Oreste Picci-oni, etc. Rossi instigated, or caused to be instigated, a cosmic-ray pro-gram of great breadth, utilizing all the then known experimental tech-niques, in each of which Rossi was himself quite expert, except for the Wilson chamber. This technique Rossi had never himself used, and he wanted someone in his group to take it up. I enthusiastically agreed, and that is how I got into the cloud chamber business in the first place. We did a great deal of reading on the subject, but most helpful, of course, was the beautiful paper by P. M. S. Blackett and R. B. Brode.[5] This paper set standards of Wilson chamber performance that we much later considerably improved on, but it was not easy to do so.

Rossi was scheduled to give a course of lectures on cosmic rays.

Unfortunately, when the course was about to begin, Bruno was stricken with a severe case of jaundice and was bedridden indefinitely. The public demanded that these lectures be given, but by whom? Despite the various more able collaborators that Bruno had to choose from, I got the tap on the shoulder. I was afraid that when they saw that the lectures were to be given by an imposter (even worse, a beginning graduate student) they would drop out like flies. But about 70% stuck with me, and after the last lecture they actually gave me a little round of applause – not enough to injure the eardrums, but better than nothing.

We wrote up his lectures at considerable length, and these served as a starting point for his treatise. Some of you may not know that Bruno has written a most refreshing popularization on the subject.[6] This is a unique source book for historians of cosmic radiation, written by a person who was in the thick of it from almost the beginning. The cover is a muon decay lifted from my thesis. I never expect in the future to be associated with a man who is so very pleasant to deal with personally and at the same time so intellectually stimulating.

Chasson: A couple of points should be noted. One is that between the dynamic meditation place in Copenhagen and the very practical place in Manchester was a very well beaten path. The other is an important personal matter of history for some of us. The Carnegie grants for cosmic-ray work included magnet money given to Bob Brode at Berkeley in the thirties. That magnet turned out to be the Calutron, which was used for some very important isotope separations. After World War II, it went back into the laboratory to do cosmic-ray work, and it was the servant of a lot of elementary particle students fostered by Bob Brode, whose name should not be forgotten at this conference.

Pais: May I make a few brief comments on some of the things I heard today? The first one is about Pauli. As I understand Pauli, he was a deeply conservative kind of physicist. He wrote a letter, as all of us know, around 1930, about the neutrino. He was asked by Fermi to speak about this at an international meeting, I think at the Marconi meeting in 1931. At that time, Bohr still held the floor about nonconservation of energy, and Pauli, although he was fresh and could really attack Bohr, was also really impressed by Bohr. By the time of the Solvay conference of 1933, Pauli knew on experimental grounds that Bohr was wrong, and then for the first time he opened his mouth. That is my first comment.

My second comment is about the question of the thirties versus the forties. Here I would like to mention one name I have not heard today, a man who did something terribly important in 1946, and that was Hans Bethe. Because before you, Viki, with French, and you Willis, with Norman Kroll, did your calculations, Hans Bethe did something that was nothing short of total chutzpah. He made a calculation on a form of lowbrow quantum electrodynamics, cut things off at the Compton wavelength of the electron, and got 1,000 megacycles. There were two reactions: Everybody was furious because Bethe had done it in an absolutely lowbrow dirty kind of a way, and everyone knew, of course, that he had to be right. The great virtue of the dirty calculation was to show that quantum electrodynamics was much better than we thought. When I hear words in a physics meeting about ontology, it makes me cringe. The point is, in the thirties, people didn't realize that quantum electrodynamics was a good theory, and therefore they tried all kinds of things – nonlinear theories, cutoff processes, and so on. It was not a conservative trend that set in; it was Bethe who calculated 1,000 megacycles, and you simply now had to understand quantum electrodynamics.

I have one final comment, and that has to do with what composes the climate of physics. We heard today about the climate of physics as seen by a graduate student at Princeton University between 1946 and 1950 (Harry Lipkin).* It just goes to show that the climate of physics is made up of many microclimates. Because I also was in Princeton during 1946 through 1950. I lived right through this period, and I can testify to the fact that, by and large, in very large parts of the community, the attitude described had absolutely nothing whatever to do with what was really going on. I am not saying that what was described is wrong; I am simply saying that when you talk about climates, and you publish proceedings of meetings of this kind for later generations, you have to realize that things are much more complicated. I remember at the Pocono meeting (not at the Shelter Island meeting – at Shelter Island things were still very confused) that Robert Oppenheimer said "Now we have field theory in hand," and I remember vividly that Rabi got up and said, "What the hell shall I measure now?" There was the sense that one was very close to the end. And then, of course, came the great disillusions. People applied it to meson theory in the usual

---

* *Ed. note*: Lipkin said his contemporaries were seeking "a new theory that would be as different . . . as quantum mechanics was from classical mechanics."

old-fashioned Born approximations or more elegantly dressed-up Born approximations, and they got nowhere. So there was a period from 1940 to 1950 that was a period of a great high. There was a lot of expectation in the middle of it. Then it calmed down again.

Notes

1 M. Barnóthy (Forró), "V. F. Hess, The Discoverer of Cosmic Rays," *Természet Tudományi Kozlöny* (publication of the Hungarian Society for Natural Sciences) *69* (1937), 200-2.

2 V. K. Bergwitz, V. F. Hess, W. Kolhörster, and E. Schweidler, "Feststellung zur Geschichte der Entdeckung und Erforschung der kosmischen Hohen-strahlung (Ultragammastrahlung)," *Physik Zeitschrift 29* (1928), 705-7.

3 R. Millikan, "History of Research in Cosmic Rays," *Nature 126* (1930), 14-30.

4 J. R. Oppenheimer, "Note on the Theory of the Interaction of Field and Matter," *Phys. Rev. 35* (1930), 461-77.

5 P. M. S. Blackett and R. B. Brode, "Energy of Cosmic Rays," *Proc. Roy. Soc. (London) A154* (1936), 564-87.

6 Bruno Rossi, *Cosmic Rays* (New York: McGraw-Hill, 1964).

17   Second round-table discussion

Spencer R. Weart (chairman), Takehiko Takabayasi, Satio Hayakawa, Charles Weiner, Bruno B. Rossi, Robert Serber, M. G. K.
Menon, and Dudley Shapere

Takabayasi: I want to speak a bit about certain early Japanese
works on elementary particle theory that are remarkable not only in
content but also in their background. They show a characteristic tendency that is distinct from the contemporary international main current
of theoretical physics, I prepared a memoir on some aspects of this
subject (Chapter 18), and I think this will provide some supplementary
material and commentary on Professor Hayakawa's presentation
(Chapter 4). Let me exemplify a few points that relate to the international situation and to methodology.

First, throughout this period Japanese researchers studied the Western work eagerly, but they were not intimately coupled to the main
international research circles, because of geographic and linguistic circumstances and also because of the intervention of the war. This marginality affected the character of their work, sometimes in a positive
sense. Yukawa's meson theory, for example, transcended the inhibitions of the contemporary Western physical thinking. Another example is the Japanese work on meson theory during World War II.
Although in the late thirties the meson theory was developed through
an interaction between the Japanese school and European and American researchers, in the years 1941-7 this interaction was practically cut
off by the war and its aftermath. In this period, important work in
particle physics was carried out in Japan. Most of it remained unknown
to the outside world, until about 1947. In my memorandum (Chapter
18) I discuss Taketani's methodology, his three-stage theory (1936,
1942), which Sakata supported and built on during the early war years

278

while Yukawa absorbed himself in meditation about the fundamental difficulties of quantum field theory. Tomonaga's approach, which resulted in the renormalization theory (1947-8), was different from that of Sakata or Yukawa: It was intended to try to solve the difficulties of meson theory by frontal attack on the field reaction. In the last stage of this work there were some interactions between this progress and advances made in the United States, but I think the historical details of the interactions are yet to be clarified.

Around 1948, communication with the United States was generally renewed. With the new progress of the postwar experimental research, the American advantage in both theory and experiment was increasingly strengthened. The Sakata model was one of the conspicuous Japanese works of this period that provoked a lot of debate, even more so when Sakata pushed it one step farther to the so-called Nagoya model and supported this escalation of model building on the basis of his stated philosophy. His pronounced viewpoint of infinite strata and infinite complexity of matter was, I think, just the opposite to the one expressed by Dirac, Heisenberg, and Yukawa. It apparently had partial relevance to the later development of particle physics.

Hayakawa: I would like to speak about quantum electrodynamics, which may be related to Schwinger's presentation (Chapter 21). I was a student of Professor Tomonaga, and I would like to make some remarks about how his quantum electrodynamics were developed in Japan right after the war; as I mentioned earlier (Chapter 4), the fundamental formalism was already completed during the war. Right after the war, I clearly remember that on one sunny spring day he called his students to his office, which was in ruins, and discussed what would be the program from that point. He mentioned that the formalism was completed and should be developed by applications to concrete problems, like the interactions of electrons and photons, and especially that we should start by the elimination of subsidiary conditions. Could it be done in a relativistically covariant way or not? So the first thing he asked us was to apply Dirac's many-time theory, to start with not the field theory, but just point electrons. And then we succeeded in something. We applied this formalistic field theory to the elimination of the subsidiary conditions. Then we extended this application to the meson further, to the interaction of the electromagnetic field with vector particles, and also to the interactions of meson and nucleon, and so forth. In late 1946, a paper by Bethe and Oppenheimer about renormalization came, and again it gave us a program: to work out this radiative correction to scattering of electrons and to see whether or not the radiative correction could give us finite results.*

At first we worked out the nonrelativistic scheme, but papers by Lewis and by Epstein came, and we had gotten nearly the same results as the Americans. We didn't know yet, of course, about the Lamb shift until the first colloquium after the summer vacation of 1947; Tomonaga brought *Newsweek*, in which the Lamb shift was reported. He told us what was found at Columbia University, and then he started writing down equations, which were the renormalization theory. If we applied the covariant theory to this phenomenon, and incorporated this with renormalization theory in the electron scattering, we could get a finite answer. So the group started to work hard and obtained results by the end of 1947. Concerning vacuum polarization, I would like to mention the work by Professor Sakata on the C meson, which stands for the cohesive force of the meson, exactly the same as the *f* field of Abraham Pais. According to the C meson theory, the vacuum polarization term should vanish. When Koba worked out this problem, there was some part left, but then they found a mistake in the calculation. The final results are well known and were quite clear. So this group obtained the results of the quantum electrodynamics in early 1948, after correcting this mistake in vacuum polarization.

Weiner: I would like to add something to Professor Hayakawa's account. Tomonaga wrote to Oppenheimer on April 2, 1948, describing the work done in Japan. Oppenheimer immediately passed this information on to the physicists who had just returned from the Pocono Manor conference, where these issues were central and where Schwinger and Feynman had presented their work. He explained: "When I returned from the Pocono conference, I found a letter from Tomonaga which seemed to me of such interest to us all that I am sending you a copy of it. Just because we were able to hear Schwinger's beautiful report, we may better be able to appreciate this independent development."[1] And then he quoted the letter from Tomonaga:

> I have taken the liberty of sending you copies of several
> papers and notes concerning the reaction of radiation field in
> scattering processes and related problems, which my collabo-
> rators and I have been investigating for last six months. I

* *Ed. note*: This work was based on W. Heitler's theory of radiation damping.

should be much obliged if you would be so kind to look them over. I should like to take this occasion to relate the circumstances of their formation and the reason why I have made up my mind to send you these manuscripts.

During the wartime, when we were perfectly isolated from the progress of physics in the world, Dr. S. Sakata, one of the main research workers on the theory of elementary particles in our country, made an attempt to overcome the divergence difficulty of the self-energy of the electron by introducing a neutral scalar field (the so-called C-meson field) which interacts with electrons, a hypothesis which we afterwards found to be identical with Dr. A. Pais' theory of f-field [Phys. Rev. 63, 227 (1946)]. In view of its promising feature my collaborator and I have applied this theory to the problem of elastic scattering of an electron in order to put it to a further test because I supposed, as you and Dr. H. A. Bethe have emphasized, that "this simple problem may afford a useful test of future theories of radiation." We have thus carried out a perturbation calculation following Dr. S. M. Dancoff [Phys. Rev. 55, 959 (1939)] and, though we at first made the same oversight as Dr. Dancoff missing some intermediate states in the calculation (the first note of Ito, Koba and Tomonaga), finally arrived at the conclusion that the new field is also capable of eliminating the divergence in the scattering cross-section in e^2-approximation (the second note of Ito, Koba and Tomonaga and the two papers of the same authors).

Shortly before we finished this work we received the striking report about the experimental evidence of the level shift of hydrogen atoms and its theoretical explanation by Dr. H. A. Bethe in terms of radiation reaction. Thereupon, I proposed a formalism to express Dr. Bethe's fundamental assumption in a more closed and plausible – as I believe – form, in which the separation of terms to be subtracted is made by a canonical transformation (the note of Tati and Tomonaga). This formalism, which we called "self-consistent subtraction method," was then applied to the scattering problem mentioned above and it was confirmed that the non-diagonal part of the mass-correction term plays a decisive role in the divergent part of the cross-section and just cancels the infinity that appeared in the usual formalism (the note of Koba and Tomonaga and the paper of the same authors).

After we had finished this work the January 15th issue of the *Physical Review* (1948) arrived, in which we found Dr. H. W. Lewis' and Dr. S. T. Epstein's works dealing with the same problem, which, as they wrote, had been undertaken by your suggestion. As the conclusion of these authors was identical with ours, we at first hesitated to make our paper public, but because it should play an introductory part of the series of papers to be published, we thought it not unreasonable to send our manuscript to the editor of our English journal *Progress of Theoretical Physics*. I hope that you and Dr. Lewis and Dr. Epstein would be good enough to acknowledge our works too.

Under the unfavorable condition after the wartime, however, it will take a long time – a year or so – before our papers will appear in print, and thus I have resolved on sending you our manuscripts. (The short notes will appear soon in the journal mentioned.)

In succession to these works we are further developing the formulation of our subtraction method in a relativistically more elegant form according to the formalism proposed by me some years ago [Progress of Theoretical Physics, 1, 27 (1946), 2, 101 (1947)] along the line mentioned in the note of Tati and Tomonaga above, and, on the other hand, we are examining the reaction of the radiation field in the collision processes of elementary particles, among which the consideration about the simplest example, the e^2-correction for Klein-Nishina Formula, is in the course of publication (the note of Koba and Takeda). I should be much obliged if you would kindly take some interest in these works.

In conclusion I wish to express my hearty thanks to you and other scientists of the United States for having bestowed so many favors upon us, such as presentation of journals and literatures, by which the scientific activity in our country will be much incited.

On another matter, I would like to comment on something Weiss-kopf mentioned. He talked about the anxiety of war and the effect on the work of individuals. Let us remember that the war started on different dates for different countries. Japan was at war in the mid-thirties. Let me quote a letter from Yoshio Nishina to Niels Bohr. In August of 1937 they were working on the 60-in. cyclotron at Riken in Tokyo. Japan was the first country outside of the United States to operate a cyclotron, the 37-in., which was completed in April 1937. The Copenhagen cyclotron was the next one that came out. In August 1937, Nishina wrote to Bohr: "If the war with China will not be finished before long, it is possible that we cannot get sufficient funds for scientific research, and that would substantially retard our progress."[2]

Later on, Ryokichi Sagane (who was the link between the Japanese cyclotron builders and the people at Berkeley, where he had spent a year learning to build cyclotrons) wrote to Lawrence that he had difficulty in getting parts, that the copper they needed was being requisitioned for the army, and that the problem of foreign exchange presented further difficulties. Let me quote him: "The regulation is now getting more strict, and it's almost hopeless to import those things which can be substituted by those made in Japan."[3] In the correspondence between Bohr and Nishina there is another ironic note about anxieties and about war. In September 1940, Bohr wrote and wished Nishina great success in building the large cyclotron, and he went on to say, "I suppose that you also realize that we here in this Institute [in Copenhagen] are hard at work with the many interesting problems opened by the wonderful discovery of nuclear fission." Bohr went on to say, "Such interests are indeed the only way sometimes to forget the great anxieties under which all people in Europe are living at present." Of course, the ironic part is that fission was directly applied to the war, with disastrous effect on Japan.[4]

I would like to comment on things that I heard from several speakers today that seem to fit around the issue of isolation, either geographically or culturally, and communication. In Japan this was true in both senses because of language and the distance from other centers of research. In California it was pretty much just geographic. The interesting thing is that under such situations (and this also happens, of course, in wartime, when there is isolation on political grounds as well) there are lots of creative things that people do to provide information and communication. In California there were a great many visiting lecturers. As Professor Serber mentioned, there was a steady stream of visitors. Millikan had in mind that if California was far from the European world, he was going to bring that world to California. Bohr and Sommerfeld and Einstein were among the visitors there in the thirties. There was a parade of people through Pasadena. In addition, there were regular meetings of journal clubs and colloquia. In Berkeley the theory group under Oppenheimer and the similar group at Stanford under Felix Bloch had weekly seminars. One week they would drive over to Stanford, and one week the group would drive over to Berkeley. There was a lot going on. The journal club at Berkeley, where the latest results in the physics journals were discussed every week, was something that both theorists and experimentalists went to. Another important thing about the stimulation of physics there was the close relationship of the theorists to the centers of experiment in the most rapidly developing field at the time. The cyclotron work at Berkeley and the various accelerator projects in Pasadena were examples of that.

The story of Japan is interesting, too, because in the late 1920s at Kyoto University, students who had heard about the new quantum mechanics translated the important papers and published a book of the translations. They also organized a student colloquium to discuss the newspapers. There was no professor at the university at the time who was really able to teach that. The students taught each other. And then one generation of students passed it down to another. Tomonaga and Yukawa were influenced by their high school teacher of dynamics, who told them about the new wave mechanics and encouraged them to study the subject.

Another thing was the effect, as mentioned in the earlier talks, of the visit of Dirac and Heisenberg in 1929. I think it is important to understand the impact of that on young people. Tomonaga was enormously impressed, and it changed many things for him to realize that such young men were making such great contributions. He was close to their age, and it gave him a great deal of encouragement to work. Tomonaga understood the lectures because he had already read the relevant papers.[5] So there are ways of overcoming isolation. I would like to hear from the speakers on the panel about their own experiences.

I want to say something about resources. It seems that in the discussion today, only implicitly did the question of resources appear. The letter I quoted earlier showed the effect that it was having on Japanese physicists in 1937. In several countries where the resources were readily available, a choice was made to use cosmic rays, because you could get the most exciting data, the highest energies possible, for relatively little cost. We ought to consider also the differential effect of scarce resources on theory as compared to experiment. Tomonaga recalled that a major problem during the war and in the immediate postwar period was that there were no scientific journals from other countries. After the war, he, and others in Japan, relied on colleagues in the United States and Europe to supply the scientific journals in order to get caught up. Beyond that, there was a problem especially important to a theorist: the pencil-and-paper problem. It became a serious problem to Tomonaga because he didn't have paper to write on. That was a scarce resource, and they were writing on the back of previously used pieces of paper and pieces of scrap paper. I would like to hear comments from people who may have experienced resource problems.

Rossi: You said that the reason for turning to cosmic rays was that cosmic rays are cheap and machines were costly. But the fact is, you could not do with the machines of that time what you could do with cosmic rays; it was an entirely different field of research. Showers and things like that could not be produced by any of the existing machines. Machines were useful for nuclear physics, but not for high-energy physics, not for particle physics. So I don't think that the financial considerations were important.

Serber: The comment that machines were expensive reminds me that early in 1939 I wrote a letter to the dean of the Graduate School at the University of Illinois asking for a grant to Don Kerst for building the betatron, and the sum that was asked for was $400.

J. Barnóthy: Dr. Rossi mentioned that cosmic-ray research was undertaken not merely because machines were costly and cosmic rays cheap but also because one could not do with the machines of that time what one could do with cosmic rays. May I add that this is still true. In cosmic radiation we find particles of 6×10^{20} eV energy, and we still do not have machines that can even approach the energies of those in the highest energy range of cosmic-ray particles.

Rossi: I agree to some extent. Not completely, because now the energy of these machines is sufficient to produce new particles, and they have opened up this new field of particle physics that was closed or restricted to cosmic rays before. But it's perfectly true that our particles of 10^{20} eV in cosmic rays cannot be produced on the earth.

Menon: Perhaps one should add that even though they exist, they are so scarce that the expenditure involved in detecting them and working on them also gets to be of significant magnitude; so I think it cuts both ways.

Weart: Dr. Rossi, if I could follow up on Charles Weiner's question, you went from an area that I suppose we would call peripheral in a certain sense (Italy) to, as you described it, a central place. Did you find that this had any effect on your work – when you left Italy and went first to Manchester, and then to the United States?

Rossi: I don't know, because I really did not feel isolated at all in Italy. For one thing, I was working in Florence, and Florence is very close to Rome, and in Rome there was Fermi and his group, with whom I had fairly frequent contact. Then I was traveling abroad. There were meetings in Switzerland very often, and I spent one summer in Germany. I felt a part of the European team, not of the Italian team. So I don't think I felt a great change, really, in that respect. In other respects, yes, but not in that respect.

Serber: As far as the theorists were concerned, the chief restriction was the inability to travel. When I went out to the West Coast, I didn't leave it for three years. Nobody was paying money to go to meetings anywhere else.

Weiner: And yet at Ann Arbor, where you went, as well as others, I guess on your way out to California, the summer school on theoretical physics did manage to get people from all over the United States and to mix them up with a good many from Europe, and it wasn't just faculty. Many of them were students. I'm not clear whether any of the students got their way paid, but I know a lot of them came, from the period when it was started in 1927 up through the thirties.

Shapere: In a paper in *Daedalus*, Steven Weinberg, writing about the more recent history of physics, remarked that one of the striking facts about the period he was concerned with was the repeated vindication of earlier methods of approach – the absence, that is, of fundamental "scientific revolutions."[6] Several remarks made at this conference have suggested that the same might be true of the period we have been considering: that, despite Bohrish cries of "not wild enough," despite challenges to the conservation-of-energy principle, despite opposition from one quarter or another to QED, despite expectations of a replacement for quantum mechanics, despite demands for a fundamental length, ideas of that sort, when all was said and done, such radical departures from established ways of thinking proved unnecessary. In comparison with the revolutionary developments of relativity and quantum mechanics, which required drastic modifications of older theories when they were applied to a new domain, here, in the study of the nucleus, elementary particles, and associated fields, the methods of quantum mechanics and quantum electrodynamics were vindicated, according to this picture, in the face of every challenge.

I would like to raise the question how adequate this picture really is. Another way of putting the question is this: What new features were introduced into nuclear physics, particle physics, and field theory in this period, and what is their weight as innovations as compared with the features of earlier quantum mechanics and quantum electrodynamics that were kept? How are we to appraise the significance of these new developments as fundamental alterations in our understanding of nature? Can the changes introduced in our period be understood simply as additions, not of a very fundamental sort, to the structure of physical theory that had been arrived at by the end of the 1920s?

To focus the problem, let me raise it with specific reference to three central ideas in the period covered by this conference. The first of these is the exchange-particle model of nuclear (and, later, of particle) interactions. From the start, with Heisenberg's 1932 paper, it seems to have been felt that this was the way one ought to go. For some time, of course, the situation wasn't clear. Heisenberg's own electron-exchange theory was faulted, as were other versions of exchange-particle theories. Even the Yukawa theory faced grave difficulties for some time. But in the final analysis, with appropriate alterations it worked out. My question is this: Is it really accurate to say that the Yukawa theory and other particle-exchange theories before and after it (but it in particular) were just analogical extensions of an established approach, without any fundamental alterations of principle? One might ask, for instance: Did its blurring of the distinction between entities and interactions constitute such a fundamental departure?

I think a full understanding of this issue is impossible without considering the reasons why people were so ready to look at particle-exchange theories. Why was that such a fundamental matter as to be a kind of guiding model for the construction of a theory? But also, on the other hand, why did people take so long in coming to accept or even to take really seriously the Yukawa theory? To what extent was it taken seriously in the first few years after its dissemination from Japan? In connection with this, I just want to read a passage from the Oppenheimer-Serber paper, which Laurie Brown was kind enough to call to my attention.[7] Here's what Oppenheimer and Serber said: "It has been suggested by Yukawa that the possibility of exchanging such particles of intermediate mass would offer a more natural explanation of the range and magnitude of the exchange forces between proton and neutron than the Fermi theory of the electron neutrino field." Then they went on to list some objections and concluded as follows: "These considerations therefore cannot be regarded as the elements of a correct theory, nor serve as any argument whatever for the existence of the particles; their valid content can at most be this; that these particles may be emitted from nuclei when sufficient energy ($>\mu c^2$) is available, and that they will ultimately prove relevant to an understanding of nuclear forces."

To continue with my question: Was the Yukawa theory considered less promising than other theories? In particular, was there something special about the Yukawa theory, as opposed to other sorts of particle-exchange theory, that involved some departure from fundamental ideas that led to resistance to it beyond its commonly cited experimental shortcomings? And in any case, whether people were aware of this or not, was there in the theory some such radical departure from previous ideas, so that the notion of its being a mere analogical extension is inadequate?

This brings me to the second major idea that I want to talk about: One possible reason, it has been suggested here, that the Yukawa theory may have been considered inferior was that, as opposed to what I'll call Heisenberg-type theories, the Yukawa theory required the introduction of a new particle. Perhaps somehow the idea of introducing a new particle, far from being a violation of a mere blind dogma, was in radical violation of some fundamental feature of prior theory. For example, possibly people felt that there should be only two fundamental particles, a positively charged one and a negatively charged one, and that this was somehow a central and basic idea not to be tampered with. I can't quite swallow that, because if that's true, what is one to make of the photon, the neutrino, and the positron? Probably many people at that time did feel that the introduction of a new particle should be resisted. But my question is this: Did they really have any good reason for this, or was it just a dogmatic quirk? And if there was good reason, was its ultimate violation serious enough to deserve being counted as a radical and fundamental innovation?

Now it might be that a full understanding of this question would force us to look back further in time than the period covered by this conference, namely to the long resistance to the acceptance of the photon, which ended only in the mid-1920s. In any case, those are two, the first two, of the ideas that I wanted to ask about. I'll leave the third one on renormalization for later. But now let me just ask the people on the panel about the first of my issues: Did the Yukawa theory introduce some fundamental departure from prior theory? Or was it just an analogical extension, nothing really radically new or revolutionary being introduced through this development?

Serber: It was an "analogical extension," but it did produce something radically new, namely a heavy boson. It seems to me that pointing this out was the intent of the statement by Oppenheimer and myself that has just been quoted by Dr. Shapere.

Weiner: The paper that was just referred to was submitted on June 1, 1937, and published two weeks later as a letter in *Physical Review*. Bob Serber referred earlier today to a paper that Oppenheimer did with Carlson on showers.[8] It was submitted in December 1936 (this was after the seminar at Caltech where Anderson and Neddermeyer presented their results, and it was before these results were definite enough). Oppenheimer and Carlson stated that one can conclude either that the theoretical estimates of the probability of these processes are inapplicable in the domain of cosmic-ray energies or that the actual penetration of these rays has to be ascribed to the presence of a component other than electrons and photons. The second alternative is necessarily radical. And then they went on to explain why, saying finally: "If these are not electrons, they are particles not previously known to physics."

Weart: Dr. Hayakawa, do you have any feeling for whether in the Japanese group Yukawa's theory was regarded as something revolutionary, something that went well beyond quantum electrodynamics, or whether it was regarded as a sort of natural development?

Hayakawa: Both ways. If you read Yukawa's paper, you will get the impression that this is a natural extension of the electromagnetic interaction. In this sense he named the meson the "heavy quantum," a counterpart of the light quantum. If we go back in history, we know about the quantization of the field. But what would be the outcome of the quantization of the field? For example, the positron can be understood by just taking the electron in the negative state and pushing it up to the positive energy state. So the number of electrons is conserved, and in the case of the neutrino, the same holds. But in the case of mesons, viewed as heavy photons, clearly we have to introduce a change in the particle number, because these are bosons. I think in this sense the fruitfulness of quantum field theory was first proved by the introduction of mesons that have finite mass.

Weart: It's an introduction of a new particle in two senses of the word: creating a particle as well as introducing it.

Shapere: What about renormalization? Did it amount simply to a salvaging of prior theory, without itself constituting a radical betrayal of that prior theory? Or was it a betrayal, either by constituting a radical new approach in physics or by constituting an admission of ultimate failure of prior theory, a mere makeshift expedient with which we could make do, but which could really not provide the kind of understanding that an ultimate physical theory should provide? Or, finally, was it even, as some absent physicists might say, an abandonment of the idea that science aims at "understanding," rather than having as its exclusive aim "getting the right numbers." What was renormalization, the status of it? We had a comment from Professor Dirac on this the other night.

Serber: The renormalization allowed you to make sense of the predictions of quantum electrodynamics and gave you a remarkable accuracy that you had no reason to expect from the original formulation of the theory and that was verified experimentally. It's true that many of us who had worked on the problem before were in a way disappointed by the fact that it had circumvented answering the questions that we hoped would be answered in the course of finding a solution to the problem – that is, the problem of the fine structure constant and why it is what it is, and why the masses are what they are. But on the other hand, this hope may well have been completely unrealistic in that electrodynamics isn't the whole world, and there are lots of other things in it, and it very well may not be possible to solve all the problems by looking only at that limited region with just electrons and photons.

Chasson: I have a historical question, not on this subject, that I would like to direct to our two Japanese colleagues. It has to do with the role of the Meteorological Research Institute under the guidance of Dr. Ishii. In Japan, in those early days of cosmic-ray investigations, that was a purely mission agency as we seem to know them in some familiar places. Wasn't that a great departure from the usual pattern of nonparticipation in pure scientific research?

Hayakawa: The head of the Meteorological Bureau was very interested in pure research. I got my first job in the Meteorological Bureau. I was interviewed by the head, and I said that I was not much interested in meteorology, but he said, "Don't worry. You can do anything you like except for playing go or chess." Perfect attitude!

Pais: I would like to try to make a comment on the questions raised by Dr. Shapere, because I think in some ways these are the questions that go really to the heart of the matter of this entire period. Let me first try to say the simple answer, and then try to explain it. In some ways, after 1925 nothing of such a revolutionary nature has ever happened again. Everything else can in some way or form be considered as application and implementation of the ideas of relativity and quantum theory. These ideas have been very successful up to a point. They have also led to very important open questions. If you take, for example, the idea of Yukawa, that was brilliant. In the deepest sense, though, it was not revolutionary. You did not need to change any of the conceptual notions of physics, although it was not so quickly assimilated. A second point that has bearing on this is the fact that the Yukawa theory was obviously a step in the right direction, but it never was the theory of nuclear forces. It was not only a question of using a new field; it was also the question whether or not you could compute something. There the blessing of quantum electrodynamics is not repeated. The blessing of quantum electrodynamics is that the fine-structure constant is a very small number. This smallness allows you to expand certain things in the theory. There was no evidence of such corresponding smallness in the Yukawa theory. If you make a big jump to 1980, look at what we think the theory of strong interactions is today: how the pion, the great contribution of Yukawa, looks like a sort of Van der Waals force of a totally secondary nature.

I would also like to comment on the question of the role of field theory from 1925 to today. It's a very curious thing. From 1925 until the early thirties there was a great upswing. Then came the difficulties, and people thought: "The difficulties are terrible! We're moving in the wrong direction!" and so on. Then, during the mid-forties, after the war, came the great stimulus of the experiments in quantum electrodynamics, and then came the renormalization theory. Renormalization is a very ingenious device. It does not change the concepts, but it develops the technique by which you can extract much more out of field theory than was thought possible by earlier generations. Then came the early fifties, and you find out that by applying the same methods to meson theory you get absolutely nowhere. That has again to do with the fact that the coupling constants are not that small. Then came a period where people essentially said: "Well, now field theory is dead. It's all over." That went on essentially until the early seventies. Then came this last decade, in which we have begun to realize that there is very much more to field theory, still much more to field theory than we thought. There are the gauge theories, quantum chromodynamics, and so on. It has been a very curious century in regard to field theories.

At this moment there is nothing that tells us that the postulates of relativity and quantum theory are inadequate. Our theories are definitely inadequate; there are too many open questions. But whether or not we have to make changes as revolutionary as was done by our grandfathers or a little bit before, that we just cannot say.

Marshak: Perhaps one can also add this: With the gauge theories, there was, in a sense, looking back now, a not very revolutionary statement, because quantum electrodynamics was successful in its renormalization program because of gauge invariance. That's a pretty deep thing. As soon as you brought in the finite mass, you not only had the large coupling constants to worry about, but you had other problems. In a way, the thing is going to be resolved through a combination of gauge theory and broken symmetry. My final sentence would be that Yukawa's contribution is going to have to be defined sometime in the future yet, to see just how the quark situation works out, and the Van der Waal's situation, before one can really answer the original question as to whether it was revolutionary or not.

Weart: Although there can be no doubt that it was an important contribution.

Nambu: In partial response to Professor Shapere and his question regarding renormalization, I think it might be useful to translate for you the Japanese word that Tomonaga used to mean *renormalization*. It's a little bit different from the word *renormalization* in English; I don't quite know the proper wording for it. It sounds more like compounding or rounding out. It's like putting things together. That's the term usually used for it, but I remember that Tomonaga also used another term. I distinctly remember that he used a word that meant renunciation, "principle of renunciation." I did not quite understand the meaning, and being a young student, I didn't have the courage to ask him (or any of the other students, for that matter). I was always left with an uneasy feeling. But he frequently made use of the word for "principle of renunciation," and I think this is also referred to in Takabayasi's chapter.

Serber: Whereas the concepts of field theory have not changed, really, the physics has, and nowadays we talk about the structure of nucleons, and we've learned about quarks. The field theories we talk about are quite different from quantum electrodynamics, and quite new elements appear, such as quark confinement.

Notes

1 Robert Oppenheimer to members of the Pocono conference, April 5, 1948. Oppenheimer papers, Manuscript Division, Library of Congress, Washington, D.C.
2 Nishina to Bohr, August 28, 1937, Bohr papers, Niels Bohr Institute, Copenhagen. See Charles Weiner, "Cyclotrons and Internationalism: Japan, Denmark and the United States, 1935-1945," in *Proceedings of the Fourteenth International Congress on the History of Science, 1974, No. 2* (Tokyo: Science Council of Japan, 1975), pp. 353-65.
3 R. Sagane to E. O. Lawrence, March 10, 1938, Lawrence papers, Bancroft Library, Berkeley.
4 Bohr to Nishina, September 14, 1940, Bohr papers, Niels Bohr Institute, Copenhagen.
5 Information on Tomonaga was provided to me by him in an interview in August 1974. His published recollections contain a great deal of valuable background.
6 Steven Weinberg, "The Search for Unity: Notes for a History of Quantum Field Theory," *Daedalus 106* (1977), 17-35.
7 J. R. Oppenheimer and R. Serber, "Note on the Nature of Cosmic-Ray Particles," *Phys. Rev. 51* (1937), 1113.
8 J. F. Carlson and J. R. Oppenheimer, "Multiplicative Showers," *Phys. Rev. 51* (1937), 220-31.

18 Some characteristic aspects of early elementary particle theory in Japan

TAKEHIKO TAKABAYASI

Born 1919, Hyogo prefecture, Japan; M. Sc., Tokyo Imperial University, 1941; extended models of elementary particles, history and philosophy of quantum theory; Nagoya University.

In the early development of particle physics (1930-50), Japan made remarkable theoretical contributions.[1] These works have a particular quality that distinguishes them from the international main current of theoretical physics. The following general points are relevant.

In the first 30 years of this century, relativity and quantum mechanics appeared as scientific revolutions. The history of particle physics, which began to be developed immediately following that period, had a somewhat different character. Although its difficulties seemed to require a further revolution in fundamental concepts, each time this was postponed, and the difficulties were overcome by some new subtle device remaining within the framework of quantum mechanics and relativity and revealing a richer world of particle physics in the high-energy regime.[2] We might say that some investigations of this kind, like the introduction of new entities or changes of model, especially suited the style of Japanese workers. Also, throughout this period Japanese research was either weakly coupled to or isolated from the main international research circles because of geographic, linguistic, and other reasons. This marginality had effects (both positive and negative) on the character and the course of Japanese work, and it also affected the international recognition of those works.

In any case, the Japanese contributions to early particle physics display some characteristic particularities, although these need not be overemphasized. In the long run, the progress of physics is international.

294

The importance of the Japanese contribution is clearly demonstrated by Hideki Yukawa's meson hypothesis of 1935.[3] In the heuristic and historical contexts, the following points are particularly important in this work: First, in contrast to Werner Heisenberg's theory of the nuclear force and nuclear structure, Yukawa started by directly postulating the nuclear "U field" as a new physical object that is not reducible to known ones, and he considered it classically.[4] Then he treated it by semiclassical arguments of the type of Oskar Klein. Second, Yukawa's theory had the character of combining, with delicate balance, the nuclear force (treated in Heisenberg's theory) and the β decay (treated by the theory of Enrico Fermi) through the agency of his new intermediate particle.[5]

In retrospect, this theory is considered as a natural step in the orthodox development of the theory, but nonetheless it was a step beyond the inhibition of Western physical thinking. Throughout the early period of particle physics the international current of thought centering around Niels Bohr followed the Scholastic tradition, according to which "one should not multiply entities unnecessarily." Thus Wolfgang Pauli did not dare to publish his neutrino idea nor develop it. Yukawa was perhaps lucky that there was no overcritical authority in Japan. J. Robert Oppenheimer made complicated responses to Yukawa's theory when Carl D. Anderson and associates discovered a new particle in 1937, and betrayed his strong ambivalence toward new particles.[6]

The first round of development of meson theory ranged from 1937 to the discovery of the π meson by Cecil Powell and associates in 1947, there being competition between Japanese and European-American physicists.[7]* Throughout this decade the original Yukawa viewpoint remained the standard one, the meson playing the roles of nuclear force quantum and β-decay intermediary in its virtual states, while its on-shell state was identified with the cosmic-ray meson; just this unifying character of the theory seemed attractive. (Thus, in particular, the lifetime of the meson presented itself as a characteristic problem whose calculation depended on those triple aspects of the theory.) Eventually this Yukawa scheme had to be decomposed and reorganized; nevertheless, in the historical setting, it worked to motivate various sorts of investigations in meson theory that, though inconclusive, were still relevant.

---

* *Ed. note*: An important role was also played by the Indian physicist Homi J. Bhabha.

A special feature of Japanese research on elementary particles derived from Yukawa's first collaborators, Shoichi Sakata and Mituo Taketani, who had studied Marxian philosophy and applied it to their research.[8] Thus Taketani formulated the three-stage theory (1936, 1942), according to which our recognition of nature proceeds by repeated cycles of phenomenological, substantialistic, and essentialistic stages, corresponding to the Hegelian triad of thesis-antithesis-synthesis. Sakata noted the particular importance of the substantialistic path lying between the two usually acknowledged approaches, the phenomenological and the essentialistic.

In particular, Sakata tackled the problems of the meson lifetime and others that had been subject to discrepancies of two orders of magnitude between theory and experiments. Instead of regarding them as due to some deeper theoretical difficulties, he took them rather as a puzzle to be resolved in a substantialistic manner. Taking a cue from the mixed-field meson theory of Møller and Rosenfeld, he thus proposed the two-meson theory.[9] By that time, Yukawa's meson had secured worldwide recognition, and there was a tacit consensus among people like Pauli, Fermi, and Yukawa that this meson should be the last new elementary particle to be admitted. Sakata's two-meson theory was so arranged that it did not impair, but only extended or supplemented, the original Yukawa scheme. Moreover, its method appeared to be just a repetition of the one undertaken by Yukawa in his original theory. Still, Sakata's theory did not get much sympathy in Japan at that time, and the extra particles it admitted were unwelcome strangers.[10]

After that time (circa 1942), the lines of investigation began to differ between Yukawa and Sakata, corresponding to their different philosophical backgrounds.[11] Yukawa began to be absorbed in the general and fundamental difficulties of quantum field theory, and he would speak about his so-called theory of *maru* (*maru* meaning circle), where he suggested the radical path of transcending the conventional mode of thinking of relativistic causality by the notion of a priori probability on an arbitrary closed hypersurface in Minkowski space.[12]

Sin-itiro Tomonaga's style was different from either Yukawa's or Sakata's.[13] He began research as a collaborator of Yoshio Nishina, who had worked at Bohr's institute. He calculated various quantum electrodynamical processes during 1933-6 and then went abroad in 1937-9 for study at Heisenberg's institute. Thus Tomonaga had more immediate access than Yukawa or Sakata to the leading Western currents of thought about physics. During his stay, Heisenberg's institute was somewhat deserted, so that Tomonaga was actually one of the few physicists who were directly influenced by Heisenberg at that time when he was young and still very original. Tomonaga became convinced of the physical reality of field reaction and the need for nonperturbative treatment of meson theory, subjects that he pursued deeply on returning to Japan.

As already stated, Yukawa would talk at that time of his theory of *maru*, motivated by his ambitious project for achieving both manifest covariance and convergence of quantum field theory. This was criticized by Tomonaga, who preferred the Dirac type of approach: that one should solve one problem at a time. Thus Tomonaga limited himself to the problem of manifest covariance and also held the more conservative viewpoint of maintaining the usual notion of causality. He altered Yukawa's mysterious notion of probability amplitude "on *maru*" to one on an arbitrary spacelike hypersurface and succeeded in formulating the "super-many-time theory."[14]

As regards the divergence problem in QED, various analyses and attempts were made in the Western countries during the thirties. Also in this connection, the idea of renormalization of mass and charge was born, but this does not by itself mean the recognition of the concept of "renormalizable field theory." It was slightly later that Tomonaga turned to the divergence of the field reaction in QED, but it was to his advantage that he did it after his experience with the field reaction in meson theory. This line of investigation and that of the super-many-time formulation mentioned earlier joined luckily together to arrive at his "self-consistent subtraction method" (1947).[15] In this process, Sakata's C-meson theory served as a catalyst.[16] To attack the divergence problem of QED, Sakata again tried the substantialistic approach, modifying the method of the mixed-field theory by Møller and Rosenfeld. Tomonaga's response to this work was one of constructive criticism: Certainly this C-meson theory eliminated the divergence of electron self-energy (at least to the second order of perturbation theory), as Sakata had claimed, but the crucial point was whether or not it also eliminated the divergences in the radiative correction to scattering. Setting this as the problem led him to the concept of renormalizable field theory.[17]

The renormalization theory was formulated in Japan and in the United States (with important contributions in Switzerland), and the two treatments have some differences as well as some parallel

Japanese particle theorists at Kyoto University on August 22, 1950. Seated left to right: G. Araki, H. Yukawa, S. Tomonaga, S. Sakata. Standing left to right: S. Ozaki, R. Utiyama, K. Husimi, T. Miyazima, M. Taketani, S. Nakamura, Z. Koba, Y. Tanikawa, M. Kobayasi, Y. Nambu (credit: Satio Hayakawa).

aspects.[18] It took place in the postwar occupation period, and the interactions and communications between the two sides were infrequent and casual. It is a problem still to be answered: To what extent might the interaction have been important?[19]

In the same postwar period, Taketani and Sakata wrote many essays about their three-stage theory, which had considerable influence among the growing après-guerre mass of young Japanese, sometimes producing adherents.[20] If one took this three-stage theory as a dogma, it could play the role of Procrustes' bed, according to which one tailors the theoretical situation at hand. It was difficult to fit the renormalization theory into this methodological scheme in any clear-cut fashion.[21] As to any philosophical methodology, Tomonaga entertained sound skepticism.[22]

The method of Tomonaga's renormalization theory may be characterized as substitution and renunciation. The theory distinguishes observable mass and charge from bare mass and charge and substitutes finite empirical values for the former. The procedure is akin, to some extent, to Heisenberg's *quantentheoretische Umdeutung* at the opening of quantum mechanics (1925), which made correspondence-theoretical substitutions to eliminate the original mechanical quantities occurring in the theory in favor of the ones directly connected to observable quantities. Furthermore, Bohr asserted that quantum mechanics implies "renunciation of the classical ideal of causality." Tomonaga's method of renormalization is similar to this situation in quantum mechanics; it steers clear of divergent physical quantities by renouncing their calculation.

I said earlier that Tomonaga worked alongside the leading Western current of theoretical physics. However, the preceding method of substitution and renunciation does not necessarily belong to the orthodox current of Western thought, which pursues logical completeness and thoroughgoingness. In fact, in the subsequent period, efforts were made in the Western countries in the direction of pushing that renunciation itself to its logical conclusion, leading to axiomatic field theory or analytic S-matrix theory. In contrast, Tomonaga's original method might be compared to the attitude of Oriental philosophy.[23] At the same time, the scrupulous style of his various investigations is reminiscent of the skillfulness of Japanese master artisans.

After the war, the leadership in particle physics was gradually taken over by experimental researchers, and the American advantage was strengthened also on the theoretical side. One of the conspicuous Japanese works was the Sakata model. Although this belongs to the mid-fifties, I want to say a few words about it, because it was much discussed.

Sakata's original viewpoint was to discriminate between "more" elementary particles and the other hadrons that should be their composites, and also to put the Gell-Mann-Nakano-Nishijima rule on a substantialistic foundation. This viewpoint itself was not necessarily very distinctive at that time, because in the period when newly discovered strange particles were a major problem, people wanted to reduce the number of "true" elementary particles anyway. The merit of Sakata's theory lay rather in its choice of three Fermi particles ("sakatons") as the fundamental particles. (In this respect it was a natural generalization of the Fermi-Yang model of 1949.)

Next, by noticing the correspondence between sakatons and leptons, Sakata and associates took the further substantialistic step of proclaiming the Nagoya model,[24] which was the simple (though radical) idea of regarding sakatons as being formed by somehow sticking a certain unknown substance, called B matter, to leptons.[25] Although the theory could not be developed much beyond this embryonic stage, Sakata remained insistent on the importance of this escalation of model building as the realization of "infinite strata" and the "inexhaustibility" of matter pronounced by Engels and by Lenin.[26] This provoked debates with respect to the philosophical doctrine itself as well as its immediate application to physics.[27]

Surveying the later development of particle physics, it turned out that these models and speculations of Sakata have proved to be sufficiently relevant: sakatons were replaced by quarks, and the Nagoya model stimulated some useful ideas (the form of the weak current, the fourth "urbaryon"), although the B matter itself could hardly be compared to colored gluons or the Higgs field. Also, people no longer have inhibitions about speculating about "subquarks" as the next stratum (whether the final one or not).

In my view, the quark is conceptually more recondite than the sakaton, in the sense that it represents an entity that has a status intermediate between the observable and the unobservable.[28] Thus, ironically, the quark theory reminds me of Chuang-tse's allegorical phrase that might be translated, "I know that a fish swimming in the river is happy, though I have no positive proof for it," " 知 魚 楽 ", a favorite of Yukawa.[29] I think also that the doctrine of infinite strata corresponds to the scientific thought of pushing experiments to higher and higher energies without limit. Speaking in more general terms, it belongs to the Western idea of an endless development of science, an idea present since the time of Francis Bacon, although it has begun to be questioned in recent times.[30]

Notes

1 Some material in English relating to the history of elementary particle physics in Japan is included in *Particle Physics in Japan 1930-50*, three volumes, ed. by L. M. Brown, M. Konuma, and Z. Maki (Research Institute for Fundamental Physics, Kyoto University. 1980).

2 This characteristic of particle physics has persisted up to the present. Besides the introduction of new particles and couplings, the devices include renormalization, parity violation, fractional charge, spontaneous breakdown of symmetry, Higgs mechanism, etc.

3 H. Yukawa, "Interaction of Elementary Particles. I," *Proc. Phys.-Math. Soc. Japan* *17* (1935), 48-57.

4 W. Heisenberg, "Structure of Atomic Nuclei," *Z. Physik 77* (1932), 1-11, and *78* (1932), 156-64. Yukawa recalls: "If I had considered the problem quantum-mechanically from the outset I should have reached something like meson-pair theory instead of my actual one." See H. Yukawa, "Reminiscence," *Shizen No. 3* (1947), 12 (in Japanese).

5 E. Fermi, "Theory of β-Rays," *Z. Physik 88* (1934), 61-171.

6 J. R. Oppenheimer and R. Serber, "Note on the Nature of Cosmic-Ray Particles," *Phys. Rev. 51* (1937), 1113; J. F. Carlson and J. R. Oppenheimer, "On Multiplicative Showers," *Phys. Rev. 51* (1937), 220-31. Ironically, Oppenheimer and Serber were right in denying that the observed cosmic-ray particle was Yukawa's particle.

7 These were hard times during World War II. A difference between Japan and the United States was that in Japan, theoretical physics followed the lead of the meson theory, whereas in the United States a strong tradition of nuclear physics formed, embracing meson physics.

8 Marxism gradually entered Japan; after the mid-twenties it had a strong impact on intellectuals, in spite of the growing oppression. At first, the main interest was in "historical materialism," but aspects of its epistemology and natural philosophy, based particularly on Engels's *Dialectics of Nature* and Lenin's *Materialism and Empiricocriticism*, which were translated into Japanese as early as 1929, were also studied.

9 Shoichi Sakata and Takesi Inoue, *Sugaku-butsuri-gakkaishi 16* (1942), 232; Shoichi Sakata and Takesi Inoue, "On the Correlations between Mesons and Yukawa Particles," *Progr. Theoret. Phys. (Kyoto) 1* (1946), 143-50. See also Y. Tanikawa, "On the Cosmic-Ray Meson and the Nuclear Meson," *Progr. Theoret. Phys. (Kyoto) 2* (1947), 220-1.

10 Yukawa's meson was experimentally found by Powell and associates, with the properties assigned it by Sakata's two-meson theory. However, international recognition for the latter work was meager, partly because it was made during wartime and was unknown outside of Japan until about 1947. This misfortune may have been a traumatic experience for Sakata, but its intrinsic success (in its major ideas) greatly strengthened his faith in his methodology.

11 See T. Takabayasi, "Philosophical Background of Yukawa Theory" (see note 1) and "Dr. Sakata and the Theory of Elementary Particles," *Kagaku 41* (1971), 86-92, and 129-33 (in Japanese).

12 P. A. M. Dirac, "Lagrangian in Quantum Mechanics," *Physikalische Zeitschrift der Sowjetunion 3* (1933), 64-72.

13 See T. Takabayasi, "Physics and Method of Dr. Tomonaga," *Kagaku 49* (1979), 780-87 (in Japanese).

14 S. Tomonaga, *Riken-iho 22* (1943), 545. At the end of this paper the author says: "This theory is nothing more than the result of applying to the usual theory a formal transformation which is almost self-evident." It is somewhat strange that the problem of a manifestly covariant formulation of quantum field theory was not explored in other countries following the works of Dirac-Fock-Podolsky (1932) and Felix Bloch (1934).

15 Yukawa once told me: "Great men are lucky; for example, Newton and Einstein. I wonder why that is so." I think there is some truth in this and that it applies to Yukawa as well as to Tomonaga.

16 S. Sakata and O. Hara, "The Self-Energy of the Electron and the Mass Difference of Nucleons," *Progr. Theoret. Phys. (Kyoto) 2* (1947), 30-1.

17 Z. Koba and S. Tomonaga, "Application of the 'Self-Consistent' Subtraction Method to the Elastic Scattering of an Electron," *Progr. Theoret. Phys. (Kyoto) 2* (1947), 218.

18 One difference is that in Japan, as mentioned, the tradition of meson theory intervened effectively. On the other hand, Sakata's *C*-meson theory was parallel to the *f*-meson theory of Abraham Pais.

19 As for the influence of the American research on the Japanese research, the stimulus given by Hans A. Bethe's theory of the Lamb shift was, of course, important. Freeman J. Dyson wrote in *Disturbing the Universe* (New York: Harper & Row, 1979) about the surprise given by Tomonaga's papers, which reached the United States early in 1948, saying: "Amid the ruin of the war, totally isolated from the rest of the world, Tomonaga was in some respects ahead of anything existing anywhere else at that time." Tomonaga once said that this postwar isolation was better in one sense: It allowed him to concentrate on his own ideas.

20 Most of them are in Japanese, but they also published purely methodological essays in *Progress of Theoretical Physics (Kyoto)*. See S. Sakata, "The Theory of the Interaction of Elementary Particles," *Progr. Theoret. Phys. (Kyoto) 2* (1947), 145-50, and M. Taketani, "Conflict between Matter and Field," *Progr. Theoret. Phys. (Kyoto) 2* (1947), 187-97.

21 About that time, Tomonaga told me, with characteristic humor: "So Mr. Taketani is perhaps considering to extend his three-stage theory to a four-stage theory."

22 Tomonaga himself attributed his critical sense about philosophical methodology to the fact that his father was a professor of philosophy at Kyoto University. In his final talk (just before his tragic throat operation toward the end of 1978), Tomonaga said (see Note 1): "Einstein must have much respected Ernst Mach but he did not obey him blindly. That is really the attitude of the true physicist. It is very dangerous to label each physicist as either Machist or anti-Machist." See also Tomonaga's posthumous book: *What the Devil Is Physics?* (Iwanami, 1979).

23 See Y. Nambu, "Footprints of Professor Tomonaga," *Kagaku 49* (1979), 754-6 (in Japanese).

24 Ziro Maki, Masami Nakagawa, Yoshio Ohnuki, and Shoichi Sakata, "A Unified Model for Elementary Particles," *Progr. Theoret. Phys. (Kyoto) 23* (1960), 1174-80.

25 Sakata here used a broadly classical picture; he intended to tuck the mass and the ability of strong interaction into this B matter; in fact, various possible difficulties must also be tucked in together, insofar as the notion of "sticking" is difficult to accord with the uncertainty principle; also, it would imply the confinement of B matter. When Sakata first announced this model at our Physics Institute of Nagoya University, I pointed out that in its mode of thinking this model is akin to the idea of the neutron entertained by Ernest Rutherford (1920) and James Chadwick (1932), as well as to the picture of the neutron that Heisenberg used in his first theory of nuclear structure (1932).

26 A world view similar to the infinite strata was also stated, for example, by the astronomer K. Charlier, according to whom each structural unit successively fits inside the next one endlessly, as in a Russian doll.

27 See, for example, *Proceedings of the International Conference on Elementary Particles, 1965*, ed. by Y. Tanikawa (Kyoto: Prog. Theor. Phys., 1966), p. 119; T. Takabayasi, "What Is Physics? – Methodological Fragments," *Butsuri-gakkaishi 22* (1967), 375-80 (in Japanese); Y. Ne'eman, "Concrete versus Abstract Theoretical Models," in *The Interaction between Science and Philosophy*, ed. Y. Elkana (New York: Humanities Press, 1975). The viewpoint opposite to infinite strata of matter was expressed notably by Dirac, by Heisenberg, and by Yukawa, not to mention Einstein.

28 T. Takabayasi, "Evolution of Elementary Particle Physics and Fundamental Concepts," *Physics Monthly 1* (1979), 43-47.

29 Curiously, this phrase resembles but also differs from Einstein's: "What does a fish know about the water in which he swims all his life?" Albert Einstein, in George Schreiber, *Portraits and Self-Portraits* (Boston: Houghton Mifflin, 1936).

30 Thus, my last question is this: "What the devil is particle physics?" (after Tomonaga's "What the devil is physics?")

Part V

A new picture

19 My work in meson physics with nuclear emulsions

CESARE MANSUETO GIULIO LATTES

Born 1924, Curitiba, Paraná, Brazil; attended University of São Paulo; cosmic rays and cosmology; Federal University of Rio de Janeiro and Centro Brasileiro de Pesquisas Físicas; on leave from University of Campinas.

At the end of World War II, I was working at the University of São Paulo, Brazil, with a slow-meson-triggered cloud chamber that I had built in collaboration with Ugo Camerini and A. Wataghin. I sent pictures obtained with this cloud chamber to Giuseppe P. S. Occhialini, who had recently left Brazil and had joined Cecil F. Powell at Bristol. On receiving from Occhialini positive prints of photomicrographs of tracks of protons and α particles, obtained in a new concentrated emulsion just produced experimentally by Ilford Ltd., I immediately wrote to him asking to work with the new plates, which obviously opened great possibilities. Occhialini and Powell arranged for a grant from the University of Bristol; I somehow managed to get to Bristol during the winter of 1946.

I was given the task of obtaining the shrinkage factor of the new emulsion (which was much more concentrated than the old ones); Occhialini and Powell were still at work on n-p scattering at around 10 MeV, using the old emulsions. I decided that the time allotted to me at the Cambridge Cockroft-Walton accelerator, which provided artificial disintegration particles as probes for the shrinkage factor, was sufficient for a study of the following reactions:

$$D \ (d, p) \ H_1^3,$$
$$Li_3^6 \ (d, p) \ Li_3^7,$$
$$Li_3^7 \ (d, p) \ Li_3^8,$$

$$\text{Be}_4^9 \ (\text{d, p 2n}) \ \text{Be}_4^8,$$
$$\text{B}_5^{10} \ (\text{d, p}) \ \text{B}_5^{11},$$
$$\text{B}_5^{11} \ (\text{d, p}) \ \text{B}_5^{12}.$$

Through analysis of the tracks, we obtained a range-energy relation for protons up to about 10 MeV that was used for several years in research where single charged particles were detected (e.g., pions and muons).[1]

In the same experiment I placed borax-loaded plates, which Ilford had prepared at my request, in the direction of the beam of neutrons from the reaction

$$\text{B}_5^{11} + \text{H}_1^2 \rightarrow \text{C}_6^{12} + \text{n}_0,$$

which gives a peak of neutrons at about 13 MeV. The idea, which worked well, was to obtain the energy and momentum of neutrons, irrespective of their direction of arrival (which was not known), through the reaction

$$\text{n}_0 + \text{B}_5^{10} \rightarrow \text{He}_2^4 + \text{He}_2^4 + \text{H}_1^3.$$

Occhialini and I decided that he should take some plates to the Pic-du-Midi in the Pyrenees for an exposure of about one month; some were loaded with borax, and some were normal plates (without borax). All were made of the new concentrated B_1-type emulsion for which a range-energy relation already existed. The normal plates were to be used for the study of low-energy cosmic rays and as a control, to see if we were detecting cosmic-ray neutrons.

When Occhialini processed the emulsion after their recovery, on the same night on which they were received in Bristol it became clear that borax-loaded emulsions had many more events than the unloaded ones; borax somehow kept the latent image from fading; normal plates had a great amount of fading. The variety of events in the borax plates, and the richness in detail, made it obvious that the neutron energy detection was but a side result. The normal events seen in the plate were such as to justify putting the full force of the laboratory into the study of normal low-energy cosmic-ray events. After a few days of scanning, a young lady, Marietta Kurz, found an unusual event: one stopping meson and, emerging from its end, a new meson of about 600 μ range, all contained in the emulsion. I should add that mesons are easily distinguished from protons in the emulsion we used because of their much larger scattering and their variation of grain density with range. A few days later, a second "double" meson was found; unfortunately, in this case the secondary did not stop in the emulsion, but one could guess, by studying its ionization (grain counting), that its extrapolated range was also about 600 μ. The first results on the double mesons were published in *Nature*.[2] By the way, the cosmic-ray neutrons (direction, energy) were also obtained in the same plates, and the results were published in the same volume of *Nature*.[3]

Having one and a half double mesons that seemed to correspond to a fundamental process (although it could have been an exothermal reaction of the type $\mu^- + X_a^b \rightarrow X_{a-2}^b + \mu^+$), the Bristol group realized that one should quickly get more events. I went to the Department of Geography of Bristol University and found that there was a meteorological station at about 18,600 ft some 20 km by road from the capital of Bolivia, La Paz. I therefore proposed to Powell and Occhialini that if they could get funds for me to fly to South America, I could take care of exposing borax-loaded plates at Chacaltaya Mountain for one month. That was done, and I left Bristol with several borax-loaded plates plus a pile of pound notes sufficient to carry me to Rio de Janeiro and back. Contrary to the recommendation of Professor Tyndall, director of the H. H. Wills Physical Laboratory, I took a Brazilian airplane, which was wise, because the British plane crashed in Dakar and killed all its passengers.

After the agreed time, I developed one plate in La Paz. The water was not appropriate, and the emulsion turned out stained. Even so, it was possible to find a complete double meson in this plate; the range of the secondary was also around 600 μ.

Back in Bristol, the plates were duly processed and scanned; about 30 double mesons were found. It was decided that I should try to get the mass ratio of the first and second mesons by doing repeated grain counting on the tracks. The results convinced us that we were dealing with a fundamental process.[4] We identified the heavier meson with the Yukawa particle and its secondary with Carl Anderson's mesotron. A neutral particle of small mass was needed to balance the momenta.

At the end of 1947, I left Bristol with a Rockefeller scholarship with the intention of trying to detect artificially produced pions at the 184-in. cyclotron that had started operation at Berkeley, California. The beam of α particles was only 380 MeV (95 MeV per nucleon), an energy insufficient for producing pions. I took my chance on the "favorable" collisions in which the internal momentum of a nucleon in the α and the momentum of the beam provided sufficient energy in the center-of-mass system. The results showed that mesons were indeed being produced. Two papers describe the method of detection and the results, the first referring to negative pions, the second to positive.[5] By making use of the range of pions and their curvature in a magnetic field, it was possible to estimate the masses to be about 300 electron masses.

Around February 1949, I was preparing to leave Berkeley to return to Brazil. At that time, Edwin McMillan, who had his 300 MeV electron synchrotron in operation, asked me to look at some plates that had been exposed to γ rays from his machine. In one night I found about a dozen pions, both positive and negative, and the next morning I delivered to McMillan the plates and maps that allowed the finding of the events. I do not know what use McMillan made of the information, but there is no doubt that they were the first artificially photoproduced pions detected.

Notes

1 C. M. G. Lattes, R. H. Fowler, and R. Cuer, "Range-Energy Relation for Protons and α-Particles in the New Ilford 'Nuclear Research' Emulsions," *Nature 159* (1947), 301-2; C. M. G. Lattes, R. H. Fowler, and R. Cuer, "A Study of the Nuclear Transmutations of Light Elements by the Photographic Method," *Proc. Phys. Soc. (London) 59* (1947), 883-900.
2 C. M. G. Lattes, H. Muirhead, G. P. S. Occhialini, and C. F. Powell, "Processes Involving Charged Mesons," *Nature 159* (1947), 694-7.
3 C. M. G. Lattes and G. P. S. Occhialini, "Determination of the Energy and Momentum of Fast Neutrons in Cosmic Rays," *Nature 159* (1947), 331-2.
4 C. M. G. Lattes, G. P. S. Occhialini, and C. F. Powell, "Observation on the Tracks of Slow Mesons in Photographic Emulsions," *Nature 160* (1947), 453-6 and 486-92; C. M. G. Lattes, G. P. S. Occhialini, and C. F. Powell, "A Determination of the Ratio of the Masses of π^- and μ^- Mesons by the Method of Grain-Counting," *Proc. Phys. Soc. London 61* (1948), 173-83.
5 Eugene Gardner and C. M. G. Lattes, "Production of Mesons by the 184-Inch Berkeley Cyclotron," *Science 107* (1948), 270-1; John Burfening, Eugene Gardner, and C. M. G. Lattes, "Positive Mesons Produced by the 184-Inch Berkeley Cyclotron," *Phys. Rev. 75* (1949), 382-7.

20　The fine structure of hydrogen

WILLIS E. LAMB, JR.

Born 1913, Los Angeles; Ph.D., University of California, 1938; Nobel Prize, 1955, for discovery of splitting of hydrogen levels (Lamb shift); Department of Physics, University of Arizona.

It will be impossible because of time limitations to give a full account of the experiments and theory dealing with the fine structure of the hydrogen atom. I have therefore decided to concentrate on telling some parts of the story that are not widely known. I spent many years writing up the essential scientific aspects of the work.[1] The opportunity provided by this Fermilab conference to tell some of the background story seems too good to miss. A lot of strange things combined together to make possible the measurement of the hydrogen fine structure.

Perhaps it should be said at the outset that I shall be mentioning the names of a lot of famous physicists. Because of the time limitations, I shall not be able to give them credit for their enormous contributions to physics or to indicate the great respect and affection I have for them. In the last category I would certainly include Robert Oppenheimer, Isidor I. Rabi, Victor Weisskopf, Hans Bethe, and Paul A. M. Dirac.

My first contact with the problem of hydrogen fine structure was quite remote and trivial. In 1887, using his interferometer, Albert Michelson had discovered that some of the spectral lines of hydrogen had a complex structure. His last years were spent at Caltech, and he liked to play chess. I was rather good at that game in my high school years in Los Angeles. I remember being introduced to him about 1929 at a chess festival of some kind in Pasadena. We talked a little about the game, but not about physics, and although I knew that he was famous for measurement of the velocity of light, I did not know of any other reasons.

Berkeley

In those days, Los Angeles High School was generally thought to be the best one in the city. However, none of my teachers there could give me any help to break into the mysterious kind of mathematics called calculus. When I went up to Berkeley in 1930 and enrolled in the College of Chemistry, I found science suddenly much harder and more interesting than it had been in high school. This college was presided over by Gilbert N. Lewis, who was dean from 1912 to 1945. (Incidentally, it may not be generally known by particle physicists that Lewis coined the word *photon* as late as 1926.[2] For him, it was a zero-mass particle that was emitted by an excited atom and that later was converted into electromagnetic radiation.) I only got to know Lewis after I became a graduate student. We gave each other some good advice, which neither of us followed. I advised him that his ideas on the electron-neutron interaction could not be correct. He advised me to become an experimentalist rather than a theorist. The reason given was that a theorist without a good idea was useless, whereas an experimentalist could always go into the laboratory and "polish up the brass."

For four undergraduate years I did very little except study chemistry, physics, and mathematics. I did not have much atomic physics, as an undergraduate, although Leonard Loeb gave a very detailed course on radioactivity and gas discharges. In some chemistry seminars I was exposed to a little atomic theory based on the vector model of the Bohr quantum theory at the level of the book by Linus Pauling and Samuel Goudsmit. I found the subject most unclear. Part of the trouble was that despite several courses in mechanics, I did not really understand angular momentum. I first heard of wave mechanics from another student about 1932, and a little later a friend pointed out someone walking ahead of us on Shattuck Avenue who was supposed to be one of the great experts on that subject. The man's name was Oppenheimer, and I had not heard it before. Not long after this, the student newspaper *Raspberry*, a western version of the Harvard *Lampoon*, ran an article, rather an admiring one for that publication, on Oppenheimer, referring to him as a "mop on stilts." By my senior year I was going regularly to Physics Department colloquia and had ample opportunity to see and hear him in action. Although Wendell Latimer made some arrangements for me to go to Harvard with a teaching fellowship in chemistry, I decided that I had to transfer to physics for graduate study, and naturally I wanted to work with Oppenheimer. In 1934 we were at the bottom of the Depression, and I would not have been very employable as a chemist. Jobs for physicists at that time were also very scarce. Fortunately, I had pretty good grades, and it seemed that I could get a teaching assistantship in physics at Berkeley at the generous stipend of $600 per year.

As a senior, and in graduate school, I was exposed to all kinds of good courses in physics: analytic mechanics from Victor Lenzen, vector and tensor analysis from William Williams, who managed to stress things that would be useful for quantum mechanics, and electromagnetic theory from Ernest O. Lawrence. I audited a two-semester graduate course on atomic physics. Robert Brode gave the first-semester lectures at the level of Harvey White's book. Although I became happier with the vector model, the discussion began with alkali atoms and seemed unnecessarily complicated. I remember my great disappointment when the lectures turned to the simplest atom, hydrogen, and the theory became even more complicated because of the interplay of spin-orbit interactions and relativity theory. This introduced me to Arnold Sommerfeld's 1916 relativistic treatment of the hydrogen fine structure, which used the old quantum theory. In the second semester I was exposed to the rest of spectroscopy by Francis Jenkins. I might mention that I applied for and received permission to skip the required optics laboratory because I found working in a darkened room with bright colored lights very much of a strain for my eyes. In electrical laboratory work I got about as far as an experiment with vacuum-tube-generated waves propagating on Lecher wires at about 5 megacycles per second, and the opportunity to look at Lissajous figures on an oscilloscope.

I soon had courses on methods of mathematical physics and quantum mechanics from Oppenheimer. I took to quantum mechanics with an enthusiasm that has lasted the rest of my life. I also listened to his lectures on statistical mechanics and developed a curiosity about the approach to thermodynamic equilibrium that is still with me. Oppenheimer's lectures were a revelation. The equations he wrote on the blackboard were not always reliable. We learned to apply correction-factor operators to allow for incorrect signs and numerical coefficients.

My predecessors as Oppenheimer graduate students at Berkeley were Harvey Hall, Frank Carlson, Melba Phillips, Leo Nedelsky, Arnold Nordsieck, and Glenn Camp. Later graduate students who came while I was still in Berkeley included Philip Morrison, Sidney Dancoff, Hartland Snyder, Robert Christie, and George Volkoff. Wendell Furry was there in a postdoctoral capacity. National Research Council fellows were Robert Serber and Edwin Uehling in 1934 and Leonard Schiff in 1937. A younger student (who did not work for Oppenheimer) was Robert Wilson, who is well known at Fermilab.

Oppenheimer's office was room 219, LeConte Hall. As were many of his students, I was given a small table in the room. Oppenheimer had no desk, but only a table in the middle of the room, heavily strewn with papers. One wall was entirely covered by a blackboard and hardly ever erased. One set of open shelves had reprints of Oppenheimer's publications. I was allowed to have a copy of most of these. In one paper, which impressed me mightily, Oppenheimer pointed out specifically for the first time how the quantum electrodynamics of Dirac, Werner Heisenberg, and Wolfgang Pauli led to self-energy divergence difficulties.[3] This paper was written before the reinterpretation of the negative energy states, and consequently the divergences were not as manageable as they were later to become. His attempt to subtract one infinite integral from another did not succeed. Despite this, Oppenheimer noted that a judicious selection of some of the integrals for the infinite energy of the 2p state of hydrogen led to a finite result. Hall carried out the necessary calculations and obtained a $2s_{1/2}$-$2p_{1/2}$ level shift of 1,700 MHz. This wasn't really a meaningful procedure, but it was certainly a suggestive one, although few people took it seriously or remembered it.

Oppenheimer's seminar on theoretical physics was held weekly, and after the arrival of Felix Bloch in 1934 it became a joint Berkeley-Stanford seminar. On average, the more numerous Berkeley group went to Stanford about one week in three or four. To this seminar, besides Oppenheimer and Bloch, came at various times, from Berkeley, Phillips, Furry, Nordsieck, Serber, Uehling, Marjorie Vold (née Young), and Schiff, and, from Stanford, William Hansen, David Webster, and Arnold Siegert. I heard many things way over my head at these meetings. Cosmic-ray showers, positron theory, and vacuum polarization were much discussed. I learned about the deviations from the Coulomb field of a point charge given by calculations of Uehling. Also mentioned was the apparent discovery of a discrepancy between the spectroscopic observations of William Houston at Caltech and the predictions of the Dirac relativistic quantum mechanics of the hydrogen atom.[4]

Ann Arbor

During 1935 I spent two months at Ann Arbor attending the summer school in theoretical physics organized by Goudsmit and George Uhlenbeck. In addition to lectures from them, I heard David Dennison, Enrico Fermi, and Bloch. Dennison lectured on molecular theory, Fermi on the physics of neutron capture and scattering, and Bloch on the theory of electrical conductivity. Uhlenbeck dealt with β-decay theory, and Goudsmit talked about atomic structure. After that summer in Michigan, I was able, for the first time, to find problems for myself that seemed to be worth doing.

Pasadena

In the spring of each year, Oppenheimer spent six weeks teaching at Caltech, and I managed to spend most of this time in Pasadena during three of my four years with him. I learned some collision theory from his lectures, and I was introduced by Paul S. Epstein to Fermi's 1932 article on the quantum theory of radiation. I heard a lecture by Dirac, but I did not come to know him until the late fifties. I did meet Houston, but I had no conversations with him about his recent observations on hydrogen fine structure. Others who came to theoretical seminars were Richard Tolman, Howard Robertson, Jesse DuMond, and Charles Lauritsen.

Stanford

Another very important part of my education came during summers, beginning in 1936, at Stanford, where Bloch let me have desk space. His arrangement for a visiting summer professor each year made it possible for me to hear lectures from George Gamow, Weisskopf, Rabi, J. H. Van Vleck, and again from Fermi. A number of summer tourists passed through. I remember particularly Edward Teller, George Placzek, George Kistiakowski, and Samuel Allison. I also saw a lot of Stanford physicists Hansen, Webster, Norris Bradbury, Paul Kirkpatrick, and Russell Varian.

Berkeley

To my regret, Oppenheimer did not care much for the kinds of problems I wanted to work on after my return from Ann Arbor. In retrospect, I think that they were important, and some still are, but they were too difficult to work on considering the current state of physics and my limited abilities. My only reason for mentioning some of them is that they did have some bearing on my state of mind in later years when the vague ideas about the work on the hydrogen fine-structure problem were being formed.

I spent some time trying to do research on various field theories of nuclear forces. At first I used the electron-neutrino field. This gave a highly singular nucleon-nucleon potential, and then only if infinite integrals were interpreted with the help of convergence procedures. I saw from this work that there would be an extra spin-dependent interaction between an atomic electron and a nucleon spin. Later, Hideki Yukawa introduced the idea that nuclear forces were mediated by a field of intermediate-mass bosons. Oppenheimer coined the words *yukon* and *dynaton* for what later was called *mesotron* and finally *meson*. I soon shifted my calculations from the electron-neutrino to Yukawa's mesotron field. This had the advantage that the integrals appearing in the calculations became far less divergent. My 1938 thesis contained two parts, one of which was called "electromagnetic properties of nuclear systems" and dealt with the spreading out of the charge distribution around a nucleon and exchange currents between nucleons.[5] At this stage, I had a number of reasons to think that there would be departures from Coulomb interactions. A further implication of this work, nonadditivity of nuclear forces, was too difficult to contemplate at the time, and even today a lot of nuclear-structure calculations may need appreciable corrections from multiparticle forces. Multiatom forces, of course, play an important role in molecular theory.

Oppenheimer did not suffer fools gladly. Sometimes I annoyed him because of my stupidity and ignorance. Mostly, he was as kind to me as I deserved. I wish I could have learned as much as he could have taught me.

Columbia

After I got my Ph.D. in 1938, Rabi, whom I had met that summer at Stanford, arranged for me to get a position as an instructor at Columbia University at the large salary of $2,400 per year. There I found Nordsieck, who had moved from Stanford a year earlier. Rabi's molecular-beam work was in full flower, staffed with Columbia personnel Polykarp Kusch, Jerome Kellogg, and two people who taught 15 hr per week at the New York City colleges, Sidney Millman and Jerrold Zacharias, moonlighting at no pay in order to be able to do research. Norman Ramsey was a promising graduate student. Another student, seen only during the night hours, Julian Schwinger, was already making major contributions to theoretical physics. Herbert Anderson was another graduate student of that time.

My teaching load was only 12 hr per week and consisted of laboratory and recitation sections for the engineering physics courses taken by freshmen and sophomores. This left me plenty of time for research and an occasional trip to a seminar at Princeton, where I often had helpful conversations with Eugene Wigner and John Wheeler.

At Columbia, my first research effort was to finish up some work that I had started at Stanford dealing with the resonance capture of neutrons by nuclei bound in solids. (Twenty years later, this paper had some importance in connection with the Mössbauer effect.)[6]

I was also involved in a controversy with Herbert Fröhlich, Walter Heitler, and Boris Kahn about possible deviations from the Coulomb field of a proton on the basis of mesotron theories of the sort considered in my thesis.[7] They were not aware of that work, but they thought that they could get a short-range repulsion of an electron by the mesotron charge distribution around the proton. I argued that their result was possible only because they were working with a perturbation theory plagued by divergent integrals.[8] We never did come to an agreement on this point. Still, the discussion was useful, because it kept me thinking about the hydrogen fine structure. Also, it taught me about the suggestion by Simon Pasternack that Houston's spectroscopic data could be interpreted in terms of a short-range repulsion between the electron and proton.[9]

In the meantime, I worked on various problems involving nuclear and molecular-beam physics. As the war came on, most of my colleagues went off to work on secret projects. My wife was classified as an enemy alien; so I was not allowed to work on radar or nuclear devices. Instead, I taught a number of advanced courses at Columbia, and also physics for the V-12 Navy program. One of the students there was Joshua Lederberg, who later received a Nobel Prize for his work in genetics.

The molecular-beam work slowed down to a halt because so many of the participants were working on other things. Up to that time, most molecular beams were detected by the process of positive-ion emission from a hot wire, a method limited to molecules containing alkali atoms and a few others. I somehow heard that one of the senior professors at Columbia, William Webb, had discovered 15 years before that metastable mercury atoms could eject electrons from solid surfaces, and it occurred to me that one might detect a beam of metastable atoms by making use of this effect. At this time, Rabi was away for long periods; when I saw him, I told him about my idea. He did not seem very enthusiastic, but he told me that there had been some work in this area by Leonard Broadway on nitrogen metastable atoms a few years before.

I refused to be discouraged and looked further into the literature to find that there were quite interesting experimental studies by Mark Oliphant and theory by Harrie Massey on the ejection of electrons when helium metastables impacted a metallic surface. During the brief time Fermi was at Columbia, there was a Puerto Rican student Amador Cobas who worked with him on a cosmic-ray project. When Fermi went off to Chicago, Cobas helped me make calculations that we thought were a bit better than the earlier ones. I hoped that once the war was over I might be able to interest Rabi in work on excited states of helium. As a matter of record, Rabi never did become enthusiastic about beams of metastable atoms. However, in later years, his student, Vernon Hughes, rather against Rabi's wishes, did much fine physics with beams of metastable helium atoms. The work with Cobas became of considerable interest to people working on surface physics, such as Homer Hagstrom at Bell Labs, and, of course, it made an essential contribution to the hydrogen fine-structure work.[10]

Near the end of 1943 it became evident to someone in authority that it was no longer necessary to withhold my clearance for war work because of my wife's status. Sometime in November, I received a telephone call from Oppenheimer in Santa Fe. He asked if I knew what he was working on. I told him that I could make a pretty good guess. He indicated that the project was very important and that he would like me to join it. Naturally, I was very much attracted to the idea, but in the end I declined the offer. At that time, I had been an instructor at Columbia for five years and could get no indication from our chairman, George Pegram, about my chances for a permanent appointment. Despite this, I thought it better to stay at Columbia. Pegram encouraged me by a promotion from instructor to associate with a small increase in salary. About that time I was invited to join a group of physicists, mostly people who had been associated with Rabi in molecular-beam work, in development of centimeter-wavelength magnetron oscillators. This work was carried on in an organization called the Columbia Radiation Laboratory (CRL), which was some kind of little brother of the much larger M.I.T. Radiation Laboratory. This laboratory was housed on one and later two floors of the physics building. There was an armed guard at the entrance day and night.

The main work at CRL involved microwave magnetron oscillators. This stimulated me many years later to develop theories of maser and laser devices.[11] (I am better known in some circles for a certain "dip" than for the "shift" of this lecture.) For present purposes, however, there were two activities that are more worthy of mention. The wavelength selected for our main effort at CRL was in the K band, centered at 1.25 cm. The older X band, at 3.2 cm wavelength, was well established from work at M.I.T. After our project was well under way, it was pointed out by Van Vleck that there was probably an absorption line at about 1.25 cm because of atmospheric water vapor and that a very real possibility existed that K-band radar would be seriously hampered by high humidity. Kellogg, Gordon Becker, and Stanley Autler began work on the absorption coefficient of water vapor as a function of wavelength. As it finally evolved, the method made use of a large copper-lined cavity, 8 ft on a side.[12] The box contained atmospheric-pressure air saturated with water vapor at a high temperature. Microwaves of the desired wavelength were fed into the box, and the spatially averaged energy density of radiation in the cavity was used to determine the desired absorption coefficient. I was able to make some useful suggestions about the experiment, based on techniques long before used in architectural acoustics.[13] I also suggested that the rotation of a large copper fan would help to further average out undesirable effects from the discrete mode structure of the cavity resonator. I believe that this led directly or indirectly to the use of such fans in many microwave ovens. The measurement of the resonance curve of pressure-broadened water vapor in 1944 was an early form of microwave spectroscopy. I had hopes that the box method would be rather generally applicable, but for various reasons it was not. The greater interest in sharp lines at low pressure, and the invention of the method of Stark modulation by E. Bright Wilson, contributed to my disappointment. Nevertheless, my thoughts were being directed in a fruitful way for the hydrogen fine-structure work.

Although my main job at CRL was to provide theory, sometimes I would think of an experimental configuration for a magnetron device that interested me. I could not hope to get anyone who was involved in the serious business of designing, building, and testing magnetrons to waste time on my ideas. I could, if I wished, build anything I wanted with the help of the superb shop facilities, as long as I carried on the fabrication myself, or with minimal disturbance of the main work of CRL. I did not know anything of metal fabrication, high-vacuum techniques, etc., but Kusch was willing to give me some instruction. Hence, I gradually learned the things I needed to know in order to make a few special kinds of magnetron devices. For example, I made one of the first continuous-wave magnetron oscillators to give a kilowatt of output at a wavelength shorter than 3 cm.

During this whole time I continued to teach. In particular, I taught a course in atomic physics in the summer session of 1945. The textbook assigned was a Dover reprint of Gerhard Herzberg's *Atomic Spectra and Atomic Structure*. I found in this book a reference to the selection rules for allowed optical transitions in atomic hydrogen. It seemed that the wavelength for a transition between the fine-structure terms for n = 2 would have a wavelength of about 3 cm. Reference to some "doubtful" observations in activated hydrogen by Otto Betz and Th. Haase were made in a footnote. I was struck by the near equality of this wavelength and that of my magnetrons; so I looked up the 1932 paper by Betz and found that, for a source of radiation, he had used a spark discharge with electrodes immersed in liquid oxygen to increase the radiative yield. This work preceded the earliest microwave spectroscopy by Claud Cleeton and Neil Williams, in 1934, who used a very primitive magnetron for a study of the ammonia inversion spectrum. Betz had a continuous spectrum, and a very small signal. He believed that he had observed the expected absorption, but his plotted data were not very convincing. I then looked for the reference to the paper by Haase. Unfortunately, 15 items of the bibliography were missing, and the Haase paper was not listed. For a while I did nothing, but eventually I sought out the older edition of Herzberg's book, which contained the desired information. It seemed that Haase had repeated Betz's work in 1935 and had found no absorption peaks. I never did find out what had really happened to Betz's work, but I speculated that the National Socialists were somehow responsible.*

---

* *Ed. note*: The literature referred to in this section is the following: Gerhard Herzberg, *Atomic Spectra and Atomic Structure*, trans. by J. W. T. Spinks (New York: Prentice-Hall, 1937; revised 1944 and reprinted by Dover Publications, New York); Otto Betz, "Ueber die Absorption kurzer elektrischer Wellen in ionisierten Gasen, ein Versuch zum Nachweis der langwelligen Strahlung des Wasserstoffatoms," *Ann. Physik 15* (1932), 321-44; T. Haase, "Ueber die Absorption von Dezimeterwellen in ionisierten Gasen und die Frage des Nachweises der Absorption langer Wellen durch angeregte Wasserstoffatome," *Ann. Physik 23* (1935), 657-76; C. E. Cleeton and N. H. Williams, "Electromagnetic Waves of 1.1 cm Wave-Length and the Absorption Spectrum of Ammonia," *Phys. Rev. 45* (1934), 234-7.

I felt that I now had something to think about! Much better sources of microwaves were available to me than had been to Betz and Haase. All I had to do was to learn how to make gas discharges that could produce activated hydrogen atoms and measure the absorption of microwaves as a function of wavelength. Such thoughts came into and out of my mind for a good part of a year. I continued to make continuous-wave magnetrons, but I moved the wavelength to the "right" value, 2.74 cm, for hydrogen. I knew about Uehling's vacuum polarization calculation, and from the experiences with "mesotron" theories of nucleon structure I thought the discrepancy from the Dirac theory would be small. Pasternack's postulate of a larger shift seemed unreasonable because of the flaw in the work of Fröhlich, Heitler, and Kahn.[7] Furthermore, the later spectroscopic observations of J. W. Drinkwater, Owen Richardson, and W. Ewart Williams indicated that there was no such discrepancy.[14] As a result, I was thinking only of small-spin-dependent or nonelectromagnetic interactions. It turned out for several reasons that 2.74 cm was not a good choice.

The magnetron was not an ideal source of microwaves. Besides being very noisy, it was hard to tune. Still, Kusch had made tunable X-band magnetrons using a "crown-of-thorns" mechanism, and perhaps this could be imitated. Klystron oscillators would have been much more convenient, but whereas we had klystrons for the X band and K band, they were not available for the wavelength that I imagined to be needed.

It seemed from Herzberg's discussion that the 2s state of hydrogen would be metastable. However, Vladimir Rojansky and Van Vleck in 1928-9 had given reasons why they did not expect this to be the case.[15] Although I had long known about Hans Bethe's article on the one- and two-electron problems in the 1933 *Handbuch der Physik* series, I had not read all of his discussion on hydrogenic intensities. I then learned from him about the metastability of the 2s state of hydrogen and its extreme sensitivity to small electric fields. I also learned about the Zeeman effect in hydrogen and realized that hydrogen metastables could be less sensitive to electric fields if they were placed in a strong magnetic field. It became clear, however, that if the atoms were allowed to move at right angles to the magnetic field, they would experience a motional electric field that could reduce the metastability.

There was a lot to do in those days. The war was over, but work continued on various projects at CRL, and we hoped that some of its research could continue into peacetime. The absent faculty members were returning. In 1945 I was given a promotion to assistant professor and even granted tenure at that rank with a salary of $5,100. Dean Pegram explained his earlier hesitancy on the basis of a belief that there would be a large oversupply of physicists when the war ended and that he had to keep his options open.

I no longer taught elementary physics, but did lots of analytic mechanics, electromagnetic theory, statistical mechanics, and quantum mechanics. Among graduate students in the postwar years were Laurie Brown, who helped to organize this symposium on the history of particle physics, and Leon Lederman, who now "works" at Fermilab.

I continued to think intermittently about how one might make a microwave measurement of the hydrogen fine structure. By July of 1946, I had some vague ideas of what to try, and I even placed an order in the shop for the parts of a crude apparatus. Because I knew how to make only things that looked like magnetrons, my apparatus looked like a magnetron. Before it was ready, I had realized that the electric field between cathode and anode would spoil the 2s metastability even in the presence of a magnetic field. Hence, I began thinking of a somewhat unconventional molecular-beam apparatus. Fortunately for me, just at this time Robert Retherford decided to return to Columbia as a graduate student. He had begun research before the war on atomic beams with Kellogg, but then he had gone off to work on vacuum tubes for Westinghouse. He was an ideal partner for the project now increasingly dear to my heart. Retherford knew a lot about vacuum techniques. He knew how to mount a very sensitive d'Arsonval galvanometer to minimize the bouncing of its spot of light caused by laboratory vibrations and the motion of the building in the wind. He had worked with FP-54 dc amplifiers, used in those days to measure exceedingly small currents. Such currents were expected to be produced when metastable atoms struck a target and ejected electrons from it. The earlier work with Kellogg had taught him how to make a Wood's gas discharge with that most beautiful of red colors indicating a good yield of atomic hydrogen. We wrote up the plan of the experiment for the October issue of the CRL quarterly progress report, which is reprinted in the last article cited in note 1.

The experiment first succeeded on Saturday, April 26, 1947, and turned out very much as expected, except for the location of the resonances. I have already given a brief description of the apparatus and the results.[1] It was obvious from the first indication of resonant transitions that discrepancies from the predictions of the Dirac theory of hydrogen were being observed and that they were far larger than then thought theoretically likely. They were, however, in good accord with Houston's earlier spectroscopic measurements of the hydrogen Balmer α spectrum and Pasternack's analysis in terms of a shift of the 2s state by about 0.03 wave number. I remember that late that night after Retherford had gone home, I went over to the laboratory to see if I could confirm the earlier results. I found that I could not manage all the knobs and data taking; so I telephoned my wife Ursula to come over to the Pupin Laboratories and help me. The two of us were able to run the apparatus and could clearly see that the galvanometer spot still moved as it had done earlier that day.

To go somewhat against my modest nature, but inspired by the writings of James Watson and Francis Crick, I will tell you that until the experiment succeeded I had no notion that a Nobel Prize might be involved. Neither were we worried that Pauling was about to scoop us. However, I awakened the following morning with the realization that a very nice experiment had worked. It made for a very good feeling. I realized that the research was of the quality deserving a Nobel Prize, but I was also aware that there were very many other considerations than scientific ones involved in an actual award. As it turned out, that did come over eight years later. The best parts of this were the messages from friends and especially a telegram from Wolfgang Pauli.

During the first week of June of 1947 there was held the Shelter Island conference, organized by Oppenheimer, with support from the National Academy of Sciences. Among those attending were Bethe, David Bohm, Gregory Breit, Karl Darrow, Herman Feshbach, Richard Feynman, Hans Kramers, Duncan McInnes, Robert Marshak, John von Neumann, Nordsieck, Abraham Pais, Pauling, Rabi, Bruno Rossi, Schwinger, Serber, Teller, Uhlenbeck, Van Vleck, Weisskopf, and Wheeler. It was at this conference that Marshak suggested that there were two kinds of mesons. I talked about the fine-structure measurements made with Retherford. The results of John Nafe and Edward Nelson and of Kusch and Henry Foley, indicating an anomaly in the electron's magnetic moment, were described by Rabi. The theoretical talk by Kramers dealt with his program that was to do for the divergences of quantum electrodynamics what Hendrik Lorentz had done for those of the classical theory of a point electron. No concrete method of calculation was described. Schwinger told about his theoretical work on the anomalous magnetic moment of the electron, which at that time was in a very early state of development.

Within a few days of my return to Columbia, I received from Bethe a copy of a manuscript that gave an amazingly short calculation of a $2s_{1/2}$-$2p_{1/2}$ shift amounting to 1,040 MHz.[16] This was done with the simplest conceivable form of nonrelativistic quantum electrodynamics. Infinities occurred, but they were canceled out in a highly plausible and physical manner. (I never asked Bethe exactly where and when he had done his calculation. It was rumored that it was done while he rode the New York Central railroad on the trip from New York to Schenectady, where he consulted for General Electric on betatrons.) I instantly understood what Bethe had done and mentally kicked myself for not being clever enough to do it first. I tried to make up for this shortcoming by embarking, with Norman Kroll, on a relativistic generalization of Bethe's result. In effect, we simply evaluated the expressions given in 1930 by Oppenheimer, with allowance for the filled negative energy electron states. In his presentation at this symposium (Chapter 3), Weisskopf has described the race of three groups of workers on this problem. He thinks that J. Bruce French and he would have won if they had not had bad advice from the experts.[17] Kroll and I were free of this handicap and published first.[18] I think that we actually got the correct result before anyone else. Of course, it turned out that our method could not be generalized to higher orders in the fine-structure constant. To do better it was necessary to use the manifestly covariant methods of Schwinger, Feynman, and Freeman Dyson. In his presentation at this symposium (Chapter 21), Schwinger has described how in 1949 he won a prize from the National Academy of Sciences for his work on covariant field theory. Kroll and I were unsuccessful contenders for this award, but we could be happy that we got the right answer for the second-order relativistic 2s-2p shift.

The hydrogen and deuterium fine-structure work continued at Columbia with Edward Dayhoff and Sol Triebwasser until the early fifties, and at its end it had attained a precision of about 0.1 MHz. The method was also extended to apply to singly ionized helium in work with Miriam Skinner and by Paul Yergin, Robert Novick, and Edgar Lipworth.[19] All the results were in quite good accord with theory.

Epilogue

For want of time, I cannot bring the story further up to date. Work on hydrogenic fine structure still continues in various places striving for higher accuracy and for higher n and higher Z. Extensions are also being made to positronium and other exotic atoms. Before closing, let me refer to some applications of the fine-structure research and give a few anecdotes. The success of quantum electrodynamic theory of fine structure and anomalous magnetic moments was a major landmark in the history of particle physics. Partly as a result of the successful renormalization calculations of quantum electrodynamics, present-day theory has moved very much ahead of the position it had in the mid-fifties.

I am sometimes asked what other uses our work might have. By measuring the rate of quenching of metastable atoms, one could accurately measure small electric fields near metal surfaces. As far as I know, this has not been done. Another much more useful application involves the production of intense beams of polarized protons, as in the work of Joseph McKibben and Gerald Ohlsen at Los Alamos.[20] The general method for this technique was spelled out in the CRL progress report that preceded the success of the fine-structure experiment.

Another way in which the fine-structure work has relations to other kinds of physics involves the quantum mechanics of the two hydrogenic states 2s and 2p. These states have nearly the same energy, but different decay rates. An applied electromagnetic field can convert one state into another. A similar situation exists, for example, in the elementary particle kaon system, with two states K_1^0 and K_2^0 having different decay modes, and slightly different masses.[21] The passage of one type of kaon through matter can convert it into the other. The probability amplitudes for the two states obey equations of a very similar nature to those for the hydrogen atom.

A fourth application involves a tenuous but logical connection between the fine-structure work and the development of masers and lasers. As described earlier, I had considered the possibility of measuring the $2s_{1/2}$-$2p_{3/2}$ absorption of microwaves in a gas discharge. A rough estimate was made of the expected absorption coefficient. This turned out to be positive if the lower atomic level had a higher population than the upper one, and negative if the population was inverted. I knew that a negative absorption coefficient meant amplification. From my earlier experiences I should have realized that it also meant that an oscillator could be made. Metastable hydrogen atoms would not have been a practical working substance for such an oscillator. However, Charles Townes and Arthur Schawlow gave an account of my considerations in their 1951 book on microwave spectroscopy. I imagine that they might thereby have received some indirect help toward their development of masers and lasers.

I shall now give some brief anecdotes, and then be finished. When Dwight Eisenhower was president of Columbia University, he was brought on a tour of the Physics Department. In the course of this, he visited the Pupin Laboratory, where Retherford and I were working. The phrase "hydrogen fine structure" did not seem to mean much to him, but he immediately associated the word "deuterium" with an allied raid into Norway during World War II. Another story involves Oppenheimer. In the sixties he asked me if I was not sorry to have missed coming to Los Alamos. I told him that I thought my decision to stay at Columbia in 1943 was the right one for me, because otherwise I would never have worked on the hydrogen fine-structure experiment. I don't think that he was very pleased by my reply.

I brought Michelson into the story because he discovered the hydrogen fine structure. Other great contributions to the area were made by Sommerfeld, with the relativistic pre-quantum-mechanical theory of 1916, and by Dirac in 1928, with the relativistic wave equation. I never met Sommerfeld, but in 1950 he sent me a handwritten note that indicated that he had heard of our measurements, and he mentioned that he was the "81-year-old greatgrandfather" of the hydrogen fine-structure theory.

In 1959 I had a walk with Dirac around the island of Mainau in Lake Constance. He told me that he had learned of the microwave measurements through a clipping from the front page of *The New York Times*. It is clear that I should have sent him the news in a more direct manner. Finally, I shall tell you that in the sixties I shared a bus ride with Dirac at a conference in Lindau. He asked me if I had enjoyed participating in the discovery of the fine-structure anomaly. I replied yes, but indicated that I would have had much more pleasure if instead I could have discovered his relativistic wave equation. After a brief pause, he said gently, "Things were simpler then."

Notes

1 W. E. Lamb, Jr., and R. C. Retherford, "Fine Structure of the Hydrogen Atom by a Microwave Method," *Phys. Rev. 72* (1947) 241-3; reprinted in *Quantum Electrodynamics*, ed. by J. Schwinger (New York: Dover, 1958), pp. 136-8; W. E. Lamb, Jr., and R. C. Retherford, "Fine Structure of the Hydrogen Atom. Part I," *Phys. Rev. 79* (1950), 549-72; W. E. Lamb, Jr., and R. C. Retherford, "Fine Structure of the Hydrogen Atom. Part II," *Phys. Rev. 81* (1951), 222-32; W. E. Lamb, Jr., "Anomalous Fine Structure of Hydrogen and Singly Ionized Helium," *Rep. Prog. Phys. 14* (1951), 19-63; W. E. Lamb, Jr., "Fine Structure of the Hydrogen Atom. Part III," *Phys. Rev. 85* (1952), 259-76; W. E. Lamb, Jr., and R. C. Retherford, "Fine

Structure of the Hydrogen Atom. Part IV," *Phys. Rev. 86* (1952), 1014-22; S. Triebwasser, E. S. Dayhoff, and W. E. Lamb, Jr., "Fine Structure of the Hydrogen Atom. Part V," *Phys. Rev. 89* (1953), 98-106; E. S. Dayhoff, S. Triebwasser, and W. E. Lamb, Jr., "Fine Structure of the Hydrogen Atom. Part VI," *Phys. Rev. 89* (1953), 106-15; W. E. Lamb, Jr., "Fine Structure of the Hydrogen Atom," in *Les Prix Nobel en 1955* (Stockholm, 1956); reprinted in *Science 123* (1956), 439-42; W. E. Lamb, Jr., "Experimental Tests of Quantum Electrodynamics," Forty-second Guthrie Lecture, in *Year Book of the Physical Society* (London, 1958), pp. 1-9; W. E. Lamb, Jr., "Some History of the Hydrogen Fine Structure Experiment," in *A Festschrift for I. I. Rabi, New York Academy of Sciences, Series II, Volume 38* (1976), 82-6.

2 G. N. Lewis, "The Conservation of Photons," *Nature 118* (1926), 874-5.

3 J. R. Oppenheimer, "Note on the Theory of the Interaction of Field and Matter," *Phys. Rev. 35* (1930), 461-77.

4 W. V. Houston, "A New Method of Analysis of the Structure of Hα and Dα," *Phys. Rev. 51* (1937), 446-9.

5 W. E. Lamb, Jr., "A Note on the Capture of Slow Neutrons in Hydrogenous Substances," *Phys. Rev. 51* (1937), 187-90; W. E. Lamb, Jr., and L. I. Schiff, "On the Electromagnetic Properties of Nuclear Systems," *Phys. Rev. 53* (1938), 651-61.

6 W. E. Lamb, Jr., "Capture of Neutrons by Atoms in a Crystal," *Phys. Rev. 55* (1939), 190-7.

7 H. Fröhlich, W. Heitler, and Boris Kahn, "Deviation from the Coulomb Law for a Proton," *Proc. Roy. Soc. (London) A171* (1939), 269-80; H. Fröhlich, H. Heitler, and Boris Kahn, "Deviation from the Coulomb Law for a Proton," *Phys. Rev. 56* (1939), 961-2.

8 W. E. Lamb, Jr., "Deviation from the Coulomb Law for a Proton," *Phys. Rev. 56* (1939), 384; W. E. Lamb, Jr., "Deviation from the Coulomb Law for a Proton," *Phys. Rev. 57* (1940), 458. I did not discover until much later that Kahn had sent a letter from Groningen in October 1940 entitled "Remark on Deviations from the Fine Structure Formula" to the Dutch journal *Physica 8* (1941), 58. This pointed out that the strong repulsion found by Fröhlich, Heitler, and Kahn would lead to leakage of vacuum electrons (Klein's paradox), which would limit the attainable 2s level shift to the order of $(r_0/a_0)^2$ Ry, which he thought would be "completely unobservable." He concluded that there are "no grounds for expecting a deviation from the [Dirac] fine structure formula." Kahn died in a Nazi annihilation camp in 1943. Further details can be found in a forward written by G. E. Uhlenbeck to a republication of Kahn's 1938 thesis, "On the Theory of the Equation of State," in *Studies in Statistical Mechanics, Vol. III*, ed. by J. de Boer and G. E. Uhlenbeck (Amsterdam: North-Holland, 1965).

9 S. Pasternack, "Note on the Fine Structure of H$_\alpha$ and D$_\alpha$," *Phys. Rev. 54* (1938), 1113.

10 A. Cobas and W. E. Lamb, Jr., "On the Extraction of Electrons from a Metal Surface by Ions and Metastable Atoms," *Phys. Rev. 65* (1944), 327-37.

11 See, for example, M. Sargent III, M. O. Scully, and W. E. Lamb, Jr., *Laser Physics*, 3rd ed. (Reading, Mass.: Addison-Wesley, 1977).

12 G. Becker and S. Autler, "Water Vapor Absorption of Electromagnetic Radiation in the Centimeter Wavelength Range," *Phys. Rev. 70* (1946), 300-7.

13 W. E. Lamb, Jr., "Theory of a Microwave Spectroscope," *Phys. Rev. 70* (1946), 308-17.

14 J. W. Drinkwater, Sir O. Richardson, and W. E. Williams, "Determinations of the Rydberg Constants e/m and the Fine Structure of H$_\alpha$ and D$_\alpha$ by Means of a Reflexion Echelon," *Proc. Roy. Soc. (London) 174* (1940), 164-88.

15 V. Rojansky and J. H. Van Vleck, "The Non-Metastability of the 2s Level in Atomic Hydrogen," *Phys. Rev. 32* (1928), 327; V. Rojansky, "On the Theory of the Stark Effect in Hydrogenic Atoms," *Phys. Rev. 33* (1929), 1-15.

16 H. A. Bethe, "The Electromagnetic Shift of Energy Levels," *Phys. Rev. 72* (1947), 339-41.

17 J. B. French and V. F. Weisskopf, "The Electromagnetic Shift of Energy Levels," *Phys. Rev. 75* (1949), 1240-8.

18 N. M. Kroll and W. E. Lamb, Jr., "On the Self-Energy of a Bound Electron," *Phys. Rev. 75* (1949), 388-98.

19 W. E. Lamb, Jr., and Miriam Skinner, "The Fine Structure of Singly Ionized Helium," *Phys. Rev. 78* (1950), 539-50. R. Novick, E. Lipworth, and P. F. Yergin, "Fine Structure of Singly Ionized Helium," *Phys. Rev. 100* (1955), 1153-73.

20 J. L. McKibben, G. P. Lawrence, and G. G. Ohlsen, "Nuclear Spin Filter," *Phys. Rev. Letters 20* (1968), 1180-2. See also W. J. Thompson and T. B. Clegg, "Physics with Polarized Nuclei," *Phys. Today 32* (Feb., 1979), 32-9.

21 See, for example, T. D. Lee and C. S. Wu, "Decays of Neutral K Mesons," *Ann. Rev. Nucl. Sci. 16* (1966), 511-90.

21   Renormalization theory of quantum electrodynamics: an individual view

JULIAN SCHWINGER

Born 1918, New York City; Ph.D., Columbia, 1939; Nobel Prize, 1965 for research on quantum electrodynamics; University of California, Los Angeles.

My task here is to tell the story, as I saw it and as I participated in it, of the development of renormalized quantum electrodynamics in the years preceding the 1950s. I am also conscious that in a meeting attended by professional historians, the emphasis must be placed on documentation, rather than mere unattested remembrance, an ideal that, like the speed of light, can be approached but never attained.

My story will be divided into four phases; preparation (1934-46); noncovariant relativistic theory (1947); first covariant relativistic theory (1947-8); second covariant relativistic theory (1949-50).

At the age of 16 I wrote, but did not publish, a paper entitled "On the Interaction of Several Electrons." It was about quantum electrodynamics. It combined the space-time-varying operator fields of the Dirac-Fock-Podolsky electrodynamics of 1932 with second-quantized operator fields for electrons, asking whether the usual formalism continues to apply when the electron interaction is the nonlocal retarded interaction of Møller.[1] In the process it made the first tentative introduction of what I would later call the interaction representation, which is no more than the extension to all operator fields of what Paul A. M. Dirac, V. A. Fock, and Boris Podolsky had done for the electromagnetic field. Let me quote one sentence from the paper: "The second term in equation (20) represents the infinite self-energy of the charges and must be discarded." The last injunction merely parrots the wisdom of my elders, to be later rejected, that the theory was fatally flawed, as witnessed by such infinite terms, which, at best, had to be discarded, or subtracted. Thus the "subtraction physics" of the 1930s.

I shall skip over the events of the next 11 years, except to note the following: In the fall of 1939 I came to Berkeley for the first time, not as a student of J. Robert Oppenheimer, but armed with a Columbia Ph.D. and a National Research Council fellowship. Our first collaboration, later that year, used quantum electrodynamics to describe the electron-positron emission from an excited oxygen nucleus, which emphasized for me the physical reality of such virtual photon processes.[2] Also important was the 1941 work on strong-coupling mesotron theory, where I gained experience in using canonical transformations for extracting the physical consequences of the theory.[3]

We now come to 1945. With the war winding down and an enormous capability in microwave technology developed, it was natural that frustrated physicists should begin to think of using their expertise in devising electron accelerators. I took a hand in that and designed parameters for an instrument I called the microtron, but that is another story. What was significant was the radiation emitted by relativistic electrons moving in circular paths under magnetic field guidance. It is an old problem, but the quantitative implications of relativistic energies had not been appreciated. In attacking this classical relativistic situation, I used the invariant proper-time formulation of action, including the electromagnetic self-action of a charge. That self-action contained a resistive part and a reactive part, to use the engineering language I had learned. The reactive part was the electromagnetic mass effect, here automatically providing an invariant supplement to the mechanical action and thereby introducing the physical mass of the charge. Incidentally, in the paper on synchrotron radiation that was published several years later, a more elementary expression of this method is used, and the reactive effect is dismissed as "an inertial effect with which we are not concerned."[4] But here was my reminder that electromagnetic self-action, physically necessary in one context, was not to be, and need not be, omitted in another context. And in arriving at a relativistically invariant result, in a subject where relativistic invariance was notoriously difficult to maintain, I had learned a simple but useful lesson: to emerge with relativistically invariant physical conclusions, use a covariantly formulated theory, and maintain covariance throughout the calculation.

Of course, the concept of electromagnetic self-action, of electromagnetic mass, had not entirely died out in that age of subtraction physics; it had gone underground, to surface occasionally. Hans Kramers must be mentioned in this connection. In a book published in 1938 he suggested that the correspondence-principle foundation of quantum electrodynamics was unsatisfactory because it was not related to a classical theory that already included the electromagnetic mass and referred to the physical electron.[5] He proposed to produce such a classical theory by eliminating the proper field of the electron, the field associated with uniform motion. Very good – if we lived in a nonrelativistic world. But it was already known from the work of Victor Weisskopf and Wendell Furry that the electromagnetic-mass problem is entirely transformed in the relativistic theory of electrons and positrons, then described in the unsymmetrical hole formulation – the relativistic electromagnetic-mass problem is beyond the reach of the correspondence principle.[6] Nevertheless, I must give Kramers very high marks for his recognition that the theory should have a structure-independent character. The relativistic counterpart of that was to be my guiding principle, and over the years it has become generalized to this commandment: Thou shalt not entangle that which is known, and reliable, with that which is unknown, and speculative. The effective-range treatment of nuclear forces, which evolved just after the war, also abides by this philosophy.[7]

The next phase opened with the famous Shelter Island conference of June 1947. Not recalling the exact dates, I looked at Willis E. Lamb and Robert Retherford's paper and learned that it was June 1 to 3;[8] then I glanced at Hans Bethe's paper and read that it was June 2 to 4.[9] Anyway, it was in June. On the train down to New York, Weisskopf and I discussed the already leaked news that Lamb and Retherford had used the wartime-developed microwave techniques to confirm Simon Pasternack's suggested upward shift of the 2s level in hydrogen.[10] We agreed that electrodynamic effects would be responsible and that a finite result would emerge from a relativistic calculation. I do not recall actually saying anything at Shelter Island, but Bethe acknowledges such remarks. As we all know, Bethe then instantly proceeded to exploit his great familiarity with hydrogenic dipole matrix elements and sum rules to compute the nonrelativistic aspects of these ideas. Owing to the comparative insensitivity of the calculation to the unknown high-energy cutoff, a better than order-of-magnitude number emerged. The agreement of that number with the observed level shift ended any doubt, if doubt there was, concerning the electrodynamic nature of the phenomenon. Yet the relativistic problem, of extracting from the theory a finite and unique prediction, remained.

The Lamb-Retherford measurement had been foreshadowed by pre-war spectroscopic observations. But the Shelter Island conference also brought a totally unanticipated announcement from Isidor I. Rabi: The hyperfine structures in hydrogen and deuterium were too large by a fraction of a percent. The significance of the small difference between these two fractions would later be explained by Aage Bohr.[11] But it was their similarity that counted first, suggesting that there was yet another flaw in the Dirac description of the electron, now referring to magnetic properties. The hypothesis that the electron had an additional magnetic moment was first explicitly published by Gregory Breit, later that year, in a curiously ambivalent way: "It is not claimed that the electron has an intrinsic moment. Aesthetic objections could be raised against such a view."[12] Perhaps that ambivalence caused Breit to falter, for he, and here I quote myself, did "not correctly draw the consequences of his empirical hypothesis." He arrived at a value of the additional magnetic moment about five times larger than what more direct experiments, not to mention the relativistic electrodynamic theory, would soon disclose. An additional magnetic moment that large would contribute about one third of the observed upward relative displacement of the 2s level of hydrogen. It was not necessary (the empirical hypothesis of an additional electron moment is easily handled correctly), but, in fact, it took the development of the relativistic electrodynamic theory to straighten out the confusion. However, I am getting ahead of my story.

At the close of the Shelter Island conference, Oppenheimer and I took a seaplane from Port Jefferson to Bridgeport, Connecticut, where civilization, as it was then understood (the railroad) could be found. As the seawater closed over the airplane cabin, I counted my last remaining seconds. But, somehow, primitive technology triumphed. A few days later I abandoned my bachelor quarters and embarked on an accompanied, nostalgic trip around the country that would occupy the whole summer. Not until September did I set out on the trail of relativistic quantum electrodynamics. But I knew what to do.

This is how I would shortly put it, in the first published report of the new electrodynamics:[13]

> Attempts to evaluate radiative corrections to electron pheno-
> mena have heretofore been beset by divergence difficulties,

attributable to self-energy and vacuum polarization effects. Electrodynamics unquestionably requires revision at ultra-relativistic energies [sic], but is presumably accurate at moderate relativistic energies. It would be desirable, therefore, to isolate those aspects of the current theory that essentially involve high energies, and are subject to modification by a more satisfactory theory, from aspects that involve only moderate energies and are thus relatively trustworthy. This goal has been achieved by transforming the Hamiltonian of current hole theory electrodynamics to exhibit explicitly the logarithmically divergent self-energy of a free electron, which arises from the virtual emission and absorption of light quanta. The electromagnetic self-energy of a free electron can be ascribed to an electromagnetic mass, which must be added to the mechanical mass of the electron. Indeed the only meaningful statements of the theory involve this combination of masses, which is the experimental mass of a free electron.

Then, skipping a bit:

It is important to note that the inclusion of the electromagnetic mass with the mechanical mass does not avoid all divergences; the polarization of the vacuum produces a logarithmically divergent term proportional to the interaction energy of the electron in an external field. However, it has long been recognized that such a term is equivalent to altering the value of the electron charge by a constant factor, only the final value being properly identified with the experimental charge. Thus the interaction between matter and radiation produces a renormalization of the electron charge and mass, all divergences being contained in the renormalization factors.

The statement beginning "However, it has long been recognized . . ." harkens back to the very beginnings of the hole theory of positrons. Allow me to translate from the French of Dirac's 1934 report to the seventh Solvay congress:[14] "In consequence of the preceding calculation it would seem that the electric charges normally observed on electrons, protons or other electrified particles are not the charges actually carried by these particles and occurring in the fundamental equations, but are slightly smaller."

One more sentence from my not-yet-written report:[13] "The simplest example of a radiative correction is that for the energy in an external magnetic field." In mid-November of 1947, I went to Washington to attend a small meeting at George Washington University and give a status report on that calculation, of the additional magnetic moment of the electron. It was not complete at the time, but I have the finished calculation, which was discovered in a pile of manuscripts on January 24, 1976, and then labeled "Original Calculation of $\alpha/2\pi$ (1947)." But the magnetic moment of the electron was not my sole concern at that time. My one distinct memory of the Washington meeting is of sitting at a big table and apparently taking notes during a lecture – Was it George Gamov explaining his ideas on the blackbody residual radiation of the big bang? I do not recall. What I do recall is that I was actually doing some simple computations, using my knowledge of the hydrogenic wave functions in momentum space, to understand the "amazingly high value," as Bethe put it, of his average excitation energy for hydrogen. With these clandestine calculations I had easily found that the logarithm of the excitation energy in Rydberg units should be approximately 211/84, or a little more than 2.5. The actual value, which requires rather extensive numerical calculations, is about 2.8.

The first report on renormalized quantum electrodynamics, excerpts of which have just been quoted, was submitted to the *Physical Review* at the end of 1947. It gave the predicted additional magnetic moment of $\alpha/2\pi$ and pointed out that not only are the hyperfine structure discrepancies accounted for but also the later more accurate atomic-moment measurements in states of sodium and gallium.[15] The report continued:

> The radiative corrections to the energy of an electron in a
> Coulomb field will produce a shift in the energy levels of hy-
> drogen-like atoms and modify the scattering of electrons in a
> Coulomb field. . . . The values yielded by our theory differ
> only slightly from those conjectured by Bethe on the basis of
> a non-relativistic calculation and are, thus, in good accord
> with experiment. Finally, the finite radiative correction to the
> elastic scattering of electrons by a Coulomb field provides a
> satisfactory termination to a subject that had been beset with
> much confusion.

Now, what is that last bit all about? Whereas the question of bound-state energies had been largely ignored, theorists had given attention to radiative corrections in scattering. In 1937, Felix Bloch and Arnold Nordsieck recognized that arbitrarily soft photons are emitted with certainty in a collision, implying that the cross section for a perfectly elastic collision is zero.[16] Yet, in a treatment that considers only soft photons, the total cross section is unchanged from its value in the absence of electromagnetic interaction. The real problem begins when hard virtual photons are reintroduced. In 1939, Sidney Dancoff performed such a relativistic calculation for both spin-0 and spin-1/2 charged particles.[17] Incidentally, on reading Dancoff's paper recently, I was somewhat astonished to see the word *renormalization*. But the context here was not mass or charge renormalization; it was the additional terms that maintain the normalization of the state vector. The confusing outcome of Dancoff's calculation was that, whereas spin 1/2 produced a divergent radiative correction, spin 0, usually associated with more severe electromagnetic self-energy problems, gave a finite correction.

The new theory removed the difficulty for spin 1/2. At about the same time, H. W. Lewis reconsidered Dancoff's spin-1/2 work and recognized that it was inconsistent in its treatment of the mechanical and the physical masses of the electron.[18] Then, on subtracting the effect of the electromagnetic mass, the divergences did cancel. But such a subtraction of two ambiguous expressions does not automatically produce an unambiguous finite residue. Lewis acknowledged that the canonical transformation method I had developed was better suited to that purpose. All this raises a question. After reporting that finite radiative corrections were attained in both bound-state and scattering calculations, why was I not specific about their precise values?

Within a month the reason would be given publicly. The American Physical Society held its 1948 New York meeting from January 29 to 31 at Columbia University. I was invited to give a paper on recent developments in quantum electrodynamics. By the way, another invited paper at that meeting was a report from the General Electric Laboratory on the observation and satisfactory spectral analysis of the visible synchrotron radiation emitted by 70-MeV electrons. On January 31 I gave my talk – twice. The only record I have of that event is a typed copy of my already submitted report, on the back page of which is written a formula for the energy shift of hydrogenic levels. One of the terms is a spin-orbit coupling, which should be the relativistic electric counterpart of the $\alpha/2\pi$ additional magnetic-moment effect. But it is smaller by a factor of 3; relativistic invariance is violated in the noncovariant theory. Oppenheimer would later record this in his report to the eighth Solvay congress.[19] But the back of the page also contains something else – the answer to the obvious question: What happens if the additional magnetic-moment coupling to the electric field is given its right value, no other change being introduced? What emerges, and therefore was known in January 1948, is precisely what other workers using noncovariant methods would later find, which is also the result eventually produced by the covariant methods. Of course, until those covariant methods were developed and applied, there could be no real conviction that the right answer had been found.

The third stage, the development of the first covariant theory, had already begun at the time of the New York meeting in January. I have mentioned that the simple idea of the interaction representation had presented itself 14 years earlier, and the space-time treatment of both electromagnetic and electron-positron fields was inevitable. I have a distinct memory of sitting on the porch of my new residence during what must have been a very late Indian summer in the fall of 1947 and with great ease and great delight arriving at invariant results in the electromagnetic-mass calculation for a free electron. I suspect this was done with an equal-time interaction. The spacelike generalization, to a plane, and then to a curved surface, took time, but all that was in place at the New York meeting. I must have made a brief reference to these covariant methods; the typed copy contains such an equation on another back page, and I know that Oppenheimer told me about Sinitiro Tomonaga after my lecture.

Tomonaga's work on a covariant Schrödinger equation had been published in Japanese in 1943; then, in 1946, it was translated into English to appear in an early issue of a new Japanese journal.[20] I have read remarks to the effect that if scientific contact had not been broken during the Pacific war, the theory that we are reviewing here would have been significantly advanced. Of course, lacking an unlimited number of parallel universes in which to act out all possible scenarios, such statements are meaningless. Nevertheless, I shall be bold enough to disagree. The preoccupation of the majority of involved physicists was not with analyzing and carefully applying the known relativistic theory of coupled electron and electromagnetic fields but with changing it. The work of Tomonaga and his collaborators, immediately after the war, centered about the idea of compensation, the introduction of the fields of unknown particles in such a way as to cancel the divergences produced by the known interactions.[21] Richard P. Feynman also advocated modifying the theory, and he would later intimate that a particular, satisfactory modification could be found.[22] My point is merely this: A formalism such as the covariant Schrödinger equation is but a shell awaiting the substance of a guiding physical principle. And the specific concept of the structure-independent renormalized relativistic electrodynamics, while always abstractly conceivable, in fact required the impetus of experiments to show that electrodynamic effects were neither infinite nor zero, but finite and small, and demanded understanding.

The first covariant formulation, in action, was exhibited at the Pocono Manor Inn conference of March 30 to April 1, 1948. I possess a copy of the notes that were taken of the 14 lectures, including those of Feynman and myself. On reading over what was written about my work, I felt no conviction that it was a reliable record of what was actually said; the intrusive hand of the reporter lies heavy on those pages. However, much the same material appeared in notes of lectures delivered several months later at the University of Michigan. Beyond the formalities of field equations, commutation relations, vacuum expectation values, and the like, the topics discussed were free electron mass, photon mass and vacuum polarization, and the electron in an external field, leading to the additional magnetic moment and the energy shifts of hydrogenic atoms. Although it is a vast improvement over the noncovariant methods, what is contained here is still quite primitive. But it introduces the essential computational device of relativistically invariant parameters, quantum counterparts of proper time. It is those parameters that appear in the various outcomes, where they greatly facilitate the separation of the renormalization terms from the actual physical effect under consideration. A logarithmically divergent, invariant electromagnetic mass for the free electron emerges in this way, as it had in the Indian summer of 1947. The photon mass would be a more vexing subject. As Oppenheimer is cited as remarking at Pocono, a covariant gauge-invariant theory could not have a nonzero photon mass, and there is no need to compute it. Yet people, notably Gregor Wentzel, would insist on doing so and end up with nonzero answers.[23] The real subtlety underlying this problem did not emerge for another decade, in the eventual explicit recognition of what others would call Schwinger terms.[24]

While the Pocono conference was in session, Tomonaga was completing a covering letter, directed to Oppenheimer, that was attached to a collection of papers describing the work that had been done in Japan, both independently and in reaction to the news from the West. In a subsequent review paper, written in response to Oppenheimer's telegraphed request, Tomonaga commented on the problem raised by the "infinity [that] is to be attributed to the vacuum polarization effect," in other words, the photon mass. Characteristically, one of the suggested remedies was compensation, the introduction of another charged particle that would produce a photon-mass term of opposite sign. In transmitting this communication to the *Physical Review*, Oppenheimer added a note about the photon mass or, as he put it, "the familiar problem of the light quantum self-energy."[25] He remarked that "as long experience and the recent discussions of Schwinger and others have shown, the very greatest care must be taken in evaluating such self-energies lest, instead of the zero value which they should have, they give non-gauge covariant, non-covariant, in general infinite results."

The Pocono conference was my first opportunity to learn what Feynman was doing with quantum electrodynamics. I had seen his work with John A. Wheeler on classical electrodynamics, and the idea of abolishing the electromagnetic field, in a fundamental sense, did not appeal to me at all.[26] Feynman had discarded the operator-field formulation, and yet as his talk proceeded, I could see points of similarity and, of course, points of difference, other than formalistic questions. We agreed in the emphasis on a manifestly covariant four-dimensional description, including the use of a four-dimensional electromagnetic gauge. It is interesting that where we differed in techniques of computation, time has seen a mutual accommodation. Feynman used not invariant parameters but noncovariant integration methods; he would later adopt invariant parametrization. Where I used two kinds of invariant functions arising from commutator and vacuum expectation value considerations, Feynman, as had E. C. G. Stueckelberg before him, used a complex combination of the two.[27] At the later stage of the second covariant theory, I would also find it to be the natural element. The mention of Stueckelberg brings me back to the remark made in connection with Tomonaga. I regret that I did not find the occasion to review the papers, but I gather that Stueckelberg had early anticipated several of the later features of the invariant perturbation theory of coupled relativistic fields. But Stueckelberg also failed to develop renormalized quantum electrodynamics prior to the experimental impetus of 1947.

The subject of vacuum polarization is a point on which, throughout this 1948 period and beyond, Feynman and I disagreed, a point not of

Murray Gell-Mann and Richard Feynman (credit: AIP, Niels Bohr Library, Marshak collection).

individual mathematical style but of fundamental physics. In his report to the eighth Solvay congress, Bethe said that "the polarization of the vacuum is consciously omitted in Feynman's theory."[28] The reasoning went this way: A modification of the electromagnetic interaction made the electromagnetic mass finite but did nothing for the apparently more severely divergent—here it is again—photon mass. Therefore, things would be simpler if all such effects (closed loops, in Feynman's graphical, acausal language) were omitted. But I knew that the virtual photon emitted by the excited oxygen nucleus created an electron-positron pair; the vacuum is polarizable. In a later paper I would use this very example to illustrate a manifestly gauge-invariant treatment of vacuum polarization.[29]

To the covariant formulation the effect on the electron spin of an external magnetic field poses no problem. The additional $\alpha/2\pi$ magnetic moment in a static field is regained, but now one also sees explicitly that this is a dynamical effect, disappearing as the invariant measure of space-time variation of the field becomes increasingly large on the relativistic scale. It was when we turned to an electrostatic field, to the relativistic justification and extension of the Bethe calculation, that an unfortunate and quite unnecessary bit of confusion entered. The problem was the joining of the relativistic calculation, where the Coulomb potential is regarded as a perturbation, to the nonrelativistic calculation, which treats the Coulomb potential exactly. Later developments would avoid that unphysical separation, but the first attacks used it. And both Feynman and I goofed – we blew it. The physical problem of bound states is not sensitive to arbitrarily soft photons – the atom defines a natural scale of frequencies. But the relativistic treatment of the Coulomb potential as a perturbation, a scattering situation, is sensitive, as in the Bloch-Nordsieck discussion. This is the so-called infrared divergence. And the nonrelativistically calculated difference between the correct and the perturbation treatments of the Coulomb field must also be sensitive, in such a way as to cancel out the infrared divergence in the complete expression. But clearly that will happen without error only if the treatment of soft photons in the relativistic and nonrelativistic parts is consistent. With our eyes on the high-energy end of the photon spectrum, both Feynman and I were careless about the low-energy end.

The following remarks are intended to clarify, not to excuse, that lapse. One provisional technique for handling the infrared problem is to pretend that the photon does have (horrors!) a nonzero mass. Actually, in a theory that otherwise is gauge-invariant, the unphysical processes thereby introduced will quietly disappear as that mass is finally set equally to zero. The relativistic perturbation calculation easily accepts a small photon mass. In the nonrelativistic dipole approximation, it is only the photon energy that makes an appearance. It is not hard to remember that the integration over photon energy is actually a momentum-space integral and take into account the altered momentum-energy relation demanded by the nonzero mass. But there is more. The nonrelativistic treatment refers only to transversely polarized photons, as is appropriate to their motion at the speed of light. But with diminishing energy, a massive photon slows down, and the longitudinal polarization begins to contribute. It is not natural to think of slow, longitudinally polarized photons, and we didn't; but one must, if the whole treatment is to be consistent.

Sometime in 1948, Weisskopf and J. B. French completed their noncovariant calculation of the bound-state energy shift, using every possible clue to maintain relativistic invariance, including the known effect of a magnetic field. Their result was similar to, but not quite identical with, what the covariant calculations of Feynman and myself had produced, which were the same, apart from Feynman's omission of the vacuum-polarization effect. Somewhat shaken, French and Weisskopf retreated to their blackboards and pondered. I, of course, believed the covariant calculation. But then I happened to chance on the almost forgotten outcome of my own noncovariant calculation using the right spin-orbit coupling. It was identical with the French-Weisskopf result! That shook me up to the point that, as Freeman Dyson in 1949 attested, I found the careless slip in the use of the photon mass.[30] This reconciled all the calculations, vacuum polarization aside.[31] And so, as far as the relativistic energy shift is concerned, although Weisskopf was not the first to find the correct result, he was the first to insist on its correctness.

From July 19 to August 7, 1948, a period of three weeks, I lectured at the University of Michigan summer school on (what else?) recent developments in quantum electrodynamics. It seems that I supplied the notes for the first part of the course, which must have been the manuscript for the paper received by the *Physical Review* on July 29.[32] The notes for the second part of the course were taken by David Park. I have read recently words to the effect that what I presented there was like a cut and polished diamond, with all the rough edges removed, brilliant and dazzling. Or, if you don't care for that simile, you can have "a marvel of polished elegance, like a difficult violin sonata played by a virtuoso – more technique than music." I gather I stand accused of presenting a finished elaborate mathematical formalism from which had been excised all the physical insights that provide signposts to its construction. To all charges I plead Not Guilty. The paper to which I have referred has a long historical and physical introduction that motivates the development and sets out the goals of relativistic renormalization theory.[32] Beyond that, the lectures presented the explicit working out of the interaction of a nonrelativistic electron with the radiation field, in the dipole approximation. The canonical transformation that isolates the electromagnetic mass is an elementary one, and the further details leading to the solution of the bound state and scattering problems were provided. This was the simple model on which the relativistic theory was erected. It was good enough for the immediate purposes but, as I have already remarked, still quite primitive. I needed no one to tell me that it was but a first step to an aesthetically satisfactory and effective relativistic theory of coupled fields. Incidentally, at about this same time the canonical transformation method was being successfully applied by E. Corinaldesi and R.

Jost to the radiative correction for the cross section of Compton scattering on a spinless charged particle.[33]

Sometime in mid-1948 I became aware that the National Academy of Sciences was offering a prize for "an outstanding contribution to our knowledge of the nature of light." Entries could be in either of two categories, of which one was a contribution published or submitted in manuscript before October 1, 1948, "which is a comprehensive contribution to a logical, consistent theory of the interaction of charged particles with an electromagnetic field including the interaction of particles moving with high relative speeds." Well! And when I noticed that Feynman was on the committee to award the prize, and therefore presumably ineligible to receive it, I decided that someone out there had me in mind. The reason I mention this "ain't the money; it's the principle of the thing."[34] I submitted the manuscripts of two completed papers and the incomplete provisional version of a third paper. The third paper began with the relativistic treatment of radiative corrections to Coulomb scattering, a topic that was experimentally remote at the time, but is now a routine aspect of interpreting high-energy experiments that employ electrons and positrons. Then the manuscript took up the topic "Radiative Corrections to Energy Levels," beginning as follows: "In situations that do not permit the treatment of the external field as a small perturbation, it is convenient to employ a representation in which the matter field spinors obey equations that correspond to a particle moving under the influence of the external potential." This is what, several years after, would be called the Furry representation.[35] The manuscript went on to study solutions of those field equations and, in the process, exhibited integral equations that were the space-time, relativistic versions of what Lippman and I would present, more symbolically, a year or so later.[36] The manuscript ended abruptly in the middle of a sentence; deadline time had arrived.

I may have been seriously distracted by the pressure of other work, for the completed third and last paper in the quantum electrodynamics series was not submitted until May 26, 1949, although a summary of the results for relativistic Coulomb scattering corrections and energy shifts was sent in at the beginning of that year.[37] I cite in this connection my only memory of the Old Stone on the Hudson meeting, held in April of 1949. On arriving, I was somewhat disconcerted to be immediately asked to report what I was thinking about, to which I replied, half facetiously and half factually, that "the Harvard group was not thinking, it was writing." But it is more probable that the delay had a psychological basis. The impetus of the experimental discoveries of 1947 was waning. The pressure to account for those results had produced a certain theoretical structure that was perfectly adequate for the original task, but demanded simplification and generalization; a new vision was required. There already were visions at large, being proclaimed in a manner somewhat akin to that of the Apostles, who used Greek logic to bring the Hebrew god to the Gentiles. I needed time to go back to the beginnings of things; not yet would I go back to the source.

My retreat began at Brookhaven National Laboratory in the summer of 1949. It is only human that my first action was one of reaction. Like the silicon chip of more recent years, the Feynman diagram was bringing computation to the masses. Yes, one can analyze experience into individual pieces of topology. But eventually one has to put it all together again. And then the piecemeal approach loses some of its attraction. Speaking technically, the summation of some infinite set of diagrams is better and more generally accomplished by solving an integral equation, and those integral equations usually have their origin in a differential equation. And so the copious notes and scratches, labeled "New Opus," and surviving from the summer of 1949, are concerned with the compact, operator expression of classes of processes. And slowly, in these pages, the integral equations and the differential equations emerge. There is another collection of scraps that at some time in the past I put into a folder and labeled "New Theory – Old Version (1949-50)," although I now believe that the reference to 1950 is erroneous – by then the New Theory in its later manifestation had arrived. There is a way to tell the difference. With the emphasis on the operator-field description of realistic interacting systems, the interaction representation had begun to lose its utility, and fields incorporating the full effects of interaction enter. The unpublished essay of the National Academy of Sciences competition had already taken a step in that direction. If fields of both types, with and without reference to interaction, appear in an equation, the historical period is that of the Old Version. The later version has no sign at all of the interaction representation. On one of these pages there is an Old Version, 1949, equation giving the first steps toward the relativistic equation for two interacting particles now known as the Bethe-Salpeter equation. Accordingly, it is not surprising to read in a footnote of a 1951 paper, presenting an operator derivation of the two-particle equation, that I had already discussed it in my Harvard lectures.[38] Before I take up what is really important in this new theory, which is the second covariant relativistic theory, the realization of the new vision that I sought, let me, for a moment, turn anecdotist.

I had been invited to the 1948 Solvay congress meeting in Brussels, but did not go, and regretted it. Accordingly, I was more than pleased to accept an invitation to present a paper at the International Congress for Nuclear Physics, Quantum Electrodynamics and Cosmic Rays, jointly sponsored by the Italian and Swiss physical societies, and to be held in Basel and Como from September 5 to 16, 1949. Stephen White, then of the *Herald Tribune*, whom I first met at the Shelter Island conference, would later point to both of these meetings as witnessing the end of the European monopoly in science and the growing dominance of American science. I wonder what he would say today, some 30 years later. My story does not concern the meeting itself, which was a great social occasion; it is about a side trip to Zürich. Rabi was in Paris, the first stop of my epic journey, and he insisted that I talk to Wolfgang Pauli, to soothe his ruffled feelings. Apparently I had transgressed, but the precise nature of my sin I do not now recall. And so we went to Pauli. He, along with F. Villars, had just completed a paper that had taken them through all the recent publications in quantum electrodynamics.[39] He sat me down and voiced his unhappiness with various aspects of my papers. To each of his complaints, I would, in effect, reply, "Yes, but I don't do it that way anymore." This refusal to be a stationary target left Pauli utterly exasperated. Nevertheless, I think we parted friends.[40]

Feynman had found his vision in a paper of Dirac that gave a correspondence-principle setting for action, the natural invariant starting point of a relativistic theory.[41] I found my vision in the same place. Working with simple mechanical systems, Feynman noticed that Dirac's asymptotic connection, between the quantum description of time evolution and the classical action, sharpened into an equality, for infinitesimal time changes.[42] The indefinite repetition of infinitesimal displacements gave a quantum description of time development in an integral form, similar to the one Norbert Wiener had earlier introduced in another context. One could easily generalize particle variables to Bose-Einstein fields and emerge with the type of functional integral that is commonly regarded today as the starting point of quantum field theory. But quantum field theory must deal with Bose-Einstein fields and Fermi-Dirac fields on a fully equivalent footing. There is nothing in these correspondence-principle-based integrals that suggests the need for anticommuting objects or supplies the meaning of integration for such variables without reference to independent knowledge of some properties of that kind of system. This was not my idea of a fundamental basis for the theory. And, as the history of physics and my own experience indicated, integral statements are best regarded as consequences of more basic differential statements. Indeed, the fundamental formulation of classical mechanics, Hamilton's principle, is a differential, a variational, principle.

There was my challenge. What is the general quantum statement of Hamilton's principle in variational form? It is not hard to find – Dirac's paper already contains some steps in this direction. Here it is.[43] Time development is represented by a transformation function relating the states of the system at two different times or, if you like, on two different spacelike surfaces. Apart from a factor of $i = (-1)^{1/2}$, the variation of this transformation function is just the corresponding matrix element, referring to those states, of the variation of the action operator – for a certain class of operator variations. It is the introduction of operator variations that cuts the umbilical cord of the correspondence principle and brings quantum mechanics to full maturity. The way is now open for Fermi-Dirac fields to appear naturally and on an equal footing with Bose-Einstein fields.

This development must have begun in late 1949 or early 1950, as indicated by a set of notes entitled "Quantum Theory of Fields, A New Formulation." These notes were taken by the current president of the California Institute of Technology, then known as Marvin Goldberger. Dated July, 1950, they refer to a field theory course that was given in the semester between January and June. First for particles, and then for fields, the notes trace how the single quantum action principle leads to operator commutation relations, equations of motion, or field equations and conservation laws. In the relativistic field context, the postulate of invariance under time reflection (remember, this is 1950) leads to two kinds of fields (two statistics) as a consequence of the more elementary analysis into two kinds of spin, integral and half-integral. This occurs because time reflection is not a canonical, a unitary, transformation, but also requires an inversion in the order of all products. That discloses the fundamental operator nature of the field, distinguishing essential commutativity from essential anticommutativity, as demanded by the spin character of the field. In a subsequent version, the existence of two kinds of fields with their characteristic operator properties is recognized at an earlier stage.[44]

Here also the non-Hermitian fields of charged particles are replaced by Hermitian fields of several components, facilitating the description of the internal degrees of freedom that would later proliferate. In this version, time reflection implies a transformation to the complex conjugate algebra, and the postulate of invariance predicts the type of spin to be associated with each statistic. An inspection of the proof shows that what is really used is the hypothesis of invariance under time and space reflection. That invariance and the spin-statistics connection are equivalent. But, with the later discovery of parity nonconservation, the common emphasis, as embodied in the so-called TCP (or is it PTC?) theorem, is to regard the spin-statistics relation as primary and the invariance under space-time reflection as a consequence.

"The Theory of Quantized Fields" is the title of a series of papers that developed and exploited the quantum action principle. The first of this series was largely written during the summer of 1950, again at the Brookhaven National Laboratory.[43] Also begun at this time was a paper that I have already mentioned as a manifestly gauge-invariant treatment of vacuum polarization.[29] But more significant here is the glimpse it gives of the new spirit, in use, but without detailed introduction. An Appendix contains a modified Dirac equation involving a so-called mass operator that is constructed from the Green's functions of electron and photon. The reader is referred to a footnote that most unhelpfully says, "The concepts employed here will be discussed at length in later publications." The purpose of the Appendix is to provide a short, but not yet the shortest, rederivation of the $\alpha/2\pi$ magnetic moment. I cannot refrain from remarking that this same year saw the first application of the Feynman-Dyson methods to a problem that had not already been solved by other procedures. This was the calculation by Robert Karplus and Norman Kroll of the α^2 modification of the electron magnetic moment.[45] They got it wrong. That error remained unnoticed until 1957, when Charles Sommerfield, as his doctoral thesis, used the mass operator technique to produce the right answer.[46]

I have earlier stated my goal of achieving an aesthetically satisfactory and effective relativistic theory of coupled fields. What I have just discussed about the two statistics is, I believe, aesthetically satisfactory. Effectiveness came with the introduction of sources.[47] The concept of source uses numerical space-time functions, totally commutative numbers for Bose-Einstein fields, totally anticommutative numbers for Fermi-Dirac fields. The latter constitute a Grassmann algebra. Often considered bizarre 30 years ago, anticommutative number systems are now the darlings of the super-symmetryists. A source enters the action operator multiplied by its associated field. Those additional action terms symbolize the interventions that constitute measurement of the system, as the test charge in electrostatics probes the electric field. The action principle expresses this succinctly. Apart from the ubiquitous i, the functional derivative of the transformation function with respect to a source is the matrix element of the associated field. That enables all operator field equations to be represented by numerical functional derivative equations. And the commutation properties of the fields at equal times, or on a spacelike surface, are implicit in the fact that the operator field equations now contain the sources, acting as driving terms. The sources serve yet a third function. Through their dynamical action, any desired initial or final state of the system can be produced from the physical ground state, the invariant vacuum state. Accordingly, it suffices to consider the transformation function connecting the vacuum states on two different spacelike surfaces, in the presence of arbitrary sources. The functional differential equations are given a less concise but more elementary form on expanding the vacuum probability amplitude as an infinite power series in the sources. The coefficient of a particular product of sources, referring to a set of space-time points, is a function of those points. I gave the name Green's function to the totality of those multipoint functions. As the equivalent of the functional differential equations, the Green's functions obey an infinite linear inhomogeneous set of coupled differential equations. The accompanying boundary conditions, implied by the reference to the vacuum state, are the generalization of those introduced by Stueckelberg and Feynman.

But the set of coupled Green's function equations is only one way of applying this flexible source method. Do you want to work directly with a perturbation expansion of the transformation function? Then use functional derivatives with respect to sources to construct the interaction term of the action operator. The transformation function for the physical, interacting system will now be produced, from the interactionless transformation function, by the effect of an exponential involving that functional derivative replacement for the field interaction term. (Confronted with a sentence like this, one appreciates why mathematics is the preferred language of theoretical physics.) The power series expansion of the exponential then generates, order by order, the desired perturbation series. Topology (the Feynman diagrams) is optional here; that is a matter of pedagogy, not physics. And for sufficiently complicated situations, it should be advantageous to have a method that supplies all relevant terms analytically, rather than by geometrical intuition. Would you rather manipulate functional integrals? Then begin with a formal solution of the functional differential equations in which an exponential function of the action (multiplied by *i*, of course, with operators replaced by functional derivatives) acts on a grand delta functional of all sources. The Fourier construction of that delta functional, using well-defined functional integration concepts, then yields the functional integral construction of the transformation function. And there are mixed procedures, with functional derivatives for one kind of source entering numerical differential equations for the other type of field.

What I have just described is all technique. Now, here is the music. It is probably a fairly widespread opinion that renormalized quantum electrodynamics is just the old quantized version of the combined Maxwell and Dirac equations, with some rules for hiding divergences. That is simply not true. A theory has two aspects. One is a set of equations relating various symbols. The other is, at some level, the physical interpretation to be associated with the symbols. In the course of the development here being described, the equations did not change, but the interpretation did. In the late 1930s, most people would not have challenged these statements: e and m, as they enter the Dirac and Maxwell equations, are the charge and the mass of the electron; an electromagnetic field operator creates or annihilates a photon; a Dirac field operator creates an electron or annihilates a positron, and its adjoint field does the inverse. And all this would be true if the two fields were uncoupled. But, in the real world, the localized excitation represented by an electromagnetic field, for example, does not just create a photon; it transfers energy, momentum, and angular momentum, and then Nature goes to work. And so, it may create a photon, or an electron-positron pair, or anything else with the right quantum numbers. The various Green's functions are the correlation functions among such localized excitations, and the study of their space-time behavior is the instrument for the identification of the physical particles and of their interactions. Renormalization, properly understood, is an aspect of the transfer of attention from the initial hypothetical world of localized excitations and interactions to the observable world of the physical particles. As such, it is logically independent of divergences. Could we construct a convergent theory of coupled fields, it would still need to be renormalized.

All that I have been saying was explicit or implicit in work performed before the end of the fifth decade, although actual publication would be delayed, sometimes indefinitely.[48] Therefore, I consider that I have not yet crossed the time barrier that defines the scope of this conference. But I feel that I cannot conclude without saying something about the influence that electrodynamics would have in other areas of physics. And I do not see how I can avoid mentioning the ultimate fate of renormalization in my hands. The only solution to this dilemma is to turn back in time.

Here is an anecdote from 1941, unattested and, unfortunately, unattestable. I had been thinking about Enrico Fermi's theory of β decay, wherein appears a very small coupling constant of order 10^{-12}. It occurred to me that the electron mass, then used as the significant mass scale, was not necessarily the relevant quantity. The neutron and proton were also involved, and possibly the nucleon mass was the appropriate unit. On introducing it, the coupling constant became of order 10^{-5}. And then I thought – perhaps the really significant mass unit is several tens of nucleon masses, for then the coupling constant could be the electromagnetic coupling constant $\alpha \cong 1/137$. One day I mentioned this bit of numerology to Oppenheimer. He stared at me, and then said coldly, "Well, it's a new idea." Indeed it was, and is.[49]

And finally, I turn to the last section of a 1949 paper by Dyson, which I think it reasonable to assume was strongly influenced by Oppenheimer.[50] In any event, here is a quotation: "What is to be looked for in a future theory is not so much a modification of the present theory which will make all infinite quantities finite, but rather a turning-round of the theory so that the finite quantities shall become primary," and then, "One may expect that in the future a consistent formulation of electrodynamics will be possible, itself free from infinities and involving only the physical constants m and e." That is just what I have accomplished in a program called "Source Theory," which is in no way limited to quantum electrodynamics.[51]

And so, if I were asked to respond to criticisms of the path I followed prior to the beginning of the sixth decade, I would answer: "I don't do it that way anymore."

Notes

1 P. A. M. Dirac, V. A. Fock, and Boris Podolsky, "On Quantum Electrodynamics," *Physikalische Zeitschrift der Sowjetunion 2* (1932), 468-79. This article is reprinted in *Selected Papers on Quantum Electrodynamics*, ed. by Julian Schwinger

(New York: Dover Publications, 1958). The preface to this collection provides a historical survey from the vantage point of 1956. P. Jordan and E. Wigner, "Ueber das Paulische Äquivalenzverbot," *Z. Physik 47* (1928), 631-51; reprinted in Schwinger (ed.), *Selected Papers on Quantum Electrodynamics* (New York: Dover, 1958). C. Møller, "Zur Theorie des Durchgangs schneller Elektronen durch Materie," *Ann. Physik 14* (1932), 531-85; Christian Møller, "Ueber den Stoss zweier Teilchen unter Berücksichtigung der Retardation der Kräfte," *Z. Physik 70* (1931), 786-95.

2 J. R. Oppenheimer and J. Schwinger, "On Pair Emission in the Proton Bombardment of Fluorine," *Phys. Rev. 56* (1939), 1066-7.

3 J. R. Oppenheimer and Julian Schwinger, "On the Interaction of Mesotrons and Nuclei," *Phys. Rev. 60* (1941), 150-2. This paper contains several references to a paper of mine "to be published soon." It appeared 29 years later as "Charged Scalar Mesotron Field," in *Quanta, Essays in Theoretical Physics Dedicated to Gregor Wentzel*, ed. by Peter Freund, Charles Goebel, and Yoichiro Nambu (Chicago: University of Chicago Press, 1970), pp. 101-38.

4 Julian Schwinger, "On the Classical Radiation of Accelerated Electrons," *Phys. Rev. 75* (1949), 1912-25; Julian Schwinger, "Electron Radiation in High Energy Accelerators," *Phys. Rev. 70* (1946), 798-9.

5 H. Kramers, *Quantentheorie des Elektrons und der Strahlung* (Leipzig: Akademische Verlagsgesellschaft, 1938); English translation by D. ter Haar (Amsterdam: North-Holland, 1957).

6 V. S. Weisskopf, "On the Self-Energy and the Electromagnetic Field of the Electron," *Phys. Rev. 56* (1939), 72-85; reprinted in Schwinger (ed.), *Selected Papers on Quantum Electrodynamics* (Note 1). The references in this article to the initial papers of 1934 do not make explicit that it was W. Furry who first appreciated the logarithmic nature of the divergence of the electromagnetic mass in the hole theory of electrons and positrons.

7 Julian Schwinger, "A Variational Principle for Scattering Problems," *Phys. Rev. 72* (1947), 742; Julian Schwinger, "On the Charge Independence of Nuclear Forces," *Phys. Rev. 78* (1950), 135-9 (and earlier references in the latter article).

8 Willis E. Lamb, Jr., and Robert C. Retherford, "Fine Structure of the Hydrogen Atom by a Microwave Method," *Phys. Rev. 72* (1947), 241-3; reprinted in Schwinger (ed.), *Selected Papers on Quantum Electrodynamics* (Note 1).

9 H. A. Bethe, "The Electromagnetic Shift of Energy Levels," *Phys. Rev. 72* (1949), 339-41; reprinted in Schwinger (ed.), *Selected Papers on Quantum Electrodynamics* (Note 1).

10 Simon Pasternack, "Note on the Fine Structure of H_α and D_α," *Phys. Rev. 54* (1938), 1113.

11 Aage Bohr, "On the Hyperfine Structure of Deuterium," *Phys. Rev. 73* (1948) 1109-11.

12 G. Breit, "Relativistic Corrections to Magnetic Moments of Nuclear Particles," *Phys. Rev. 71* (1947), 400-2 ["However Breit has not correctly drawn the consequences of his empirical hypothesis. The effects of a nuclear magnetic field and a constant magnetic field do not involve different combinations of μ and $\delta\mu$." (from the work cited in Note 13, footnote 2)].

13 Julian Schwinger, "On Quantum-Electrodynamics and the Magnetic Moment of the Electron," *Phys. Rev. 73* (1948), 416-17; reprinted in Schwinger (ed.), *Selected Papers on Quantum Electrodynamics* (Note 1).

14 P. A. M. Dirac, "Théorie du Positron," in *Rapport du 7ᵉ Conseil Solvay de Physique, Structure et Propriétés des Noyaux Atomiques, 1933* (Paris: Gauthier-Villars,

1934), pp. 203-12; reprinted in Schwinger (ed.), *Selected Papers on Quantum Electrodynamics* (Note 1).

15 H. M. Foley and P. Kusch, "On the Intrinsic Moment of the Electron," *Phys. Rev. 73* (1948), 412; reprinted in Schwinger (ed.), *Selected Papers on Quantum Electrodynamics* (Note 1).

16 F. Bloch and A. Nordsieck, "Note on the Radiation Field of the Electron," *Phys. Rev. 52* (1937), 54-9; reprinted in Schwinger (ed.), *Selected Papers on Quantum Electrodynamics* (Note 1).

17 S. M. Dancoff, "On Radiative Corrections for Electron Scattering," *Phys. Rev. 55* (1939), 959-63.

18 H. W. Lewis, "On the Reactive Terms in Quantum Electrodynamics," *Phys. Rev. 73* (1948), 173-6.

19 J. R. Oppenheimer, "Electron Theory," *Rapport du 8ᵉ Conseil Solvay de Physique, Particules Élémentaires, 1948*, ed. by R. Stoops (Brussels: Secrétaires du Conseil, 1950), pp. 269-81; reprinted in Schwinger (ed.), *Selected Papers on Quantum Electrodynamics* (Note 1).

20 S. Tomonaga, "On a Relativistically Invariant Formulation of the Quantum Theory of Wave Fields," *Progr. Theoret. Phys. (Kyoto) 1* (1946), 27-42; reprinted in Schwinger (ed.), *Selected Papers on Quantum Electrodynamics* (Note 1).

21 D. Ito, Z. Koba, and S. Tomonaga, "Correction due to the Reaction of 'Cohesive Force Field' for the Elastic Scattering of an Electron," *Progr. Theoret. Phys. (Kyoto) 2* (1947), 216-17.

22 R. P. Feynman, "Space-Time Approach to Quantum Electrodynamics," *Phys. Rev. 76* (1949), 769-89; reprinted in Schwinger (ed.), *Selected Papers on Quantum Electrodynamics* (Note 1).

23 Gregor Wentzel, "New Aspects of the Photon Self-Energy Problem," *Phys. Rev. 74* (1948), 1070-5.

24 Julian Schwinger, "Field Theory Commutators," *Phys. Rev. Letters 3* (1959), 296-7.

25 J. R. Oppenheimer, "Note on the Above Letter," *Phys. Rev. 74* (1948), 225.

26 John Archibald Wheeler and Richard Phillips Feynman, "Interaction with the Absorber as the Mechanism of Radiation," *Rev. Mod. Phys. 17* (1945), 157-81.

27 E. C. G. Stueckelberg, "La mécanique due point matériel en théorie de relativité et en théorie des quanta," *Helv. Phys. Acta 15* (1942), 23-37.

28 For the eighth Solvay conference, see Note 19. *Ed. note*: Professor H. A. Bethe was invited to the 1948 Solvay conference but was absent. Although it is stated that he transmitted a report, it is not in the printed proceedings.

29 Julian Schwinger, "On Gauge Invariance and Vacuum Polarization," *Phys. Rev. 82* (1951), 664-79; reprinted in Schwinger (ed.), *Selected Papers on Quantum Electrodynamics* (Note 1).

30 F. J. Dyson, "The S Matrix in Quantum Electrodynamics," *Phys. Rev. 75* (1949), 1736-55; reprinted in Schwinger (ed.), *Selected Papers on Quantum Electrodynamics* (Note 1); see footnote 8 in this paper.

31 Julian Schwinger, "On Radiative Corrections to Electron Scattering," *Phys. Rev. 75* (1949), 898-9; reprinted in Schwinger (ed.), *Selected Papers on Quantum Electrodynamics* (Note 1). Footnote 5 in the paper refers to this story and supplies some references.

32 Julian Schwinger, "Quantum Electrodynamics. I. A Covariant Formulation," *Phys. Rev. 74* (1948), 1439-61.

33 E. Corinaldesi and R. Jost, "Die höheren strahlungstheoretischen Näherungen zum Compton-Effekt," *Helv. Phys. Acta 21* (1948), 183-6. The attack on the spin-1/2 problem was begun by D. Feldman and J. Schwinger, "Radiative Correction to the

Klein-Nishina Formula," *Phys. Rev. 75* (1949), 338. *Ed. note*: The problem was solved by L. M. Brown and R. P. Feynman, "Radiative Corrections to Compton Scattering," *Phys. Rev. 85* (1952), 231-44.

34 Arthur Roberts, "It ain't the money; it's the principle of the thing" was composed in celebration of the Nobel Prize award to I. I. Rabi in 1944.

35 W. H. Furry, "On Bound States and Scattering in Positron Theory," *Phys. Rev. 81* (1951), 115-24.

36 B. A. Lippman and Julian Schwinger, "Variational Principles for Scattering Processes. I," *Phys. Rev. 79* (1950), 469-80.

37 Julian Schwinger, "Quantum Electrodynamics. III. The Electromagnetic Properties of the Electron-Radiative Corrections to Scattering," *Phys. Rev. 76* (1949), 790-817; reprinted in Schwinger (ed.), *Selected Papers on Quantum Electrodynamics* (Note 1). Julian Schwinger, "On Radiative Corrections to Electron Scattering," *Phys. Rev. 75* (1949), 898-9.

38 Murray Gell-Mann and Francis Low, "Bound States in Quantum Field Theory" (footnote 6), *Phys. Rev. 84* (1951), 350-4. Indeed, my own publication of these matters was submitted before this paper: paper 31 in Schwinger (ed.), *Selected Papers on Quantum Electrodynamics* (Note 1).

39 W. Pauli and F. Villars, "On the Invariant Regularization in Relativistic Quantum Theory," *Rev. Mod. Phys. 21* (1949), 434-44; reprinted in Schwinger (ed.), *Selected Papers on Quantum Electrodynamics* (Note 1).

40 The nature of charge renormalization was not clearly understood for some time. Excerpt of a letter of April 13, 1948, from A. Pais to S. Tomonaga: "In fact it seems one of the most puzzling problems how to 'renormalize' the charge of the electron and of the proton in such a way as to make the experimental values for these quantities equal to each other." It was during this visit, I believe, that I communicated to Pauli the remark that charge renormalization is a property of the electromagnetic field alone, leading to a universal renormalization factor (relating the physical charge to the hypothetical charge) that is less than unity. See G. Källen, *Quantum Electrodynamics* (New York: Springer-Verlag, 1972), p. 215, footnote 1.

41 P. A. M. Dirac, "The Lagrangian in Quantum Mechanics," *Physikalische Zeitschrift der Sowjetunion 3* (1933), 64-72; reprinted in Schwinger (ed.), *Selected Papers on Quantum Electrodynamics* (Note 1).

42 R. P. Feynman, "Space-Time Approach to Non-Relativistic Quantum Mechanics," *Rev. Mod. Phys. 20* (1948), 367-87; reprinted in Schwinger (ed.), *Selected Papers on Quantum Electrodynamics* (Note 1).

43 Julian Schwinger, "The Theory of Quantized Fields. I," *Phys. Rev. 82* (1951), 914-27; reprinted in Schwinger (ed.), *Selected Papers on Quantum Electrodynamics* (Note 1).

44 Julian Schwinger, "The Theory of Quantized Fields. II," *Phys. Rev. 91* (1953), 713-28; reprinted in Schwinger (ed.), *Selected Papers on Quantum Electrodynamics* (Note 1).

45 Robert Karplus and Norman N. Kroll, "Fourth Order Corrections in Quantum Electrodynamics and the Magnetic Moment of the Electron," *Phys. Rev. 77* (1950), 536-49.

46 Charles M. Sommerfield, "Magnetic Dipole Moment of the Electron," *Phys. Rev. 107* (1957), 328-9.

47 Julian Schwinger, "On the Green's Functions of Quantized Fields. I," *Proc. Natl. Acad. Sci. U.S. 37* (1951), 452-9; reprinted in Schwinger (ed.), *Selected Papers on Quantum Electrodynamics* (Note 1).

48 For a compact survey of the whole development, see "A Report on Quantum

Electrodynamics," paper 160 in *Selected Papers of Julian Schwinger* (Boston: D. Reidel, 1979), hereafter referred to as *SP*.

49 This is the anticipation of the unification of electromagnetism with the weak interactions, and of the not yet experimentally verified heavy-boson intermediary of the charge-exchange weak interactions, explicitly proposed in *SP* 82 (Note 48).

50 F. J. Dyson, "The S Matrix in Quantum Electrodynamics," *Phys. Rev. 75* (1949), 1736-55; reprinted in Schwinger (ed.), *Selected Papers on Quantum Electrodynamics* (Note 1).

51 See *SP* 135, 137, 147, 151, etc. (Note 48).

22   Two shakers of physics: memorial lecture for Sin-itiro Tomonaga[1]

JULIAN SCHWINGER

I am deeply honored to have the privilege of addressing you. It is natural that I should do so, as the Nobel Prize partner whose work on quantum electrodynamics was most akin in spirit to that of Sin-itiro Tomonaga. But not until I began preparing this memorial did I become completely aware of how much our scientific lives had in common. I shall mention those aspects in due time. More immediately provocative is the curious similarity hidden in our names. The Japanese character (the *kanji*) *shin* 振 has, among other meanings, those of "to wave" or "to shake." The beginning of my Germanic name, Schwing, means "to swing" or "to shake." Hence my title: "Two shakers of physics."

One cannot speak of Tomonaga without reference to Hideki Yukawa and, of course, Yoshio Nishina. It is a remarkable coincidence that both Japanese Nobel prize winners in physics were born in Tokyo, both had their families move to Kyoto, both were sons of professors at Kyoto University, both attended the Third High School in Kyoto and both attended and graduated from Kyoto University with degrees in physics. In their third and final year at the university, both learned the new quantum mechanics together. (Tomonaga would later remark, about this independent study, that he was happy not to be bothered by the professors.) Both graduated in 1929 into a world that seemed to have no place for them. (Yukawa later said that "the depression made scholars.") Accordingly, both stayed on as unpaid assistants to Professor Hidehiko Tamaki; Yukawa would eventually succeed him. In 1931, Nishina came on the scene. He gave a series of lectures at Kyoto University on quantum mechanics. Shoichi Sakata, then a student, later reported that Yukawa and Tomonaga asked the most questions afterward.

Nishina was a graduate in electrical engineering of Tokyo University. In 1917 he joined the recently founded Institute of Physical and Chemical Research, the Rikagaku Kenkyusho – Riken. A private institution, Riken was supported financially in various ways, including the holding of patents on the manufacture of sake. After several years at Riken, Nishina was sent abroad for further study, a pilgrimage that would last for eight years. He stopped at the Cavendish Laboratory in Cambridge, England, and at the University of Göttingen in Germany; then, finally, he went to Denmark and Niels Bohr in Copenhagen. He would stay there for six years, and out of that period came the famous Klein-Nishina formula. Nishina returned to Japan in December 1928 to begin building the Nishina group. It would, among other contributions, establish Japan in the forefront of research on nuclear and cosmic-ray physics – *soryushiron*.

There was a branch of Riken at Kyoto in 1931 when Nishina, the embodiment of the *Kopenhagener Geist*, came to lecture and to be impressed by Tomonaga. The acceptance of Nishina's offer of a research position brought Tomonaga to Tokyo in 1932. (Three years earlier he had traveled to Tokyo to hear lectures at Riken given by Werner Heisenberg and Paul A. M. Dirac.) The year 1932 was a traumatic one for physics. The neutron was discovered; the positron was discovered. The first collaborative efforts of Nishina and Tomonaga dealt with the neutron and the problem of nuclear forces. Although there were no formal publications, this work was reported at the 1932 autumn and 1933 spring meetings that were regularly held by the Riken staff. Then, at the 1933 autumn meeting, the subject became the positron. It was the beginning of a joint research program that would see the publication of a number of papers concerned with various aspects of electron-positron pair creation and annihilation. Tomonaga's contributions to quantum electrodynamics had begun.

Although these papers were visible evidence of interest in quantum electrodynamics, we are indebted to Tomonaga for telling us, in his Nobel address, of an unseen but more important step: He read the 1932 paper of Dirac that attempted to find a new basis for electrodynamics. Dirac argued that "the role of the field is to provide a means for making observations of a system of particles," and therefore "we cannot suppose the field to be a dynamical system on the same footing as the particles and thus something to be observed in the same way as the particles." The attempt to demote the dynamical status of the electromagnetic field, or, in the more extreme later proposal of John

Wheeler and Richard Feynman, to eliminate it entirely, was a false trail, contrary to the fundamental quantum duality between particle and wave, or field. Nevertheless, Dirac's paper was to be very influential. Tomonaga said:

> This paper of Dirac's attracted my interest because of the novelty of its philosophy and the beauty of its form. Nishina also showed a great interest in this paper and suggested that I investigate the possibility of predicting some new phenomena by this theory. Then I started computations to see whether the Klein-Nishina formula could be derived from this theory or whether any modification of the formula might result. I found out immediately, however, without performing the calculation through to the end, that it would yield the same answer as the previous theory. The new theory of Dirac's was in fact mathematically equivalent to the older Heisenberg-Pauli theory and I realized during the calculation that one could pass from one to the other by a unitary transformation. The equivalence of these two theories was also discovered by Rosenfeld and Dirac-Fock-Podolsky and was soon published in their papers.

I graduated from a high school that was named for Townsend Harris, the first American consul in Japan. Soon after, in 1934, I wrote but did not publish my first research paper. It was on quantum electrodynamics. Several years before, the Danish physicist Christian Møller had proposed a relativistic interaction between two electrons, produced through the retarded intervention of the electromagnetic field. It had been known since 1927 that electrons could also be described by a field, one that had no classical macroscopic counterpart. And the dynamical description of this field was understood, when the electrons interacted instantaneously. I asked how things would be when the retarded interaction of Møller was introduced. To answer the question, I used the Dirac-Fock-Podolsky formulation. But because I was dealing entirely with fields, it was natural to introduce for the electron field, as well, the analogue of the unitary transformation that Tomonaga had already recognized as being applied to the electromagnetic field in Dirac's original version. Here was the first tentative use of what Tomonaga, in 1943, would correctly characterize as "a formal transformation which is almost self-evident" and I, years later, would call the interaction representation. No, neither of us, in the 1930s, had reached what would eventually be named the Tomonaga-Schwinger equation, but each of us held a piece that, in combination, would lead to that equation: Tomonaga appreciated the relativistic form of the theory, but was thinking in particle language; I used a field theory, but had not understood the need for a fully relativistic form. Had we met then, would history have been different?

The reports of the spring and autumn 1936 meetings of the Riken staff show something new: Tomonaga had resumed his interest in nuclear physics. In 1937 he went to Germany, to Heisenberg's Institute at Leipzig. He would stay for two years, working on nuclear physics and on the theory of mesons, to use the modern term. Tomonaga had come with a project in mind: treat Niels Bohr's liquid-drop model of the nucleus, and the way an impinging neutron heats it up, by using the macroscopic concepts of heat conduction and viscosity. This work was published in 1938. It was also the major part of the thesis submitted to Tokyo University in 1939 for the degree of Doctor of Science – *Rigakuhakushi*. Heisenberg's interest in cosmic rays then turned Tomonaga's attention to Yukawa's meson.

The not yet understood fact, that the meson of nuclear forces and the cosmic-ray meson observed at sea level are not the same particle, was beginning to thoroughly confuse matters at this time. Tomonaga wondered whether the problem of the meson lifetime could be overcome by including an indirect process in which the meson turns into a pair of nucleons (proton and neutron) that annihilate to produce the final electron and neutrino. The integral over all nucleon pairs, resulting from the perturbation calculation, was – infinite. Tomonaga kept a diary of his impressions during this German period. It poignantly records his emotional reactions to the difficulties he encountered. Here are some excerpts (translated by Fumiko Tanihara):

> It has been cold and drizzling since morning and I have devoted the whole day to physics in vain. As it got dark I went to the park. The sky was gray with a bit of the yellow of twilight in it. I could see the silhouetted white birch grove glowing vaguely in the dark. My view was partly obscured by my tired eyes; my nose prickled from the cold and upon returning home I had a nosebleed. After supper I took up my physics again, but at last I gave up. Ill-starred work indeed!

Then:

Recently I have felt very sad without any reason, so I went to a film. . . . Returning home I read a book on physics. I don't understand it very well. Meanwhile I suffer. . . . Why isn't nature clearer and more directly comprehensible?

Again:

As I went on with the calculation, I found the integral diverged – was infinite. After lunch I went for a walk. The air was astringently cold and the pond in Johanna Park was half frozen, with ducks swimming where there was no ice. I could see a flock of other birds. The flower beds were covered with chestnut leaves against the frost. . . . Walking in the park . . . I was no longer interested in the existence of neutron, neutrino. . . .

And, finally:

I complained in emotional words to Professor Nishina about the slump in my work, whereupon I got his letter in reply this morning. After reading it my eyes were filled with tears. . . . He says: only fortune decides your progress in achievements. All of us stand on the dividing line from which the future is invisible. We need not be too anxious about the results, even though they may turn out quite different from what you expect. By-and-by you may meet a new chance for success. . . .

Toward the close of Tomonaga's stay in Leipzig, Heisenberg suggested a possible physical answer to the clear inapplicability of perturbation methods in meson physics. It involved the self-reaction of the strong meson field surrounding a nucleon. Heisenberg did a classical calculation, showing that the scattering of mesons by nucleons might thereby be strongly reduced, which would be more in conformity with the experimental results. About this idea, Tomonaga later remarked:

Heisenberg, in this paper published in 1939, emphasized that the field reaction would be crucial in meson-nucleon scattering. Just at that time I was studying at Leipzig, and I still remember vividly how Heisenberg enthusiastically explained this idea to me and handed me galley proofs of his forthcoming paper. Influenced by Heisenberg, I came to believe that the problem of field reactions far from being meaningless was one which required a frontal attack.

Sin-itiro Tomonaga in 1953 at Tokyo University of Education (credit: Akio Honma).

Indeed, Tomonaga wanted to stay on for another year to work on the quantum mechanical version of Heisenberg's classical calculation. The growing clouds of war made this inadvisable, however, and Tomonaga returned to Japan by ship. As it happened, Yukawa, who had come to Europe to attend a Solvay Congress, which unfortunately was canceled, sailed on that same ship. When the ship docked at New York, Yukawa disembarked and, beginning at Columbia University, where I first met him, made his way across the United States, visiting various universities. But Tomonaga, after a day's sightseeing in New York that included the Japanese pavilion at the World's Fair, continued with the ship through the Panama Canal and on to Japan. About this, Tomonaga said: "When I was in Germany I had wanted to stay another year in Europe, but once I was aboard a Japanese ship I became eager to arrive in Japan." He also remarked about his one-day excursion in

New York that "I found that I was speaking German rather than English, even though I had not spoken fluent German when I was in Germany."

Tomonaga had returned to Japan with some ideas concerning the quantum treatment of Heisenberg's proposal that attention to strong field reactions was decisive for understanding the meson-nucleon system. But soon after he began work, he became aware, through an abstract of a paper published in 1939, that Gregor Wentzel was also attacking this problem of strong coupling. Here is where the scientific orbits of Tomonaga and myself again crossed. At about the time that Tomonaga returned to Japan, I went to California to work with J. Robert Oppenheimer. Our first collaboration was a quantum electrodynamic calculation of the electron-positron pair emitted by an excited oxygen nucleus. And then we turned to meson physics. Heisenberg had suggested that meson-nucleon scattering would be strongly suppressed by field reaction effects. There also existed another proposal to the same end – that the nucleon possessed excited states, isobars, that would produce almost canceling contributions to the meson scattering process. We showed, classically, that the two explanations of suppressed scattering were one and the same: The effect of the strong field reaction, of the strong coupling, was to produce isobars, bound states of the meson about the nucleon. The problem of giving these ideas a correct quantum framework naturally arose. And then we became aware, through the published paper, of Wentzel's quantum considerations on a simple model of the strong coupling of meson and nucleon. I took on the quantum challenge myself. Not liking the way Wentzel had handled it, I redid his calculation in my own style and, in the process, found that Wentzel had made a mistake. In the short note that Oppenheimer and I eventually published, this work of mine is referred to as "to be published soon." And it was published, 29 years later, in a collection of essays dedicated to Wentzel. Recently, while surveying Tomonaga's papers, I came upon his delayed publication of what he had done along the same lines. I then scribbled a note: "It is as though I were looking at my own long unpublished paper." I believe that both Tomonaga and I gained from this episode added experience in using canonical (unitary) transformations to extract the physical consequences of a theory.

I must not leave the year 1939 without mentioning a work that would loom large in Tomonaga's later activities. But, to set the stage, I turn back to 1937. In that year, Felix Bloch and Arnold Nordsieck considered another kind of strong coupling–that between an electric charge and arbitrarily soft (extremely low frequency) light quanta. They recognized that in a collision, say between an electron and a nucleus, arbitrarily soft quanta will surely be emitted; a perfectly elastic collision cannot occur. Yet, if only soft photons, those of low energy, are considered, the whole scattering process goes on as though the electrodynamic interactions were ineffective. Once this was understood, it was clear that the real problem of electrodynamic field reaction begins when arbitrarily hard (unlimited high-energy) photons are reintroduced. In 1939 Sidney Dancoff performed such a relativistic scattering calculation both for electrons, which have spin 1/2, and for charged particles without spin. The spin-0 calculation gave a finite correction to the scattering, but for spin 1/2, the correction was infinite. This was confusing. And to explain why that was so, we must talk about electromagnetic mass.

It was already part of classical physics that the electric field surrounding an electrically charged body carries energy and contributes mass to the system. That mass varies inversely as a characteristic dimension of the body and therefore is infinite for a point charge. The magnetic field that accompanies a moving charge implies an additional momentum, an additional, electromagnetic, mass. It is very hard, at this level, to make those two masses coincide, as they must, in a relativistically invariant theory. The introduction of relativistic quantum mechanics, of quantum field theory, changes the situation completely. For the spin-1/2 electron-positron system, obeying Fermi-Dirac statistics, the electromagnetic mass, while still infinite, is only weakly, logarithmically, so. In contrast, the electromagnetic mass for a spin-0 particle, which obeys Bose-Einstein statistics, is more singular than the classical one. Thus, Dancoff's results were in contradiction to the expectation that spin 0 should exhibit more severe electromagnetic corrections.

Tomonaga's name had been absent from the Riken reports for the years from 1937 to 1939, when he was in Germany. It reappeared for the 1940 spring meeting under the title "On the Absorption and Decay of Slow Mesons." There the simple and important point was made that when cosmic-ray mesons are stopped in matter, the repulsion of the nuclear Coulomb field prevents positive mesons from being absorbed by the nucleus, whereas negative mesons will preferentially be absorbed before decaying. This was published as a *Physical Review* letter in 1940. Subsequent experiments showed that no such asymmetry existed in very light elements; the cosmic-ray meson does not interact strongly with nuclear particles. The Riken reports from autumn of 1940 to autumn of 1942 traced stages in the development of Tomonaga's strong- and intermediate-coupling meson theories. In particular, under the heading "Field Reaction and Multiple Production" there was discussed a coupled set of equations corresponding to various particle numbers that is the basis of an approximation scheme now generally called the Tamm-Dancoff approximation. This series of reports on meson theory was presented to the Meson symposium (*Chukanshi Toronkai*) that was initiated in September 1943, where also was heard the suggestion of Sakata's group that the cosmic-ray meson was not the meson responsible for nuclear forces.

But meanwhile there occurred the last of the Riken meetings held during the war, that of spring 1943. Tomonaga provided the following abstract with the title "Relativistically Invariant Formulation of Quantum Field Theory":

> In the present formulation of quantum fields as a generalization of ordinary quantum mechanics such non-relativistic concepts as probability amplitude, canonical commutation relations and Schrödinger equation are used. Namely these concepts are defined referring to a particular Lorentz frame in space-time. This unsatisfactory feature has been pointed out by many people and also Yukawa emphasized it recently. I made a relativistic generalization of these concepts in quantum mechanics such that they do not refer to any particular coordinate frame and reformulate the quantum theory of fields in a relativistically invariant manner.

In the previous year Yukawa had commented on the unsatisfactory nature of quantum field theory, pointing both to its lack of an explicit, manifestly covariant form and to the problem of divergences – infinities. He wished to solve both problems at the same time. To that end, he applied Dirac's decade-earlier suggestion of a generalized transformation function by proposing that the quantum field probability amplitude should refer to a closed surface in space-time. From the graphic presentation of such a surface as a circle, the proposal became known as the theory of *maru*. Tomonaga's reaction was to take one problem at a time, and he first proceeded to "reformulate the quantum theory of fields in a relativistically invariant manner." And in doing so he rejected Yukawa's more radical proposal in favor of retaining the customary concept of causality – the relation between cause and effect. What was Tomonaga's reformulation?

The abstract I have cited was that of a paper published in the *Bulletin of the Institute, Riken-Iho*. But its contents did not become known outside of Japan until it was translated into English to appear in the second issue, that of August-September, 1946, of the new journal *Progress of Theoretical Physics*. It would, however, be some time before this issue became generally available in the United States. Incidentally, in this 1946 paper Tomonaga gave his address as Physics Department, Tokyo Bunrika University. While retaining his connection with Riken, he had, in 1941, joined the faculty of this university that later, in 1949, became part of the Tokyo University of Education.

Tomonaga began his paper by pointing out that the standard commutation relations of quantum field theory, referring to two points of space at the same time, are not covariantly formulated; in a relatively moving frame of reference the two points will be assigned different times. This is equally true of the Schrödinger equation for time evolution, which uses a common time variable for different spatial points. He then remarked that there is no difficulty in exhibiting commutation relations for arbitrary space-time points when a noninteracting field is considered. The unitary transformation to which we have already referred, now applied to all the fields, provides them with the equations of motion of noninteracting fields, whereas in the transformed Schrödinger equation, only the interaction terms remain. About this, Tomonaga says: "In our formulation, the theory is divided into two sections. . . . One section gives the laws of behavior of the fields when they are left alone, and the other gives the laws determining the deviation from this behavior due to the interactions. This way of separating the theory can be carried out relativistically." Certainly commutation relations referring to arbitrary space-time points are four-dimensional in character. But what about the transformed Schrödinger equation, which still retains its single time variable? It demands generalization.

Tomonaga was confident that he had the answer, for, as he put it later, "I was recalling Dirac's many-time theory which had enchanted me ten years before." In the theory of Dirac, and then of Dirac-Fock-Podolsky, each particle is assigned its own time variable. But, in a field theory, the role of the particle is played by the small volume elements of space. Therefore, assign to each spatial volume element an independent time coordinate. Thus the "super-many-time theory." Let me be more precise about that idea. At a common value of the time, distinct spatial volume elements constitute independent physical systems, for no physical influence is instantaneous. But more than that, no physical influence can travel faster than the speed of light. Therefore, any two space-time regions that cannot be connected, even by light signals, are physically independent; they are said to be in spacelike relationship. A three-dimensional domain such that any pair of points is in spacelike relationship constitutes a spacelike surface in the four-dimensional world. All of space at a common time is but a particular coordinate description of a plane spacelike surface. Therefore the Schrödinger equation, in which time advances by a common amount everywhere in space, should be regarded as describing the normal displacement of a plane spacelike surface. Its immediate generalization is to the change from one arbitrary spacelike surface to an infinitesimally neighboring one, which change can be localized in the neighborhood of a given space-time point. Such is the nature of the generalized Schrödinger equation that Tomonaga constructed in 1943, and to which I came toward the end of 1947.

By this time the dislocation produced by the war became dominant. Much later, Tomonaga recalled that "I myself temporarily stopped working on particle physics after 1943 and was involved in electronics research. Nevertheless the research on magnetrons and on ultra-short-wave circuits was basically a continuation of quantum mechanics." Tatsuoki Miyazima had this remembrance:

> One day our boss Dr. Nishina took me to see several engineers at the Naval Technical Research Institute. They had been engaged in the research and development of powerful split anode magnetrons, and they seemed to have come to a concrete conclusion about the phenomena taking place in the electron cloud. . . . Since they were engineers their way of thinking was characteristic of engineers and it was quite natural that they spoke in an engineer's way, but unfortunately it was . . . completely foreign to me at the beginning. . . .
> Every time I met them, I used to report to Tomonaga how I could not understand them, but he must have understood something . . . because, after a month or so, he showed me his idea . . . [of] applying the idea of secular perturbation, well-known in celestial mechanics and quantum theory, to the motion of the electrons in the cloud. . . . I remember that the moment he told me I said, "This is it." Further investigation

actually showed that the generation of electromagnetic oscillations in split anode magnetrons can be essentially understood by applying his idea.

When Tomonaga approached the problem of ultrashortwave circuits, which is to say, the behavior of microwaves in waveguides and cavity resonators, he found the engineers still using the old language of impedance. He thought this artificial, because there no longer are unique definitions of current and voltage. Instead, being a physicist, Tomonaga began with the electromagnetic field equations of James C. Maxwell. But he quickly recognized that those equations contain much more information than is needed to describe a microwave circuit. One usually wants to know only a few things about a typical waveguide junction: If a wave of given amplitude moves into a particular arm, what are the amplitudes of the waves coming out of the various arms, including the initial one? The array of all such relations forms a matrix, even then familiar to physicists as the scattering matrix. I mention here the amusing episode of the German submarine that arrived bearing a dispatch stamped *Streng Geheim* (top secret). When delivered to Tomonaga, it turned out to be Heisenberg's paper on the scattering matrix. Copies of this top-secret document were soon circulating among the physicists. Tomonaga preferred to speak of the scattering matrix as the characteristic matrix, in this waveguide context. He derived properties of that matrix, such as its unitary character, and showed how various experimental arrangements could be described in terms of the characteristic matrix of the junction. In the paper published after the war, he remarked, concerning the utility of this approach, that "The final decision, however, whether or not the new concept is here preferable to impedance should of course be given not only by a theoretical physicist but also by general electro-engineers." But perhaps my experience is not irrelevant here.

During the war, I also worked on the electromagnetic problems of microwaves and waveguides. I also began with the physicist's approach, including the use of the scattering matrix. But long before this three-year episode was ended, I was speaking the language of the engineers. I should like to think that those years of distraction for Tomonaga and myself were not without their useful lessons. The waveguide investigations showed the utility of organizing a theory to isolate those inner structural aspects that are not probed under the given experimental circumstances. That lesson was soon applied in the effective-range description of nuclear forces. And it was this viewpoint that would lead to the quantum electrodynamic concept of self-consistent subtraction or renormalization.

Tomonaga already understood the importance of describing relativistic situations covariantly – without specialization to any particular coordinate system. At about this time I began to learn that lesson pragmatically, in the context of solving a physical problem. As the war in Europe approached its end, the American physicists responsible for creating a massive microwave technology began to dream of high-energy electron accelerators. One of the practical questions involved was posed by the strong radiation emitted by relativistic electrons swinging in circular orbits. In studying what is now called synchrotron radiation, I used the reaction of the field created by the electron's motion. One part of that reaction describes the energy and momentum lost by the electron to the radiation. The other part is an added inertial effect characterized by an electromagnetic mass. I have mentioned the relativistic difficulty that electromagnetic mass usually creates. But, in the covariant method I was using, based on action and proper time, a perfectly invariant form emerged. Moral: To end with an invariant result, use a covariant method and maintain covariance to the end of the calculation. And, in the appearance of an invariant electromagnetic mass that simply added to the mechanical mass to form the physical mass of the electron, neither piece being separately distinguishable under ordinary physical circumstances, I was seeing again the advantage of isolating unobservable structural aspects of the theory. Looking back at it, the basic ingredients of the coming quantum electrodynamic revolution were then in place. Lacking was an experimental impetus to combine them and take them seriously.

Suddenly the Pacific war was over. Amid total desolation, Tomonaga reestablished his seminar. But meanwhile, something had been brewing in Sakata's Nagoya group. It goes back to a theory of Møller and Leon Rosenfeld, who tried to overcome the nuclear force difficulties of meson theory by proposing a mixed field theory, with both pseudoscalar and vector mesons of equal mass. I like to think that my modification of this theory, in which the vector meson is more massive, was the prediction of the later discovered ρ meson. Somewhat analogously, Sakata proposed that the massless vector photon is accompanied by a massive scalar meson called the cohesive or C meson. About this, Tomonaga said:

In 1946, Sakata proposed a promising method of eliminating the divergence of the electron mass by introducing the idea of a field of cohesive force. It was the idea that there exists an unknown field, of the type of the meson field, which interacts with the electron in addition to the electromagnetic field. Sakata named this field the cohesive force field, because the apparent electromagnetic mass due to the interaction of this field and the electron, though infinite, is negative and therefore the existence of this field could stabilize the electron in some sense. Sakata pointed out the possibility that the electromagnetic mass and the negative new mass cancel each other and that the infinity could be eliminated by suitably choosing the coupling constant between this field and the electron. Thus the difficulty which had troubled people for a long time seemed to disappear insofar as the mass was concerned.

Let me break in here and remark that this solution of the mass divergence problem is, in fact, illusory. In 1950, Toichiro Kinoshita showed that the necessary relation between the two coupling constants will no longer cancel the divergences when the discussion is extended beyond the lowest order of approximation. Nevertheless, the C-meson hypothesis served usefully as one of the catalysts that led to the introduction of the self-consistent subtraction method. How that came about is described in Tomonaga's next sentence: "Then what concerned me most was whether the infinities appearing in the electron scattering process could also be removed by the idea of a plus-minus cancellation."

I have already referred to the 1939 calculation of Dancoff, on radiative corrections to electron scattering, that gave an infinite result. Tomonaga and his collaborators proceeded to calculate the additional effect of the cohesive force field. It encouragingly gave divergent results of the opposite sign, but they did not precisely cancel Dancoff's infinite terms. This conclusion was reported in a letter of November 1, 1947, submitted to *Progress of Theoretical Physics*, and also presented at a symposium on elementary particles held in Kyoto that same month. Meanwhile, parallel calculations of the electromagnetic effect were going on, repeating Dancoff's calculations, which had not been reported in detail. But then Tomonaga suggested a new and much more efficient method of calculation. It was to use the covariant formulation of quantum electrodynamics and subject it to a unitary transformation that immediately isolated the electromagnetic mass term. Tomonaga said:

> Owing to this new, more lucid method, we noticed that
> among the various terms appearing in both Dancoff's and our
> previous calculation, one term had been overlooked. There
> was only one missing term, but it was crucial to the final con-
> clusion. Indeed, if we corrected this error, the infinities ap-
> pearing in the scattering process of an electron due to the
> electromagnetic and cohesive force fields cancelled com-
> pletely, except for the divergence of vacuum polarization
> type.

A letter of December 30, 1947, corrected the previous erroneous announcement.

But what is meant by "the divergence of vacuum polarization type"? From the beginning of Dirac's theory of positrons it had been recognized that, in a sense, the vacuum behaved as a polarizable medium; the presence of an electromagnetic field induced a charge distribution acting to oppose the inducing field. As a consequence, the charges of particles would appear to be reduced, although the actual calculation gave a divergent result. Nevertheless, the effect could be absorbed into a redefinition, a renormalization, of the charge. At this stage, then, Tomonaga had achieved a finite correction to the scattering of electrons by combining two distinct ideas: the renormalization of charge and the compensation mechanism of the *C*-meson field.

But meanwhile, another line of thought had been developing. In this connection, let me quote from a paper, published at about this time, by Mituo Taketani:

> The present state of theoretical physics is confronted with dif-
> ficulties of extremely ambiguous nature. These difficulties can
> be glossed over but no one believes that a definite solution
> has been attained. The reason for this is that, on one hand,
> present theoretical physics itself has logical difficulties, while,
> on the other hand, there is no decisive experiment whereby
> to determine this theory uniquely.

In June of 1947 those decisive experiments were made known, in the United States.

For three days at the beginning of June, some 20 physicists gathered at Shelter Island, located in a bay near the tip of Long Island, New

York. There we heard the details of the experiment by which Willis E. Lamb, Jr., and Robert Retherford had used the new microwave techniques to confirm the previously suspected upward displacement of the 2s level of hydrogen. Actually, rumors of this had already spread, and on the train to New York, Victor F. Weisskopf and I had agreed that electrodynamic effects were involved and that a relativistic calculation would give a finite prediction. But there was also a totally unexpected disclosure, by Isidor I. Rabi: The hyperfine structures in hydrogen and deuterium were larger than anticipated by a fraction of a percent. Here was another flaw in the Dirac electron theory, now referring to magnetic rather than electric properties.

Weisskopf and I had described at Shelter Island our idea that the relativistic electron-positron theory, then called the hole theory, would produce a finite electrodynamic energy shift. But it was Hans Bethe who quickly appreciated that a first estimate of this effect could be found without entering into the complications of a relativistic calculation. In a *Physical Review* article received on June 27, he said:

> Schwinger and Weisskopf, and Oppenheimer have suggested that a possible explanation might be the shift of energy levels by the interaction of the electron with the radiation field. This shift comes out infinite in all existing theories, and has therefore always been ignored. However, it is possible to identify the most strongly (linearly) divergent term in the level shift with an electromagnetic mass effect which must exist for a bound as well as a free electron. This effect should properly be regarded as already included in the observed *mass* of the electron, and we must therefore subtract from the theoretical expression, the corresponding expression for a free electron of the same average kinetic energy. The result then diverges only logarithmically (instead of linearly) in nonrelativistic theory: Accordingly, it may be expected that in the hole theory, in which the *main* term (self-energy of the electron) diverges only logarithmically, the result will be *convergent* after subtraction of the free electron expression. This would set an effective upper limit of the order of mc^2 to the frequencies of light which effectively contribute to the shift of the level of a bound electron. I have not carried out the relativistic calculations, but I shall assume that such an effective relativistic limit exists.

The outcome of Bethe's calculation agreed so well with the then not very accurately measured level shift that there could be no doubt of its electrodynamic nature. Nevertheless, the relativistic problem, of producing a finite and unique theoretical prediction, still remained.

The news of the Lamb-Retherford measurement and of Bethe's nonrelativistic calculation reached Japan in an unconventional way. Tomonaga said:

> The first information concerning the Lamb shift was obtained not through the *Physical Review,* but through the popular science column of a weekly U.S. magazine. This information about the Lamb shift prompted us to begin a calculation more exact than Bethe's tentative one.

He went on:

> In fact, the contact transformation method . . . could be applied to this case, clarifying Bethe's calculation and justifying his idea. Therefore the method of covariant contact transformations, by which we did Dancoff's calculation over again would also be useful for the problem of performing the relativistic calculation for the Lamb shift.

Incidentally, in speaking of contact transformations, Tomonaga was using another name for canonical or unitary transformations. Tomonaga announced his relativistic program at the previously mentioned Kyoto symposium of November 24 to 25, 1947. He gave it a name that appears in the title of a letter accompanying the one of December 30 that points out Dancoff's error. This title is "Application of the Self-Consistent Subtraction Method to the Elastic Scattering of an Electron." And so, at the end of 1947, Tomonaga was in full possession of the concepts of charge and mass renormalization.

Meanwhile, immediately following the Shelter Island conference, I found myself with a brand new wife, and for two months we wandered around the United States. Then it was time to go to work again. I also clarified for myself Bethe's nonrelativistic calculation by applying a unitary transformation that isolated the electromagnetic mass. This was the model for a relativistic calculation, based on the conventional hole-theory formulation of quantum electrodynamics. But here I held an unfair advantage over Tomonaga, for, owing to the communication problems of the time, I knew that there were two kinds of experimental effects to be explained: the electric one of Lamb and the magnetic one of Rabi. Accordingly, I carried out a calculation of the energy shift in a homogeneous magnetic field, which is the prediction of an additional magnetic moment of the electron, and also considered the Coulomb field of a nucleus in applications to scattering and to the energy shift of bound states. The results were described in a letter to the *Physical Review*, received on December 30, 1947, the very same date as Tomonaga's proposal of the self-consistent subtraction method. The predicted additional magnetic moment accounted for the hyperfine structure measurements and also for later, more accurate, atomic moment measurements. Concerning scattering, I said that "the finite radiative correction to the elastic scattering of electrons by a Coulomb field provides a satisfactory termination to a subject that has been beset with much confusion." Considering the absence of experimental data, this was perhaps all that needed to be said. But when it came to energy shifts, what I wrote was that "The values yielded by our theory differ only slightly from those conjectured by Bethe on the basis of a non-relativistic calculation, and are, thus, in good accord with experiment." Why did I not quote a precise number?

The answer to that was given in a lecture before the American Physical Society at the end of January 1948. Quite simply something was wrong. The coupling of the electron spin to the electric field was numerically different from what the additional magnetic moment would imply; relativistic invariance was violated in this noncovariant calculation. One could, of course, adjust that spin coupling to have the right value, and, in fact, the correct energy shift is obtained in this way. But there was no conviction in such a procedure. The need for a covariant formulation could no longer be ignored. At the time of this meeting, the covariant theory had already been constructed and applied to obtain an invariant expression for the electron electromagnetic mass. I mentioned this briefly. After the talk, Oppenheimer told me about Tomonaga's prior work.

A progress report on the covariant calculations, using the technique of invariant parameters, was presented at the Pocono Manor Inn conference held March 30 to April 1, 1948. At that very time, Tomonaga was writing a letter to Oppenheimer that would accompany a collection of manuscripts describing the work of his group. In response, Oppenheimer sent a telegram: "Grateful for your letter and papers. Found most interesting and valuable mostly paralleling much work done here. Strongly suggest you write a summary account of present state and views for prompt publication *Physical Review*. Glad to arrange. . . ." On May 28, 1948, Oppenheimer acknowledged the receipt of Tomonaga's letter entitled "On Infinite Field Reactions in Quantum Field Theory." He wrote:

> Your very good letter came two days ago and today your manuscript arrived. I have sent it on at once to the *Physical Review* with the request that they publish it as promptly as possible. . . . I also sent a brief note . . . which . . . may be of some interest to you in the prosecution of the higher order calculations. Particularly in the identification of light quantum self energies, it proves important to apply your relativistic methods throughout. We shall try to get an account of Schwinger's work on this and other subjects to you in the very near future.

He ended the letter expressing the "hope that before long you will spend some time with us at the Institute where we should all welcome you so warmly."

The point of Oppenheimer's added note is this: In examining the radiative correction to the Klein-Nishina formula, Tomonaga and his collaborators had encountered a divergence additional to those involved in mass and charge renormalization. It could be identified as a photon mass. But unlike the electromagnetic mass of the electron, which can be amalgamated, as Tomonaga put it, into an already existing mass, there is no photon mass in the Maxwell equations. Tomonaga noted the possibility of a compensation, a cancellation, analogous to the idea of Sakata. In response, Oppenheimer essentially quoted my observation that a gauge-invariant relativistic theory cannot have a photon mass and, further, that a sufficiently careful treatment would yield the required zero value. But Tomonaga was not convinced. In a paper submitted about this time, he spoke of the "somewhat quibbling way" in which it was argued that the photon mass must vanish. And he was right, for the real subtlety underlying the photon mass problem did not surface for another 10 years, in the eventual recognition of what others would call "Schwinger terms."

But even the concept of charge renormalization was troubling to some physicists. Abraham Pais, on April 13, 1948, wrote a letter to Tomonaga in which, after commenting on his own work parallel to that of Sakata, he remarked: "It seems one of the most puzzling problems how to 'renormalize' the charge of the electron and of the proton in such a way as to make the experimental values for these quantities equal to each other." Perhaps I was the first to fully appreciate that charge renormalization is a property of the electromagnetic field alone, which results in a renormalization, a fractional reduction of charge, that is the same for all. But while I'm congratulating myself, I must also mention a terrible mistake I made. Of course, I wasn't entirely alone – Feynman did it too. It occurred in the relativistic calculation of energy values for bound states. The effect of high-energy photons was treated covariantly; that of low-energy photons in the conventional way. These two parts had to be joined together, and a subtlety involved in relating the respective four- and three-dimensional treatments was overlooked for several months. But sometime around September 1948 it was straightened out, and apart from some uncertainty about the inclusion of vacuum polarization effects, all groups, Japanese and American, agreed on the answer. As I have mentioned, it was the result I had reached many months before by correcting the obvious relativistic error of my first noncovariant calculation.

In that same month, September 1948, Yukawa, accepting an invitation of Oppenheimer, went to the Institute for Advance Study at Princeton, New Jersey. The letters that he wrote back to Japan were circulated in a new informal journal called *Elementary Particle Physics Research – Soryushiron Kenkyu*. Volume 0 of that journal also contained the communications of Oppenheimer and Pais to which I have referred and a letter of Heisenberg to Tomonaga, inquiring whether Heisenberg's paper, sent during the war, had arrived. In writing to Tomonaga on October 15, 1948, Yukawa said, in part, "Yesterday I met Oppenheimer, who came back from the Solvay Conference. He thinks very highly of your work. Here, many people are interested in Schwinger's and your work and I think that this is the main reason why the demand for the *Progress of Theoretical Physics* is high. I am very happy about this."

During the period of intense activity in quantum electrodynamics, Tomonaga was also involved in cosmic-ray research. The results of a collaboration with Satio Hayakawa were published in 1949 under the title "Cosmic Ray(s) Underground." By now, the two mesons had been recognized and named: π and μ. This paper discussed the generation of, and the subsequent effects produced by, the deep-penetrating μ meson. Among other activities in that year of 1949, Tomonaga published a book on quantum mechanics that would be quite influential, and he accepted Oppenheimer's invitation to visit the Institute for

Advanced Study. During the year he spent there, he turned in a new direction, one that would also interest me a number of years later. It is the quantum many-body problem. The resulting publication of 1950 is entitled "Remarks on Bloch's Method of Sound Waves Applied to Many-Fermion Problems." Five years later he would generalize this in a study of quantum collective motion.

But the years of enormous scientific productivity were coming to a close, owing to the mounting pressures of other obligations. In 1951, Nishina died, and Tomonaga accepted his administrative burdens. Tomonaga's attention turned toward improving the circumstances and facilities available to younger scientists, including the establishment of new institutes and laboratories. In 1956 he became president of the Tokyo University of Education, which post he held for six years. Then, for another six years, he was president of the Science Council of Japan, and also, in 1964, he assumed the presidency of the Nishina Memorial Foundation. I deeply regretted that he was unable to be with us in Stockholm on December 10, 1965, to accept his Nobel Prize. The lecture that I have often quoted here was delivered May 6, 1966.

Following his retirement in 1970, he began to write another volume of his book on quantum mechanics, which, unfortunately, was not completed. However, two other books, one left in an unfinished state, were published. To some extent, these books are directed to the general public rather than the professional scientist. And here, again, Tomonaga and I found a common path. I have recently completed a series of television programs that attempt to explain relativity to the general public. I very much hope that this series, which was expertly produced by the British Broadcasting Corporation, will eventually be shown in Japan.

On July 8, 1979, our story came to a close. But Sin-itiro Tomonaga lives on in the minds and hearts of the many people whose lives he touched, and graced.

Acknowledgments

I wish to thank Laurie Brown for providing access (before publication) to a number of sources, including the following: *Particle Physics in Japan, 1930-1950*, edited by L. M. Brown, M. Konuma, and Z. Maki (Research Institute for Fundamental Physics, University of Kyoto, 1981), containing an extensive discussion with Tomonaga; translations from Tomonaga's Japanese writings by Fumiko Tanihara and Noriko Eguchi; Laurie M. Brown, "Yukawa's Prediction of the Meson," *Centaurus 25* (1981), 71-132.

Note

1 This chapter was originally prepared as a memorial lecture for Sin-itiro Tomonaga and was delivered July 8, 1980, at the Nishina Memorial Foundation, Tokyo.

23 Particle physics in rapid transition: 1947-1952[1]

ROBERT E. MARSHAK

Born 1916, New York City; Ph.D., Cornell, 1939; former president, City University of New York; theoretical physicist and organizer of series "Rochester Conferences"; Virginia Polytechnic Institute and State University.

When invited to give the final presentation at this international symposium on the history of particle physics, I was asked to say something about the three Shelter Island conferences, organized in 1947 to 1949 by J. Robert Oppenheimer, and the first couple of Rochester conferences, which I organized. I have tried to fulfill that request by limiting myself to the period 1947 to 1952, but my title is also intended to convey a sense of the rapid developments in this field as a consequence of the completion of the first "high-energy" accelerators at Berkeley during 1947 to 1948.* That is not to say that cosmic-ray experiments immediately took a back seat during those years – quite the contrary. The stream of initial discoveries by cosmic-ray physicists, followed by their rapid exploration in accelerator experiments, became a source of tremendous excitement and theoretical inspiration.

I make no pretense at completeness, but I do hope that in covering some of the highlights of this five-year period I shall not be unduly influenced by my personal interests and involvement.

The Yukawa meson (pion) and the second-generation lepton (muon)

Let us recall that the neutron was discovered only in 1932, that the neutrino was postulated by Wolfgang Pauli somewhat earlier, and

* *Ed. note*: That is, the 340-MeV synchrocyclotron and the 300-MeV electron synchrotron.

that Enrico Fermi's theory of β decay was published in 1934. Hideki Yukawa's hypothesis of a boson of intermediate mass mediating the strong short-range nuclear force, and also explaining β decay, was put forward in 1935. Yukawa first proposed a scalar meson, but because this gave a wrong sign to the nuclear force, he proposed instead a vector meson.* When a particle of several hundred electron masses was discovered in the cosmic radiation in 1937, Yukawa's hypothesis seemed to be spectacularly confirmed; however, the confirmation was more apparent than real.

Regarding the question whether or not the observed cosmic-ray meson at sea level was the nuclear meson, the type of quantum mechanical calculation then current (weak-coupling approximation) predicted its scattering cross section to be about 10^{-26} cm^2 per nucleon, at an energy of the order of 1 GeV; however, cosmic-ray experiments at sea level using cloud chambers with metal plates seemed to give cross sections lower by at least a factor of 100.[2] This discrepancy was troubling, and the difficulty of reconciling the properties of a nuclear meson with the small observed scattering cross section of a sea-level meson was pointed out as early as 1939 by L. W. Nordheim and M. H. Hebb.[3]

During the next few years, several independent attempts were made to solve this dilemma. Victor F. Weisskopf and I tried to show that the replacement of a single boson by a pair of spin-1/2 mesons in Yukawa's theory could reduce the scattering cross section and still maintain the strength of the nuclear force predictions; however, the weak-coupling nature of the calculations did not convince anyone, least of all ourselves.[4] Pauli and Sidney M. Dancoff proposed a more plausible resolution of the dilemma, namely that the nuclear meson was pseudoscalar and that its coupling with the nucleon could be so arranged that the scattering cross section would go as a^2 (where a is the size of the nucleon) in a "strong coupling" type of calculation; because a could be chosen much smaller than the Compton wavelength of the meson (i.e., $a\mu \ll 1$, where μ is the mass of the meson), the small scattering cross section could easily be explained.[5] The strong-coupling explanation of the small scattering cross section did not turn out to be correct, but the idea of strongly coupling a pseudoscalar meson to a "finite-source" nucleon led later, in the hands of Pauli and others, to the prediction of the (3,3) nucleon isobar, which will be described later.

---

* *Ed. note*: The terms *scalar* and *vector* here refer, respectively, to spin 0 and spin 1.

A third explanation offered for the relatively weak interaction of the sea-level meson was being advanced in Japan in 1943, unbeknown to the rest of the world, by Shoichi Sakata and Takesi Inoue. Apparently, a "meson club," led by Yukawa and Sin-itiro Tomonaga, was meeting regularly during World War II; Sakata and Inoue delivered a paper there in 1943, proposing to explain the aforementioned dilemma by postulating two mesons, of which the heavier would be the nuclear meson (with spin 0 and/or 1), decaying into the cosmic-ray meson seen at sea level (with spin 1/2) and having a weaker coupling (by a factor of 10) with the nucleon.[6] Another member of the meson club, Yasutaka Tanikawa, worked out a two-meson theory with a spin-0 assignment to the sea-level meson. Because these Japanese authors were trying to explain a discrepancy of a factor 100 in the scattering cross section, they predicted a lifetime for the decay of the nuclear force meson into the cosmic-ray sea-level meson of the order of 10^{-21} sec. This means that the light meson of the Japanese authors was not the second-generation lepton (muon) at all, but a meson that was less strongly coupled than the nuclear meson by only a factor of 10. This is not to denigrate for one moment the very imaginative hypothesis proposed by Sakata and associates to explain the small scattering cross section of the sea-level meson, but I do think that it is fair to say that (apart from communication difficulties occasioned by World War II) it would have been difficult to decide between the strong-coupling explanation of the small scattering cross section and that of Sakata and associates until 1947, when the results of the experiments of Marcello Conversi and associates and C. M. G. Lattes and associates were announced.[7] I should point out that because of the war the English version of the Sakata and Inoue paper was not published until 1946 and that of the Tanikawa paper until 1947 and that these papers did not reach the United States until December 1947, about six months after the first Shelter Island conference.

The drama of the first Shelter Island conference (held June 2-4, 1947) can now be understood, and the crucial importance of the paper by Conversi and associates will become apparent. The war ended in August 1945, and it took many of us almost a year to discharge our obligations to the various war laboratories where we had worked (in terms of writing reports, etc.). By the beginning of 1947, Oppenheimer had become director of the Institute for Advanced Study at Princeton, and he turned his thoughts to civilian science, in particular, theoretical physics. He persuaded the National Academy of Sciences to sponsor a small conference (25 participants) in an isolated location (Shelter Island, off Long Island) for three days during the first week of June. His intention was to concentrate on theoretical physics, but he did invite Isidor Isaac Rabi and Bruno Rossi as "experimental consultants." Oppenheimer had the good judgment to start the discussion at the Shelter Island conference with an attempt to understand two experiments that seemed to contradict the then existing theory: the Lamb shift and the experiment of Conversi and associates. The discussion of the Lamb shift led to the surprisingly satisfactory nonrelativistic calculation by Hans Bethe soon after the conference and to the relativistic quantum electrodynamics theories of Julian Schwinger and Richard P. Feynman, which have been thoroughly discussed at this symposium. I shall therefore limit myself to the other major unexplained experiment discussed at the first Shelter Island conference.

The paper of Conversi and associates (published February 1, 1947) disclosed that a substantial fraction of negative sea-level mesons decayed in a carbon plate, but were absorbed in an iron plate. All positively charged mesons decayed in both carbon and iron plates. According to the theory of Tomonaga and Toshima Araki, nuclear force mesons carrying positive charge should always decay (in agreement with experiment), and those carrying negative charge should never decay (in disagreement with experiment).[8] Analysis of this experiment by Fermi, Weisskopf, and Edward Teller led to the startling conclusion that "the time of capture from the lowest orbit of carbon is not less than the time of natural decay, that is, about 10^{-6} second. This is in disagreement with the previous estimate by a factor of about 10^{12}. Changes in the spin of the mesotron or the interaction form may reduce this disagreement to 10^{10}."[9] Cognizant of the Italian experiment and its analysis, Oppenheimer sent to each participant in the first Shelter Island conference a provocative memorandum entitled "The Foundations of Quantum Mechanics: Outline of Topics for Discussion." This memorandum is sufficiently interesting to quote a goodly portion of it:

> It was long ago pointed out by Nordheim that there is an apparent difficulty in reconciling on the basis of usual quantum mechanical formalism the high rate of production of mesons in the upper atmosphere with the small interactions which these mesons subsequently manifest in traversing matter.

Participants at the Shelter Island conference, June 1947: 1, I. I. Rabi; 2, L. Pauling; 3, J. Van Vleck; 4, W. E. Lamb, Jr.; 5, G. Breit; 6, D. Mac Innes (Nat. Acad. of Sci. representative); 7, K. K. Darrow; 8, G. E. Uhlenbeck; 9, J. Schwinger; 10, E. Teller; 11, B. Rossi; 12, A. Nordsieck; 13, J. von Neumann; 14, J. A. Wheeler; 15, H. A. Bethe; 16, R. Serber; 17, R. E. Marshak; 18, A. Pais; 19, J. R. Oppenheimer; 20, D. Bohm; 21, R. P. Feynman; 22, V. F. Weisskopf; 23, H. Feshbach (credit: Duncan Mac Innes, Nat. Acad. of Sci.).

To date no completely satisfactory understanding of this discrepancy exists, nor is it clear to what extent it indicates a breakdown in the customary formalism of quantum mechanics. It would appear profitable to discuss this and related questions in some detail.

We might start this discussion by an outline of the current status of theories of multiple production. . . . However, no reasonable formulation of theories along this line will satisfactorily account for the smallness of the subsequent interaction of mesons with nuclear matter. . . . There are two reasons for these apparent difficulties. One is that in all current theory there is a formal correspondence between the creation of a particle and the absorption of an anti-particle. The other is that multiple processes are in these theories attributable to the higher order effects of coupling terms which are of quite low order, first or second, in the meson wave fields. The question that we should attempt to answer is whether, perhaps along the lines of an S matrix formulation, both these conditions must be abandoned to accord with the experimental facts.

The discussion of the Italian experiment became very animated at the first Shelter Island conference, but there was very little inclination to support Oppenheimer's suggestion that one should consider surrendering microscopic reversibility. At one point, Weisskopf suggested a possible way of overcoming the apparent lack of reversibility (between "creation" and "absorption"), namely, to postulate that the primary cosmic-ray proton converts a normal nucleon in an "air" nucleus into an excited nucleon, capable of emitting mesons; the lifetime of the "meson-pregnant" state could be chosen sufficiently long to account for the subsequent weak interaction between mesons and nucleons.[10] This hypothesis seemed rather inelegant, and I proposed an alternative solution of the difficulty, to wit, that two kinds of mesons exist in nature, possessing different masses: The heavy meson is produced with large cross section in the upper atmosphere and is responsible for nuclear forces, whereas the light meson is a decay product of the heavy meson and is the meson normally observed to interact weakly with matter at sea level. My recollection is that the two-meson hypothesis was well received; I particularly recall Feynman's enthusiasm for the idea, and even Weisskopf said that he preferred this hypothesis to his own.

When I left the Shelter Island conference, I decided to publish a brief note on the two-meson hypothesis and to use the observed decay rate from the lowest orbit in the carbon atom to estimate the lifetime for the decay of the heavy (nuclear) meson into the light (sea-level) meson. Before I could do this, I had to attend a meeting at Lake Geneva, Wisconsin, in my capacity as chairman of the Federation of Atomic Scientists; this was in mid-June, and Philip Morrison informed me at the meeting that the latest issue of *Nature* carried an article by Lattes and associates showing two pictures (in Ilford plates) of a meson coming to rest in the emulsion and decaying into a lighter meson of fixed range.[7] (The issue carrying the Bristol discovery was dated May 24, 1947, but it did not reach the United States until several weeks later, after the Shelter Island conference, because journals were not sent airmail.) As soon as I returned to Rochester, I read that article and was immediately convinced that the two beautiful π-μ decays (as they came to be known) were small in statistics, but clear cut in their support of the two-meson hypothesis. I decided to enlist Bethe's help in writing the paper because of his extensive knowledge of the cosmic-ray data. Our paper was sent to the *Physical Review* during July.[11] (There was no reference in our paper to Bruno Pontecorvo's important observation concerning the Italian experiment, because his paper was published only in August, and preprints were uncommon in those days.[12])

There is no point going into the details of the paper by Bethe and myself, except to note that its most important result was the prediction, on the basis of the observed rate of decay of the light cosmic-ray meson in the Italian experiment, of the lifetime for the decay of the heavy nuclear meson into the light meson, which we estimated to be 10^{-8} sec. (The difference between our estimate and the 10^{-21} sec that Sakata and Inoue found is basically the factor 10^{12} discrepancy pointed out by Fermi, Teller, and Weisskopf.[9]) "For the sake of a model," as we put it, we assumed that the heavy meson had spin 1/2 and the light meson spin 0, rather than spin 0 or 1 for the former meson and spin 1/2 for the latter, as Sakata and Inoue had assumed. It was clear from the outset (S is the spin) that the choice $S(\pi) = 1/2$, $S(\mu) = 0$ or $S(\pi) = 0$, $S(\mu) = 1/2$ would have very little effect on the lifetime for π-μ decay. The essential point was that $S(\mu) < 1$, the condition established by Robert F. Christy and Shuichi Kusaka from an analysis of the burst production in cosmic rays.[13] I have no doubt in my mind that had the Italian experiment existed when Sakata and Inoue first developed their

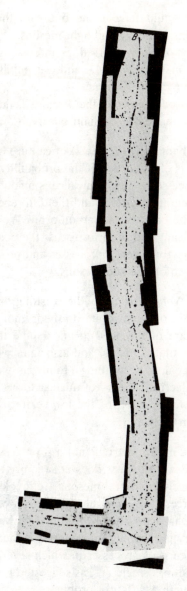

One of the first observations of the decay of a pion into a muon, showing the existence of two different mesons in the cosmic rays. (Figure 1 of C. M. Lattes, H. Muirhead, G. P. S. Occhialini, and C. F. Powell, "Processes Involving Charged Mesons." Reprinted by permission from *Nature*, Vol. *159*, No. 4047 (May 24, 1947), p. 695. Copyright © 1947–MacMillan Journals Limited.) The poster announcing this symposium was based on this mosaic of photomicrographs.

two-level meson theory, they would have come out with the right decay lifetime and the right spins of heavy and light mesons.

Pontecorvo's paper, to which I have referred, is interesting for a variety of reasons. It was sent on June 21, 1947, and was published in the August 1, 1947, issue of the *Physical Review*. He was reacting to the paper by Conversi and associates, and the Fermi, Teller, and Weisskopf analysis of it, and he made this brilliant observation:

> We notice that the probability ($\sim 10^6 \text{ sec}^{-1}$) of capture of a bound negative meson is of the order of the probability of ordinary K capture processes, when allowance is made for the difference in the disintegration energy and the difference in the volumes of the K shell and of the meson orbit. We assume that this is significant and wish to discuss the possibility of a fundamental analogy between β processes and processes of emission or absorption of charged mesons.

He obviously had not seen the *Nature* article by the Bristol group, and understandably he was not aware of our two-meson theory nor that of Sakata and Inoue. He was aware of the need to understand the production process for the mesons that Conversi and associates were observing at sea level, and he appealed, of all things, to my early meson-pair (sometimes called heavy-electron) theory of nuclear forces, which could account, somewhat unconvincingly, for the low scattering cross section.[14] Specifically, he said in his paper:

> The hypothesis that the meson decay is not a β process, while the meson absorption is a β process, does not require that hypothetical particles such as neutral mesons are invoked to account for nuclear forces. In fact, a heavy electron pair theory of nuclear forces was successfully developed by Marshak. Moreover, a pair theory is capable of accounting, at least in principle, for the existence of processes in which several pairs of mesons are produced in a single act, as suggested by Heisenberg in connection with a different problem.

This was a curious twist, but it did not diminish the importance of Pontecorvo's statement that the muon is a heavy electron (i.e., the second-generation lepton). Indeed, reading Pontecorvo's paper persuaded me that the choice of spin 0 for the nuclear meson and of spin 1/2 for the sea-level meson was the highly likely choice in the two-meson theory (before the paper of Sakata and Inoue arrived in the United States in December 1947).

In November 1947, I visited Cecil F. Powell's laboratory in Bristol. Armed with the basic ideas of the two-meson theory and the likelihood that the nuclear meson (pion) was the boson and that its decay product (the muon) was a fermion, I could point out a series of interesting consequences that could be tested by experiment (with nuclear emulsions and with other techniques), namely, that negative pions should be captured by light, as well as heavy, nuclei, in contrast to negative muons, which should be captured only by heavy nuclei; the nuclear capture of a π^- should convert an appreciable fraction of its rest energy into star energy, whereas the nuclear capture of a μ^- should convert only a negligible fraction of its rest mass into star energy, etc.* I do not recall speaking to a single theorist in Powell's laboratory (a recollection confirmed by Dr. M. G. K. Menon at this symposium), but it did not seem to matter, because there were still so many basic discoveries to make by means of the nuclear emulsion technique, and so many improvements in this technique still possible (e.g., development of electron-sensitive plates, proper correction for fading, etc.). Several years later, Louis Leprince-Ringuet, himself a distinguished cosmic-ray physicist, could say about Bristol: "En Europe, il y a pour les émulsions, Bristol, le grand soleil, et puis un tout petit nombre de petits satellites dont la dimension, même en faisant la somme, reste très inferieure a celle de Bristol."[15]

After Bristol, during that European trip in November 1947, I visited Patrick M. S. Blackett's laboratory in Manchester, where I was briefed by his colleagues George D. Rochester and Clifford C. Butler. (Blackett by that time had transferred his interests to the creation of the new field of paleomagnetism.) Rochester and Butler showed me their first two photographs of V particles – the first serious evidence for the existence of particles heavier than pions (with masses of about 1,000 times the electron mass) decaying into two charged particles.[16] Here were two excellent examples of what later came to be known as "strange particles," and I should have "spread the good word" when I returned to the United States. But, unfortunately, Rochester and Butler also showed me some other cloud chamber photographs in which they said they were observing muons directly produced in the metal plates inside

---

* *Ed. note*: By "star" is meant a nuclear disintegration into several particles whose tracks in nuclear emulsion thus form a starlike figure.

the cloud chamber. They insisted that they could distinguish between the pion and muon masses and that they were seeing the direct production of muons. I told them that this was in direct contradiction with my two-meson theory (in which I naturally had great confidence) and that I could not accept their interpretation of the photographs. Because I was confident that they were wrong about the direct production of muons, I was hesitant (quite unfairly, it turned out) to accept their interpretation of the V particles. I now believe that the source of the difficulty was the incorrect calibration that Rochester and Butler had received from the Bristol laboratory that the pion-to-muon mass ratio was 1.7, instead of the correct value 1.32 (the high value was connected with a fading problem that was soon resolved). I would guess that with the substantially smaller mass ratio, Rochester and Butler would not have argued for the direct production of muons.*

There was one more flap connected with the direct production of muons – at Berkeley – before the two-meson theory could settle down and rest on its laurels. In early 1948, Lattes came over from Bristol to show the physicists using the Berkeley synchrocyclotron how to detect mesons with the Ilford nuclear emulsion plates. With the 340-MeV α particles produced by the Berkeley synchrocyclotron, there was no problem seeing the copious production of pions. But then one of the groups began to see muons in the Ilford plates that seemed to come from the target, apparently giving evidence for the direct production of muons in α-nucleus collisions. This was reported at the second "Shelter Island conference" (actually held in the Poconos, March 30 to April 2, 1948), and my arguments against the direct production of muons in nuclear collisions apparently made little impression on Feynman, who reported on this conference in the June 1948 issue of *Physics Today*. On this point, Feynman commented in his report as follows:

> One surprising thing was discussed in considerable detail. Apparently light mesotrons are also made at Berkeley by direct collision of the alpha particles with nuclei of the target. The reason this is surprising is that previous experience with cosmic rays showed that light mesotrons interact only very slightly with nuclei. For example, they are captured by heavy nuclei only with difficulty, and by light also. All the light mesotrons were thought to result from decay of heavy ones.

---

* *Author's note added in proof*: This statement has subsequently been confirmed in a letter to me from Rochester.

But if the Berkeley experiments are correctly interpreted, this is not true. They apparently are easily knocked out of nuclei. But then why do they have such difficulty in going back in?

Within several months the Berkeley effect vanished when it was ascertained that the "apparent" direct production of muons was actually due to the direct production of pions, producing muons that were returned to the target area under the influence of the synchrocyclotron magnetic field. I mention this incident not to embarrass a physicist whom I deeply respect but to illustrate a point of some interest to historians of science that authors of theories develop a "mind set" about their own theories that is not always shared by colleagues (Feynman will know whereof I speak if I remind him of the stormy weather he encountered in connection with his formulation of QED at the third Shelter Island conference in 1949).

In concluding this section, it should be acknowledged that the determination that there were two kinds of mesons and that the heavier one was strongly interactive with nucleons, whereas the lighter one was only weakly interactive (like the electron), was a cosmic-ray experimental triumph. The two-meson theory may have added some drama to the sequence of experiments at the Bristol laboratory following the discovery of the π-μ decays, and it was clear from my visit in November 1947 (and its failure in the subsequent years to enlist any significant theoretical help) that the Bristol group regarded its program as following a clear direction without the need for any theoretical encumbrance. Some sense of the low esteem in which theory was held in this laboratory is communicated by quoting from a review article written by Powell on "Mesons" and published in 1950[17]:

> In spite of the progress represented by the new discoveries,
> the question remains, however, whether, in spite of the great
> stimulus which they have provided for the progress of theory
> and experiment, the basic features of Yukawa's original ideas
> are essentially correct. Even if it is a valuable analogy to
> identify the π mesons with the "heavy quanta" of the nuclear
> field, there appears to be no place in any formalism, devel-
> oped hitherto, for the μ mesons. These particles have a mass
> of 215 m_e; they have a very weak interaction with nucleons,
> so that they are able to penetrate hundreds of nuclei without
> disintegrating them; they have half-integral spin and decay
> with the emission of an electron and, probably, two neutri-

nos; they have a lifetime of 2.1×10^{-6} sec., and the negative
particles when arrested in elements with Z greater than 15
interact with the nuclei before decay and lead to a nuclear
transmutation with the emission of a neutrino and a neutron.
The great inadequacy of our present theoretical ideas is well
illustrated by the fact that there has scarcely been an attempt
to account for the existence of these particles and the details
of their properties.

From the present vantage point, the Bristol attitude was probably not
detrimental to the progress of particle physics, because pion theory
could not give quantitative predictions, and there were so many basic
cosmic-ray experiments that remained to be done with the emulsion
technique. The situation changed rapidly as the new accelerators
swung into action, as we shall see.

From penetrating cosmic-ray mesons to the (3,3) pion-nucleon resonance

The next topic, which will be treated much more briefly, illustrates the important role that theory can play in converting a semi-
quantitative cosmic-ray experiment into the formulation of a major
theoretical model that can only be tested by accurate accelerator experiments. This illustration is fairly typical of what transpired during
the five-year period 1947 to 1952 in the history of particle physics:
Cosmic-ray experimentation would uncover some new qualitative features at high energies (or equivalently small distances), theorists would
articulate these results into a set of model options, and the accelerator
physicists would then help to pin down whether or not one of these
models could give a quantitative fit to the experimental facts.

In the present instance, the starting point is the upper limit of the
scattering cross section by nucleons for sea-level mesons, which, as
noted earlier, was found to be about 10^{-28} cm^2.[2] The suggestion of
Pauli and Dancoff that a pseudoscalar meson was strongly coupled (in
the p state) to a finite-source nucleon (of radius a) was one of the three
approaches mentioned in an earlier section to explain the dilemma of
this small cross section for a presumed nuclear force meson.[5] By the
fall of 1944, Pauli was giving lectures at the Massachusetts Institute of
Technology on the meson theory of nuclear forces; these were published in a little book in 1946.[18]

In his book, Pauli emphasized the strong-coupling approach to meson theory and argued that the basic results could be derived by means

of a very simple semiclassical treatment, which should work increasingly well as the strong-coupling limit is approached. In expounding on this approach, Pauli explained that

> the assumption of large spin inertia leads to the so-called "strong coupling" case and results in the existence of excited states (isobars) of the nucleon with higher values of the spin. Effects involving the isospin, similar to those mentioned here, result from the interaction of charged meson fields with the "isospin" of the nucleon. The same constant a^{-1} will, in fact, determine also a "charge inertia" and in these theories excited states of the nucleon with higher values of the charge will exist.

By following through on these ideas, Pauli derived a cross section σ for the scattering of mesons by nucleons, which in the strong coupling limit ($a\mu \ll 1$), with energy of the incident meson not too large (energy $E < 1/a$), becomes $\sigma = 8a^2$. This confirmed Werner Heisenberg's early conjecture that it is possible to achieve a reduced scattering cross section of mesons by nucleons in the strong-coupling limit and that the cross section will then depend on the size of the nucleon source, which can be made appreciably smaller than the meson Compton wavelength.[5]

Having demonstrated the possibility of explaining the penetrating character of the sea-level cosmic-ray mesons, Pauli proceeded to examine further consequences of the strong-coupling meson theory. For example, he could easily derive a formula for the excitation energy of the nucleon isobars that were expected as a consequence of the presumed large "spin inertia" and "isospin inertia" of the nucleon. On the basis of the so-called symmetric pseudoscalar theory, he derived a formula for the excitation energy (ΔE) of the nucleon isobars having the simple form

$$\Delta E = \frac{3[J(J+1) - \frac{3}{4}]a\mu^2}{2f^2}. \tag{1}$$

In equation (1), a is the size of the nucleon, μ is the mass of the pion, f is the coupling constant, and J is the spin of the isobaric state; the restriction on equation (1) is that $|n| \leq J$, with ($n + 1/2$) the charge of the isobaric state. It is seen from equation (1) that $J = 1/2$ corresponds to the ordinary proton ($n = 1/2$) and neutron ($n = -1/2$) states of the nucleon, whereas $J = 3/2$ would predict equal excitation energies for the charge states Q = 2, 1, 0, −1. This is the famous (3,3) nucleon isobar (see below).

For sufficiently large coupling constant f, equation (1) predicts bound nucleon isobars (i.e., with excitation energy less than the sum of the meson and nucleon masses), and a search for stable nucleon isobars was actually carried out on the Rochester synchrocyclotron when it began to operate in 1949, but without success. The limiting energy of the Rochester machine made it impossible to look for "metastable" nucleon isobars, where the excitation energy of the nucleon isobar would exceed the sum of the pion and nucleon masses.[19]

The first strong indication of a metastable nucleon isobar with spin J = 3/2 and isospin I = 3/2, the so-called (3,3) isobar, came from the photo-pion production experiments with the Cornell and Berkeley electron synchrotrons. Figure 23.1 shows the experimental points for photoproduction of π^0 and π^+ and the comparison with the theoretical prediction of Keith A. Brueckner and Kenneth M. Watson, assuming a J = 3/2, I = 3/2 pion-nucleon resonance.[20] The Brueckner-Watson (3,3) pion-nucleon resonance is, of course, nothing more than the metastable J = 3/2, I = 3/2 nucleon isobar predicted by Pauli on the basis of the symmetric pseudoscalar theory.

Further strong support for the existence of the (3,3) pion-nucleon resonance came from the pion-nucleon experiments reported by the Chicago group at the second Rochester conference held in January 1952.[21] This group found the result that $\sigma(\pi^+ + p \rightarrow \pi^+ + p):\sigma(\pi^- + p \rightarrow n + \pi^0):\sigma(\pi^- + p \rightarrow \pi^- + p)$ were, within experimental error, in the ratio 9:2:1. In trying to explain this striking result, Fermi, speaking for the Chicago group, said at the second Rochester conference:

> If one assumes charge independence, i.e. that the isospin is a good quantum number, the two possible isospins, namely I = 3/2 and I = 1/2 scatter independently. If moreover, one assumes that the isospin I = 1/2 does not scatter at all one gets just the ratio 9:2:1; on the other hand, to assume that the I = 3/2 does not scatter at all would lead to the ratio 0:1:2. This conclusion is independent of angular momentum, spin correlation, or anything else. One can therefore interpret the experimental results by postulating the existence of a broad resonance level I = 3/2 in the band of energy 100-200 MeV, with the consequence that practically all the scattering comes through I = 3/2 in this energy region.

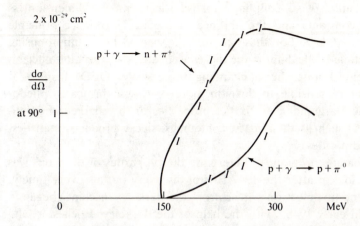

Figure 23.1. Experimental points from Cornell and Berkeley; the solid curves are from the theory of Brueckner and Watson. From Hans Bethe, Lectures on Meson Physics (Los Alamos Notes) (1952), 54.

This argument, confirmed in many ways in later years, put the concept of the isospin invariance of the pion-nucleon interaction into the fore-front of theoretical thinking in particle physics. It also became the paradigm for establishing the existence of innumerable hadronic reso-nances in succeeding years.

I hope that the sketchy remarks in this section have given some indication of the interplay between cosmic-ray experiment, theory, and accelerator experiments in contributing to the progress of particle physics.

Pion phenomenology

I have already alluded to the skepticism prevalent in the Bristol laboratory concerning the predictive power of the theory of the pion-nucleon interaction or even of the theory of muon interac-tion with other known particles. This skepticism was justified, and in this and the next section I shall comment briefly on the theoretical responses during those early years (in a sense, meson physics was born in the year 1947, when π-μ decays were first discovered). It was clear from the outset that any attempt to develop a renormaliz-able theory of the pion-nucleon interaction – along the lines of QED – would founder on the much greater strength of the pion-nucleon interaction compared to the electromagnetic (making it difficult to justify a weak-coupling expansion in terms of a dimensionless coupling constant) and the absence of a gauge invariance principle like that in QED. For these reasons, it seemed highly unpromising to do detailed calculations on the consequences of the pion-nucleon interaction, despite the outstanding successes of QED. Instead, it seemed more sensible to develop more or less sophisticated phenomenological arguments to fix the properties of the charged and neutral pions, such as their masses, lifetimes, decay products, statistics, spins, and parities.

A systematic program to determine the properties of the pions was not only logical after the 1947 Bristol discovery but also very timely, because the first electron synchrotron (the Berkeley machine) began to operate in early 1948. With the help of the Berkeley, Rochester, and Columbia synchrocylotrons (the Rochester machine in 1949 and the Columbia one in 1950), this program was essentially accomplished during 1951 to 1952. The full story is told in two books on meson physics that appeared in 1952 (where all the references are given), and I shall here only mention briefly a few of the highlights.[19,22] The copious production of charged pions by the Berkeley machine permitted an early and fairly precise determination of the mass. Nature's kindness in enforcing the condition $m(\pi^+) - m(\pi^0) > m(n) - m(p)$ meant that an accurate determination could also be made of the neutral pion mass by studying the energy spectrum of the two decay photons from the π^0 produced in the reaction $\pi^- + p \rightarrow n + \pi^0$. The lifetime for the dominant π^+ decay process, namely $\pi^+ \rightarrow \mu^+ + \nu$, was also measured at Berkeley. As far as the lifetime for $\pi^0 \rightarrow 2\gamma$ was concerned, only an upper limit could be found, which was pushed down to 10^{-14} sec by a cosmic-ray experiment with nuclear emulsions showing a pion production event of great multiplicity.[23] The Rochester machine had sufficient intensity of slow negative pions to be able to demonstrate that the absorption of π^- by complex nuclei releases large amounts of energy, thereby affirming that the charged pion is a boson (the observation of two γ rays from the decay of π^0 had already settled the boson character of the neutral pion).

The spin of the charged pion was measured directly by the simple trick of applying the principle of detailed balancing to the reaction π^+ + d (deuteron) $\rightarrow$ p + p and its inverse.[24] Thus, if we denote the cross section for the first process by $\sigma_{1\rightarrow2}$ and for the inverse $\sigma_{2\rightarrow1}$, the two cross sections are related by detailed balancing (in the center-of-mass system) as follows:

$$\frac{d\sigma_{2\to1}}{d\Omega} = \frac{3(2S+1)q^2}{2p^2}\frac{d\sigma_{1\to2}}{d\Omega} \tag{2}$$

where S is the spin of the π^+ meson, p is the momentum of either proton, and q is the momentum of the π^+ meson or the deuteron. This experiment was carried out in collaborative fashion: The cross section $\sigma_{2\to1}$ was measured at Berkeley (because the Berkeley machine could produce π^+'s directly in proton-proton collisions and the Rochester machine could not), whereas the cross section $\sigma_{1\to2}$ was measured at Rochester at the appropriate energy (because π^+'s of the right energy could be produced by the Rochester machine in proton-nucleus collisions). Using equation (2) to compare the two measured cross sections fixed the spin S as zero for π^+.[25] A more accurate measurement of the cross section for the first reaction could be carried out on the newly operating Columbia machine, and the Rochester result was confirmed.[26]

The parity determinations were made on the Berkeley machine by studying all possible end products from the two slow negative pion reactions $\pi^- + p$ and $\pi^- + d$, as follows:

$$\pi^- + p \to n + \gamma \tag{3a}$$
$$\to n + \pi^0 \tag{3b}$$
$$\pi^- + d \to 2n \tag{4a}$$
$$\to 2n + \gamma \tag{4b}$$
$$\to 2n + \pi^0. \tag{4c}$$

The comparability of the reactions (3a) and (3b), as well as the comparability of the reactions (4a) and (4b) [but the absence of (4c)], fixed the parities of π^- and π^0 as both negative. Figure 23.2 shows the curve obtained for the γ-ray spectrum resulting from reactions (3a) and (3b) and demonstrates the great power of accelerator experiments vis-à-vis cosmic-ray experiments attempting to do similar things.[27]

Putting all of this together, it had been established during 1951 to 1952, primarily with the help of the three synchrocyclotrons, that the charged and neutral pions were both pseudoscalar particles with almost equal masses. Their dominant decay modes were known, as well as the lifetime for the charged pion decay; only an upper limit for the neutral pion lifetime was known, but the method that was later to be used for an actual determination had already been put forward.[28] The closeness of the charged and neutral pion masses and their pseudoscalar character were consistent with the symmetric pseudoscalar theory that gave rise to the (3,3) pion-nucleon resonance. All of this encouraged serious

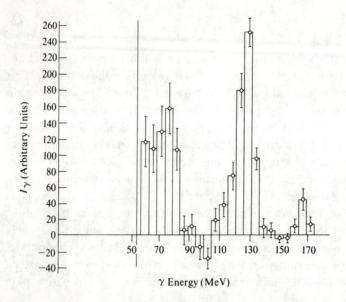

Figure 23.2. Measured γ-ray spectrum from the absorption of slow
π mesons in hydrogen. $\gamma_0 = 1/(1-\beta_0^2)^{1/2}$, where β_0 is the velocity of
the π^0 meson; ω is the γ ray energy. From Robert E. Marshak in
Meson Physics (McGraw-Hill 1952), 133; Berkeley experiment, W.
K. H. Panofsky, R. L. Aamodt, and J. Hadley in *Phys. Rev. 81*
(1951), 565.

theoretical work on the strong pion-nucleon interaction in the years
that followed.

Intimations of the universal weak interaction

Pontecorvo was really the first person to suggest that the negative cosmic-ray sea-level meson was a "heavy electron," having a weak
interaction with the proton, comparable to the electron, and giving rise
to neutrinos in the process $\mu^- + p \rightarrow n + \nu$, analogous to the reaction
$e^- + p \rightarrow n + \nu$. The suggestion was made without the knowledge that
μ^- was itself the decay product of the heavier Yukawa meson. At the
time that Pontecorvo proposed the μ-e analogy, the decay mechanism
for the muon had not been established, and, indeed, Pontecorvo even
suggested the possibility that $\mu \rightarrow e + \gamma$ was the decay mode.[12]

It was not until the pion and the muon were understood to be
distinct particles, and a determined effort was made by the cosmic-ray
and accelerator experimentalists to measure the electron spectrum

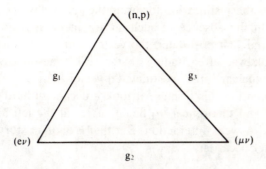

Figure 23.3. The Puppi triangle, showing interaction strengths g_1, g_2, g_3 between the known fermion pairs. From G. Puppi in *Nuovo Cim.* 5 (1948), 587.

from muon decay, that the three-body decay mode $\mu \to e + \nu_1 + \nu_2$ was finally pinned down (ν_1 and ν_2 were neutral particles, nonphotonic in character, and with small masses, consistent with zero). With this piece in place, quite a few persons noticed that the three processes $n \to p + e^- + \bar{\nu}$, $\mu^- \to e^- + \nu + \bar{\nu}$, and $\mu^- + p \to n + \nu$ all had comparable strengths.[29] Indeed, if one wrote down effective four-fermion interactions for these three processes with three different coupling constants g_1, g_2, and g_3, respectively, one found $g_1 \sim g_2 \sim g_3 \sim 10^{-49}$ erg-cm^3. One could even draw the Puppi triangle (Figure 23.3), and one began to see "intimations of the universality of the weak interaction."

I advisedly use the word *intimations*, because there was still a long road to travel before the universal (V-A) theory of weak interactions was proposed.[30] During the period 1947 to 1949, when the first thinking concerning a possible universal weak interaction surfaced, it was known that Gamow-Teller selection rules (in addition to the Fermi selection rules) were operating in β decay, and it was thought that the Gamow-Teller selection rules required a tensor (and not an axial vector) term in the effective four-fermion interaction. A serious measurement of the Michel parameter in muon decay had not as yet been made, and, of course, there was not the slightest smell of parity violation. During that period, Fermi himself was willing to be guided by experiment and certainly was not pushing for his original vector formulation of β decay.[31] It would be proper to argue, as some people have done, that Fermi had an uncanny insight into the correct structure of the weak interaction if he had invoked, let us say, the principle of gauge invariance as the justification for using the analogy with electromagnetic theory. In the absence of such an argument, it is no more surprising that Fermi first postulated the vector four-fermion interaction than that Yukawa, after finding that a scalar meson gave the wrong sign of the nuclear force, postulated a vector meson to explain his force. I only mention this because it took a decade of hard work, both experimental and theoretical, to finally arrive at the left-handed neutrino and the V-A weak interaction. But that is another story, best reserved for another occasion.

Strange particles and the second-generation quark

It is fitting to conclude this presentation with another major triumph of cosmic-ray experimentation during the period 1947 to 1952. I refer, of course, to the discovery of the strange particles. I noted earlier that the first strong indication of the existence of particles in the cosmic radiation heavier than the pion was obtained by Rochester and Butler.[16] By 1952, experiments with cloud chambers and nuclear emulsions had produced a list of "strange particles," whose masses were intermediate between the pion and nucleon masses, that were produced directly in nuclear collisions, whose lifetimes (of the charged strange particles) were shorter than the lifetime of the charged pion, and whose decay modes were most varied, including decays into only strongly interacting pions. By 1952, a list of the strange particles would look something like that in Table 23.1.

At the second Rochester conference, held in January 1952, the discussion of the strange particles was the hottest item on the agenda. Oppenheimer introduced the subject in his characteristically succinct and evocative fashion:

> Three alternative explanations have been proposed which may be schematized as (1) the live parent, (2) the heavy brother, and (3) selection rules. The first turned out to be the correct explanation of the μ meson. This was the conjecture of Marshak and Bethe that there was a step after production in which the strongly interacting particle could turn into something else that did not interact strongly with nuclear matter. This explanation does not, however, work out well for the V particles.

I might say that the strange particles were referred to, throughout the discussion of this subject at the second Rochester conference, as V

Table 23.1. *Tentative list of V particles*

Particle	Symbol	Reference	Mass	Average lifetime	Mode of decay	Nuclear inter-action	Directly produced in nuclear inter-actions	Remarks
Neutral V_2	V_2^0	Rochester et al., Manchester, 1947	~800	~10^{-9} sec	$V_2^0 \to \pi^+ + \pi^-$	?	Yes	Mode of decay still under discussion
τ meson	$\tau^\pm$	Brown et al., Bristol, 1949	969	≥10^{-9} sec	$\tau^\pm \to \pi^\pm + \pi^+ + \pi^-$	?	Yes	Two-body decay; sign of charge unknown
χ meson	χ	Menon & O'Ceallaigh, Bristol, 1952	~1,100	~10^{-9} sec	$\chi \to \pi^\pm + ?^0$	?	Probably yes	
Charged V	$V^\pm$	Rochester et al., Manchester, 1947	~1,000	~10^{-9} sec	$V^\pm \to ?^\pm + ?^0$	?	Yes	May be identical with K mesons or χ mesons, or may include both
Charged S	S	Bridge & Annis, Cambridge, Mass., 1951	~1,400	~10^{-9} sec	$S \to$ meson(π, μ) $+ ?^0 + (?^0)$	?	Yes	Probably identical with charged V particles
K meson	K^+	O'Ceallaigh, Bristol, 1951	~1,300	≥10^{-9} sec	$K^+ \to \mu^+ + ?^0 + ?^0$	?	Yes	Nature of two neutral particles not yet determined
Neutral V_1	V_1^0	Armenteros et al., Manchester, 1951	~2,200	~10^{-9} sec	$V_1^0 \to p^+ + \pi^-$	?		Mode of decay still under discussion

particles or "megalomorphs" – the latter because they had so much structure. Oppenheimer then called on Abraham Pais to present an account of his "ordering principle for megalomorphian zoology" to explain the long lifetimes of the V, a combination of associated production and a multiplicative quantum number. I quote from his remarks:

> The following is an attempt to decouple the production of these heavy particles from their decay, talking for the moment only of the observed species. This involves an approach not previously mentioned at this conference. An ordering is attempted by assigning to the protons, neutrons, V's, π's, and v's, a "mass number" as indicated in [Table 23.2].
>
> We begin by introducing the Yukawa interaction but avoid all details specifying the coupling. This coupling is then simply written $(N_0 N_0 \pi_0)$. This suggests a matrix of interactions of the form $(N_i N_j \pi_k)$, with i, j, and k equal to zero or one and with the possibility of higher indices in the future. A selection rule is now introduced which does not follow from present theories, but which is not in contradiction with known facts. This rule is $(N_i N_j \pi_k)$ can only be strong if $i + j + k$ is even.[32]

Pais certainly paved the way for the new era of the "additive strangeness quantum number" that commenced in 1953. His listing (Table 23.2) is also remarkably prescient with regard to the number of generations listed. It also looks as if Pais was trying to say something about a quark-lepton symmetry principle!

Concluding remarks

It seems highly appropriate to conclude my presentation with Pais's theory of the strange particles. I began with the first Shelter Island conference in 1947, where, basically, the "second-generation lepton" was born, and I am concluding with the second Rochester conference, where, with a certain amount of poetic license, we can say that the "second-generation quark" was born. Cosmic-ray physics can take pride in the fact that it was responsible for both seminal achievements.

The five years in the history of particle physics that I have covered in this chapter were unique in several respects. The cosmic-ray experimentalists, the theoretical particle physicists, and the high-energy-accelerator experimentalists joined together, on fairly equal terms, to gain an understanding of the subnuclear structure of matter. After

Table 23.2. *Elementary particle families*

N_0 (nucleons $(p = N_0^+, n = N_0^0)$	π_0 (pions) $(\pi^0 = \pi_0^0, \pi^\pm = \pi_0^\pm)$	e
N_1 (heavy V particle) $(V_1^0 = N_1^0; V^\pm = N_1^\pm)$	π_1 (light v particle) $(v_2^0 = \pi_1^0; v^\pm = \pi_1^\pm)$	μ
$(N_2?)$ ("resonance" in π-nucleon scattering)	?	?

1952, the ascendency of accelerator experiments over cosmic-ray experiments increased so rapidly that at the sixth Rochester conference (held in April 1956) Robert B. Leighton was led to remark that "next year those people still studying strange particles using cosmic rays had better hold a rump session of the Rochester Conference somewhere else." This was the price of progress, but it did mean that elegant experiments carried out by small teams of investigators in relatively poor countries could no longer compete as effectively with the "big machines." This trend was accentuated by the increasing concentration of particle physics experimentation in the very large accelerator laboratories throughout the world.

The Rochester conference was founded in 1950 to give "equal time" to cosmic-ray experiments, accelerator experiments, and particle theory, to the discoveries of individual scientists and teams of workers, to the interesting speculations of the young neophyte and the accomplished virtuoso, to particle physicists of all nations independent of size, economic wealth, or political complexion. After 30 years, the Rochester conference has inevitably mirrored some of the changing dynamics in the conduct of particle physics experimentation, but I do believe that the original concept of providing a democratic forum for the full exchange of ideas from whatever responsible source has been maintained. The texture of those five glorious years 1947 to 1952, during which particle physics experienced its birth pangs, will never be recaptured, but the maturation of our science has brought and will continue to bring its own intellectual joys.

Notes

1 Work supported by the U.S. Department of Energy.

2 J. G. Wilson, "The Scattering of Mesotrons in Metal Plates," *Proc. Roy. Soc. (London) A174* (1940), 73-85.

3 L. W. Nordheim and M. H. Hebb, "On the Production of the Hard Component of the Cosmic Radiation," *Phys. Rev. 56* (1939), 494-501.

4 R. E. Marshak and V. F. Weisskopf, "On the Scattering of Mesons of Spin $\hbar/2$ by Atomic Nuclei," *Phys. Rev. 59* (1941), 130-5.

5 W. Pauli and S. M. Dancoff, "The Pseudoscalar Meson Field with Strong Coupling," *Phys. Rev. 62* (1942), 85-108; W. Heisenberg ["Zur Theorie der explosionsartigen Schauer in der kosmischen Strahlung. II," *Z. Physik 113* (1939), 61-86] first worked out the strong-coupling theory for vector mesons. See also J. R. Oppenheimer and J. Schwinger, "On the Interaction of Mesotrons and Nuclei," *Phys. Rev. 60* (1941), 150-2.

6 Articles by S. Sakata and Y. Tanikawa in *Report of the Symposium on Meson Theory* (1943); published in English: Shoichi Sakata and Takesi Inoue, "On the Correlations between Mesons and Yukawa Particles," *Progr. Theoret. Phys. (Kyoto) 1* (1946), 143-50; Y. Tanikawa, "On the Cosmic-Ray Meson and the Nuclear Meson," *Progr. Theoret. Phys. (Kyoto) 2* (1947), 220-1.

7 M. Conversi, E. Pancini, and O. Piccioni, "On the Disintegration of Negative Mesons," *Phys. Rev. 71* (1947), 209-10; C. M. G. Lattes, H. Muirhead, G. P. S. Occhialini, and C. F. Powell, "Processes Involving Charged Mesons," *Nature 159* (1947), 694-7.

8 S. Tomonaga and G. Araki, "Effect of the Nuclear Coulomb Field on the Capture of Slow Mesons," *Phys. Rev. 58* (1940), 90-1.

9 E. Fermi, E. Teller, and V. Weisskopf, "The Decay of Negative Mesotrons in Matter," *Phys. Rev. 71* (1947), 314-15.

10 I urged Weisskopf to publish a note on his hypothesis, and he did: V. F. Weisskopf, "On the Production Process of Mesons," *Phys. Rev. 72* (1947), 510.

11 R. E. Marshak and H. A. Bethe, "On the Two-Meson Hypothesis," *Phys. Rev. 72* (1947), 506-9.

12 B. Pontecorvo, "Nuclear Capture of Mesons and the Meson Decay," *Phys. Rev. 72* (1947), 246-7.

13 R. F. Christy and S. Kusaka, "Burst Production by Mesotrons," *Phys. Rev. 59* (1941), 414-21.

14 R. E. Marshak, "Heavy Electron Pair Theory of Nuclear Forces," *Phys. Rev. 57* (1940), 1101-6.

15 L. Leprince-Ringuet, in *Proceedings of the Third Annual Rochester Conference, December 18-20, 1952*, ed. by H. P. Noyes, M. Camac, and W. D. Walker (New York: Interscience, 1953), pp. 47-50.

16 G. D. Rochester and C. C. Butler, "Evidence for the Existence of New Unstable Elementary Particles," *Nature 160* (1947), 855-7.

17 C. F. Powell, "Mesons," *Reports on Progress in Physics 13* (1950), 350-424.

18 W. Pauli, *Meson Theory of Nuclear Forces* (New York: Interscience, 1946; 2nd ed., 1948).

19 See R. E. Marshak, *Meson Physics* (New York: McGraw-Hill, 1952), p. 360; it is easy to give a semiclassical derivation of the width of a "metastable" nuclear isobar.

20 H. A. Bethe, *Lectures on High Energy Phenomena* (Los Alamos: 1953). K. A. Brueckner and K. M. Watson, "Phenomenological Relationship between Photomeson Production and Meson-Nucleon Scattering," *Phys. Rev. 86* (1952), 923-8.

21 E. Fermi, "Discussion of Pion-Nucleon Scattering," in "Proceedings of Second Rochester Conference, Jan. 1952" (unpublished), ed. by A. M. L. Messiah and H. P. Noyes, pp. 24-6. See also H. L. Anderson, E. Fermi, R. Martin, and D. E. Nagle, "Angular Distribution of Pions Scattered by Hydrogen," *Phys. Rev. 91* (1953), 155-68.

22 A. M. Thorndike, *Mesons* (New York: McGraw-Hill, 1952).

23 M. F. Kaplan, B. Peters, and H. L. Bradt, "Evidence for Multiple Meson and γ-Ray Production in Cosmic-Ray Stars," *Phys. Rev. 76* (1950), 1735.

24 R. E. Marshak, "Meson Reactions in Hydrogen and Deuterium," *Phys. Rev. 82* (1951), 313; W. B. Cheston, "On the Reactions $\pi^+ + d \leftrightarrows p + p^*$," *Phys. Rev. 83* (1951), 1118-22.

25 D. C. Clark, A. Roberts, and Richard Wilson, "Cross Section for the Reaction $\pi^+ + d \rightarrow p + p$, and the Spin of the π^+ Meson," *Phys. Rev. 83* (1951), 649.

26 R. Durbin, H. Loar, and J. Steinberger, "The Spin of the Pion via the Reaction $\pi^+ + d \leftrightarrows p + p$," *Phys. Rev. 83* (1951), 646-8.

27 Wolfgang K. H. Panofsky, R. Lee Aamodt, and James Hadley, "The Gamma-Ray Spectrum Resulting from Capture of Negative π-Mesons in Hydrogen and Deuterium," *Phys. Rev. 81* (1951), 565-74.

28 H. Primakoff, "Photo-Production of Neutral Mesons in Nuclear Electric Fields and the Mean Life of the Neutral Meson," *Phys. Rev. 81* (1951), 899.

29 O. Klein, "Mesons and Nucleons," *Nature 161* (1948), 897-9; J. Tiomno and John A. Wheeler, "Energy Spectrum of Electrons from Meson Decay," *Rev. Mod. Phys. 21* (1949), 144-52; T. D. Lee, M. Rosenbluth, and C. N. Yang, "Interaction of Mesons with Nucleons and Light Particles," *Phys. Rev. 75* (1949), 905; G. Puppi, "Sui mesoni dei raggi cosmici," *Nuovo Cimento 5* (1948), 587-8.

30 E. C. G. Sudarshan and R. E. Marshak, "The Nature of the Four-Fermion Interaction," *Proceedings of the Padua-Venice Conference on Mesons and Recently Discovered Particles* (1957), V-14 to V-22; R. P. Feynman and M. Gell-Mann, "Theory of the Fermi Interaction," *Phys. Rev. 109* (1958), 193-8; J. J. Sakurai, "Mass Reversal and Weak Interactions," *Nuovo Cimento 7* (1958), 649-60.

31 Enrico Fermi, *Elementary Particles* (New Haven: Yale University Press, 1951), p. 40.

32 A. Pais, "An Ordering Principle for Megalomorphian Zoology," in "Proceedings of the Second Rochester Conference, Jan. 1952" (unpublished), ed. by A. M. L. Messiah and H. P. Noyes, pp. 87-93.

Name index

Adair, Robert K., 216
Adams, Raymond V., 151
Ageno, Mario, 192, 243
Alger, Horatio, 131
Allison, Samuel, 315
Alvarez, Luis, 239, 253
Amaldi, Edoardo, 226, 238, 243, 245–6, 272
Anderson, Carl D., 6, 9, 14–19, 22, 52, 65, 84, 89, 90, 114–15, 128, 131–2, 152, 157, 178, 206, 208, 211–12, 251, 263–4, 270, 289, 295, 309
Anderson, Herbert L., 268, 317
Aoki, Hiroo, 88
Araki, Gentaro, 100
Araki, Toshima, 85, 96, 162, 222, 229, 246, 248, 379
Arnold, H. H., 124
Armenteros, R., 179–80
Astier, A., 179–80
Auger, Pierre, 113–14, 116, 227
Autler, Stanley, 319

Bacon, Francis, 300
Ballario, Carlo, 243
Barnóthy, J., 285
Barnóthy (Forró), M., 261
Barry, J. Griffiths, 215
Becker, Gordon, 319
Becker, H., 142
Becquerel, Henri, 6
Bernardini, Gilberto, 19, 20, 165, 192, 223, 226, 230–1, 237–9, 243, 263, 265, 267, 271–3
Bethe, Hans, 14–16, 18, 24, 26–7, 63, 69, 70, 75, 97, 101, 117, 128, 145, 156, 195, 208–9, 249, 276, 279, 281, 311, 321, 323–4, 331, 334, 339, 370–1, 379, 382, 396

Betz, Otto, 320–1
Bhabha, Homi J., 16, 90, 93, 117, 156, 211, 217
Birus, Karl, 99
Blackett, P.M.S., 9, 19, 87, 89, 90, 95, 115–16, 141, 146, 148, 156, 184–6, 208, 213, 251, 264, 274, 385
Bloch, Felix, 72, 208, 211–12, 270, 284, 314–15, 335, 360
Bohm, David, 323
Bohr, Aage, 332
Bohr, Margrethe, 184
Bohr, Niels, 8, 16, 23–4, 56, 58, 61, 112, 184, 206, 264–5, 275, 283, 295, 299, 355, 357
Born, Max, 21, 41
Bose, Satyandra Nath, 45
Bostick, Winston, 188
Bothe, W., 8, 112–13, 116, 128, 142, 155, 261
Bowen, Ira S., 120–1, 124–5, 127, 206, 214
Bradbury, Norris, 315
Breit, Gregory, 210, 216, 323, 332
Bridge, H. S., 274
Broadway, Leonard, 318
Brode, Robert B., 274–5, 313
Broglie, Louis de, 43, 58
Brown, Laurie M., 89, 287, 322, 374
Brueckner, Keith A., 390
Butler, Clifford C., 20, 151–2, 251, 385–6, 396

Cacciapuoti, Bernardo N., 192, 243
Camerini, Ugo, 307
Cameron, G. Harvey, 112, 121–2
Camp, Glenn, 314
Campbell, N. R., 23

Subject index